Methods in Molecular Biology

Series Editor
John M. Walker
School of Life and Medical Sciences
University of Hertfordshire
Hatfield, Hertfordshire, UK

For further volumes:
http://www.springer.com/series/7651

For over 35 years, biological scientists have come to rely on the research protocols and methodologies in the critically acclaimed *Methods in Molecular Biology* series. The series was the first to introduce the step-by-step protocols approach that has become the standard in all biomedical protocol publishing. Each protocol is provided in readily-reproducible step-by-step fashion, opening with an introductory overview, a list of the materials and reagents needed to complete the experiment, and followed by a detailed procedure that is supported with a helpful notes section offering tips and tricks of the trade as well as troubleshooting advice. These hallmark features were introduced by series editor Dr. John Walker and constitute the key ingredient in each and every volume of the *Methods in Molecular Biology* series. Tested and trusted, comprehensive and reliable, all protocols from the series are indexed in PubMed.

Statistical Genomics

Edited by

Brooke L. Fridley and Xuefeng Wang

Department of Biostatistics and Bioinformatics, H. Lee Moffitt Cancer Center & Research Institute, Tampa, FL, USA

Humana Press

Editors
Brooke L. Fridley
Department of Biostatistics and Bioinformatics
H. Lee Moffitt Cancer Center & Research Institute
Tampa, FL, USA

Xuefeng Wang
Department of Biostatistics and Bioinformatics
H. Lee Moffitt Cancer Center & Research Institute
Tampa, FL, USA

ISSN 1064-3745 ISSN 1940-6029 (electronic)
Methods in Molecular Biology
ISBN 978-1-0716-2985-7 ISBN 978-1-0716-2986-4 (eBook)
https://doi.org/10.1007/978-1-0716-2986-4

This Humana imprint is published by the registered company Springer Science+Business Media, LLC, part of Springer Nature.
The registered company address is: 1 New York Plaza, New York, NY 10004, U.S.A.

Preface

The term "statistical genomics" was popularized after 2001 when the initial analysis of the human genome sequence was completed, even though the sequencing cost per human genome was estimated at $100 million in that year. As the genome sequencing cost has drastically decreased and the generation of other types of molecular data have become more widely available, the field has since evolved and expanded tremendously over the past 20 years. The focus of its investigation has shifted towards developing and deploying efficient analytical approaches that are more *integrative*, *incorporative*, *interactive*, and *interdisciplinary*. Thus, this book includes chapters that focus on integrating genomics with other "omics" data, such as transcriptomics, epigenomics, proteomics, metabolomics, and metagenomics. Large-scale initiatives such as UK Biobank and other international genetics consortia have granted the community more open-access resources for downstream analyses. Meanwhile, concepts such as summary statistics and genetic risk score modeling have greatly facilitated post-GWAS research. The ever-growing collection of shared data resources is the main impetus for building user-friendly and interactive database and computational tools to allow efficient genomic annotation, collaboration and sharing of results, and secondary data analyses. In the past two decades, statistical modeling itself has also gone beyond conventional statistical inference and has embraced the ever-escalating computing power and big data technologies of today. As will be demonstrated in several chapters, the multifaceted analytical challenges in the heterogenous omics data have nurtured interdisciplinary models that often encompass innovative integration of computational biology, machine learning, and bioinformatics techniques. We also hope that by covering these diverse and timely topics in this book, we provide researchers firsthand insights into promising future directions and priorities in the context of pan-omics and the precision medicine era.

Tampa, FL, USA

Brooke L. Fridley
Xuefeng Wang

Contents

Contributors

ASH A. ALIZADEH • *Division of Oncology, Department of Medicine, Stanford University, Stanford, CA, USA; Institute for Stem Cell Biology and Regenerative Medicine, Stanford University, Stanford, CA, USA; Division of Hematology, Department of Medicine, Stanford Cancer Institute, Stanford University, Stanford, CA, USA; Stanford Cancer Institute, Stanford University, Stanford, CA, USA*

RHONDA BACHER • *Department of Biostatistics, University of Florida, Gainesville, FL, USA*

HAIMENG BAI • *Department of Population and Quantitative Health Sciences, Cleveland Institute for Computational Biology, Case Western Reserve University, Cleveland, OH, USA; Department of Nutrition, Case Western Reserve University School of Medicine, Cleveland, OH, USA*

WILLIAM S. BUSH • *Department of Population and Quantitative Health Sciences, Cleveland Institute for Computational Biology, Case Western Reserve University, Cleveland, OH, USA*

XUEYUAN CAO • *College of Nursing, University of Tennessee Health Science Center, Memphis, TN, USA*

PRABHAKAR CHALISE • *Department of Biostatistics & Data Science, University of Kansas Medical Center, Kansas City, KS, USA*

XIAORU DONG • *Department of Biostatistics, University of Florida, Gainesville, FL, USA*

ABDELRAHMAN H. ELSAYED • *Department of Biostatistics, St. Jude Children's Research Hospital, Memphis, TN, USA; Department of Pathology, St. Jude Children's Research Hospital, Memphis, TN, USA*

SILI FAN • *Graduate Group of Biostatistics, University of California, Davis, CA, USA*

BROOKE L. FRIDLEY • *Department of Biostatistics and Bioinformatics, H. Lee Moffitt Cancer Center & Research Institute, Tampa, FL, USA*

ANDREW J. GENTLES • *Department of Biomedical Data Science, Stanford University, Stanford, CA, USA; Stanford Center for Biomedical Informatics Research, Department of Medicine, Stanford University, Stanford, CA, USA; Stanford Cancer Institute, Stanford University, Stanford, CA, USA*

COLEMAN HARRIS • *Department of Biostatistics, Vanderbilt University Medical Center, Nashville, TN, USA*

YOUNGCHUL KIM • *Department of Biostatistics and Bioinformatics, H. Lee Moffitt Cancer Center and Research Institute, Tampa, FL, USA*

DEUKWOO KWON • *Department of Population Health Science & Policy, Icahn School of Medicine at Mount Sinai, New York, NY, USA*

HONGZHE LI • *Department of Biostatistics, Epidemiology and Informatics, University of Pennsylvania, Philadelphia, PA, USA*

QIAN LI • *Department of Biostatistics, St. Jude Children's Research Hospital, Memphis, TN, USA*

BOGDAN A. LUCA • *Department of Biomedical Data Science, Stanford University, Stanford, CA, USA; Stanford Center for Biomedical Informatics Research, Department of Medicine, Stanford University, Stanford, CA, USA*

SIYUAN MA • *Department of Biostatistics, Epidemiology and Informatics, University of Pennsylvania, Philadelphia, PA, USA*

QIANXING MO • *Department of Biostatistics & Bioinformatics, H. Lee Moffitt Cancer Center & Research Institute, Tampa, FL, USA*

AARON M. NEWMAN • *Department of Biomedical Data Science, Stanford University, Stanford, CA, USA; Institute for Stem Cell Biology and Regenerative Medicine, Stanford University, Stanford, CA, USA; Stanford Cancer Institute, Stanford University, Stanford, CA, USA*

OSCAR OSPINA • *Department of Biostatistics and Bioinformatics, Moffitt Cancer Center, Tampa, FL, USA*

STANLEY B. POUNDS • *Department of Biostatistics, St. Jude Children's Research Hospital, Memphis, TN, USA*

XIAOYU SONG • *Department of Population Health Science and Policy, Icahn School of Medicine at Mount Sinai, New York, NY, USA; Tisch Cancer Institute, Icahn School of Medicine at Mount Sinai, New York, NY, USA; Institute for Healthcare Delivery Science, Icahn School of Medicine at Mount Sinai, New York, NY, USA*

ALEX SOUPIR • *Department of Biostatistics and Bioinformatics, Moffitt Cancer Center, Tampa, FL, USA*

CHLOÉ B. STEEN • *Department of Biomedical Data Science, Stanford University, Stanford, CA, USA; Department of Medical Genetics, Oslo University Hospital, Oslo, Norway; Division of Oncology, Department of Medicine, Stanford University, Stanford, CA, USA*

MINGXIANG TENG • *Department of Biostatistics and Bioinformatics, Moffitt Cancer Center, Tampa, FL, USA*

ANDREW E. TESCHENDORFF • *CAS Key Laboratory of Computational Biology, Shanghai Institute of Nutrition and Health, University of Chinese Academy of Sciences, Chinese Academy of Sciences, Shanghai, China; UCL Cancer Institute, Paul O'Gorman Building, University College London, London, UK*

SIMON VANDEKAR • *Department of Biostatistics, Vanderbilt University Medical Center, Nashville, TN, USA*

MAGGIE HAITIAN WANG • *JC School of Public Health and Primary Care, the Chinese University of Hong Kong (CUHK), Shatin, Hong Kong; CUHK Shenzhen Institute, Shenzhen, China*

XUEFENG WANG • *Department of Biostatistics and Bioinformatics, H. Lee Moffitt Cancer Center & Research Institute, Tampa, FL, USA*

YINGYING WEI • *Department of Statistics, the Chinese University of Hong Kong (CUHK), Shatin, Hong Kong*

CHRISTOPHER M. WILSON • *Department of Biostatistics and Bioinformatics, Moffitt Cancer Center, Tampa, FL, USA*

JULIA WROBEL • *Department of Biostatistics and Informatics, University of Colorado Anschutz Medical Campus, Aurora, CO, USA*

XIAOXUAN XIA • *JC School of Public Health and Primary Care, the Chinese University of Hong Kong (CUHK), Shatin, Hong Kong; Department of Statistics, the Chinese University of Hong Kong (CUHK), Shatin, Hong Kong*

SIJIE YAO • *Department of Biostatistics and Bioinformatics, H. Lee Moffitt Cancer Center & Research Institute, Tampa, FL, USA*

XUEYI ZHANG • *Department of Population and Quantitative Health Sciences, Cleveland Institute for Computational Biology, Case Western Reserve University, Cleveland, OH, USA*

YEXIAN ZHANG • *CUHK Shenzhen Institute, Shenzhen, China*
TIANYU ZHU • *CAS Key Laboratory of Computational Biology, Shanghai Institute of Nutrition and Health, University of Chinese Academy of Sciences, Chinese Academy of Sciences, Shanghai, China*

Chapter 1

Multi-omics Data Deconvolution and Integration: New Methods, Insights, and Translational Implications

Xuefeng Wang and Brooke L. Fridley

Abstract

In the current era of multi-omics, new sequencing and molecular profiling technologies have facilitated our quest for a deeper and broader understanding of the variations and dynamic regulations in human genomes. However, analyzing and integrating data generated from diverse platforms, modalities, and large-scale heterogeneous samples to extract functional and clinically valuable information remains a significant challenge. Here, we first discuss recent advances in methods and algorithms for analyzing data at the genome, transcriptome, proteome, metabolome, and microbiome levels, followed by emerging methods for leveraging single-cell sequencing and spatial transcriptomic data. We also highlight the mechanistic insights that these advances can bring to the field, as well as the current challenges and outlooks relating to their translational and reproducible adoption at the population level. It is evident that novel statistical methods, which were inspired by new assays, will enable the associated molecular profiling pipelines and experimental designs to continuously improve our understanding of the human genome and the downstream consequences in the transcriptome, epigenome, proteome, metabolome, regulome, and microbiome.

Key words Multi-omics study, Data integration, Visualization, Translational biomarkers

1 Introduction

Since the initial completion of the Human Genome Project two decades ago, the study of genomics has evolved and expanded tremendously because of the rapid decrease in the cost of genome sequencing and the increased availability of various types of molecular data. At the same time, statistical modeling has advanced far beyond the traditional statistical inference framework to take advantage of today's big data technologies, novel machine learning algorithms, and ever-increasing computing capacity. The field of statistical genomics has thus become more integrative, incorporative, interactive, and interdisciplinary than ever before.

Brooke L. Fridley and Xuefeng Wang (eds.), *Statistical Genomics*,
Methods in Molecular Biology, vol. 2629, https://doi.org/10.1007/978-1-0716-2986-4_1,

The approaches developed in statistical genomics have now been extended to a variety of omic data types rather than genomics alone, including data on the transcriptome, methylome, proteome, metabolome, and microbiome. While integrative omic data analysis holds promise for building a more potent framework for discovering new biomarkers and potential molecular mechanisms, many challenges arise in analyzing each type of omics data set. Since many molecular measures are extremely tissue- and cell-type-specific, the noise (and the confounding effect) caused by cellular heterogeneity has been increasingly recognized as a major barrier in interpreting and translating omic data into meaningful insights. The recent development of single-cell and spatial biology technologies now make it possible to quantify omic features at unprecedented resolutions, which could mark the field's next major turning point. Nevertheless, the cost and complexity of these technologies have largely limited their widespread adoption at this time.

Meanwhile, enormous amounts of bulk-cell and bulk-tissue omics data have accumulated in the public domain, making large-scale secondary analyses possible and necessary. Integrating an external data source as an initial discovery or independent validation dataset is now a common practice in molecular studies. Against this background, many "deconvolution" methods have been developed to counteract the influence of the sample heterogeneity. As will be discussed below, using reference-based or reference-free methods, the inputted omic profiles can be decomposed according to their constituent cell types (for solving relative proportions), gene hallmark signatures, or latent factors (for discovering underlying clusters). In this chapter looking into statistical genomic methods, we review and recap some important achievements in multi-omics data analysis as elaborated in other chapters, with a special focus on emerging data deconvolution and integration techniques.

2 Emerging Methods for Multi-omics Assays: From Deconvolution to Integration

While new molecular profiling technologies are more mature, the transcriptome remains the most established and well-studied modality among all multi-omics assays. Transcriptomic profiling quantitatively measures the gene expression level at a genome-wide scale. Compared to the upstream DNA mutational data and the downstream proteomic analyses, the gene expression profile provides a more direct knowledge of gene regulation and functional information, thus making it a mainstay in both molecular biology and translational science. Due to the rapid development and reduced cost of next-generation sequencing (NGS) techniques, RNA sequencing (RNAseq) has become more widely adopted than microarrays in research settings. In this article, we highlight two key challenges in RNAseq data analysis. First, the experimental and analytical steps involved in the processing of RNAseq data can introduce significant batch-effect variability among and within different datasets. Such batch effect is usually stronger than the data generated from the same platform of gene expression arrays. Therefore, a careful diagnosis of the batch effect and any other implicit biases is highly recommended before any downstream modeling steps. If raw data is available, it is always advisable to have read count data renormalized and processed using the same approach or pipeline. If raw data is not available, batch effect correction methods such as ComBat [1, 2] should be applied to attenuate any bias present in the data. Second, it is important to consider the fact that gene expression data generated from bulk tissue or blood samples can be markedly confounded by the heterogeneous cellular composition of each sample. To tackle this issue, a series of transcriptome deconvolution methods (or digital cytometry), such as CIBERSORT [3, 4], have been developed to estimate the relative abundance of cell-type-specific signatures and have gained wide popularity in the field. Most of the deconvolution algorithms rely on the reference matrix, also known as the gene expression profile/reference matrix (GEP matrix), to solve for the individual gene signature score. The GEP matrix, estimated based on the single-cell RNA-sequencing data (ideally generated from the same tissue type), was found to provide better dissection of tissue heterogeneity and thus more accurate estimates of gene-based signatures [4, 5]. Following a similar idea, a new tool called EcoTyper was recently designed to further characterize transcriptionally defined cell states and their co-association patterns, also known as ecotypes [6, 7]. As discussed in greater detail in Chapter 4, EcoTyper provides a suite of functions that accomplish two major tasks: (i) the identification of cell states and ecotypes from bulk or single-cell expression data and (ii) the recovery of previously defined cell states and ecotypes from gene expression data. Overall, as exemplified above, gene transcriptome data represent a particularly well-established molecular modality with a plethora of well-developed analytical tools.

In addition to these novel bioinformatic tools, transcriptomic analysis has been revolutionized by the wide adoption of single-cell RNAseq (scRNAseq) technologies and recent advances in spatially-resolved transcriptomics (ST). Because scRNAseq provides gene expression profiles at individual cell-level resolution, it is capable of dissecting tissue heterogeneity. While many steps can be adapted from the bulk RNAseq analysis pipeline, scRNAseq data present several unique analytical challenges that require special consideration, such as sparsity caused by widespread dropout events, heterogeneous distributions of expression values, and high technical noise inherent in the technology [8]. Chapter 6 provides a useful compilation of data workflow, statistical methods, and practical guidance for the analysis of data from scRNAseq gene expression experiments. It was demonstrated that the upstream raw scRNAseq data requires a special cascade of quality control, normalization, and batch effect correction procedures, whereas the downstream steps involve unique dimensionality reduction (for efficient clustering), cell-type annotation, differential expression, and trajectory inference steps. All steps in the proposed scRNAseq analysis pipeline can be implemented by open-source R/Bioconductor packages, such as "Seurat" [9]. By adding the spatial resolution to the gene expression data, the analysis of ST data presents not only some similar but also additional analytical challenges. Chapter 7 highlights some of the current algorithms in ST data processing, from visualization and clustering to deconvolution (e.g., regions of interest (ROI) phenotyping), particularly focusing on the barcode-based ST techniques. The usage of the clustering analysis methods with and without explicitly considering spatial information was noted. An important consideration is that most current ST data do not yet provide single-cell resolution as in scRNAseq and only have aggregated expression data (i.e., mini-bulk data). Thus, bulk-tissue deconvolution methods as described above can still be applied to ROIs to identify the predominant cell types. The chapter concluded that more development is needed in the field to better understand spatial cell-cell communication and spatial heterogeneity patterns. As an alternative approach to investigate cell-to-cell spatial relationships, Chapter 8 discussed a series of statistical methods and applications for processing data from multiplexed imaging technologies such as immunofluorescence (mIF) and immunohistochemistry (mIHC), as well as imaging-based methods for deconvolving the tumor immune microenvironment (TIME). As the granularity, data volume, and type of modality of cell-level omics/imaging data continue to increase, we envision that paradigms such as deep neural networks, multitask learning, and transfer learning might be a promising option for building an integrative framework that incorporates and integrates data across different ROIs, samples, and platforms.

Similar to the challenge faced in the transcriptomic data, DNA methylation (DNAme), a major component of epigenome data, is also highly tissue- and cell-type-specific. Thus, the DNAme measurement from bulk tissues can be strongly confounded by cellular composition and heterogeneity level. Because it is more stable both biologically and chemically, the tumor DNAme profile is well recognized as an important translational biomarker. A series of reference-based deconvolution methods for analyzing DNAme data have been proposed, following a similar concept used in deconvolving bulk-tissue gene expression data [10]. The basic idea of the reference-base algorithm is to assume that bulk DNAme profiles are the weighted sum of reference profiles, and solving for these weights will thus reveal relative fractions for each cell-type group included in the reference panels. Chapter 3 describes why the existing reference methods can be suboptimal for solid tissues as compared to blood, because the experimental purification of individual cell types for generating an orthogonal DNAme reference profile is not feasible. To tackle this issue, a method called EpiSCORE [11, 12] was developed to generate high-resolution cell-type deconvolution of bulk DNA methylomes by leveraging the existing single-cell expression data. EpiSCORE uses a predictive semi-Bayesian model to translate the reference information from scRNAseq data to generate a more accurate DNAme reference. The chapter covers a step-by-step tutorial from the construction of the DNAme reference matrix to the cell-type deconvolution, using data from skin and brain as examples. If DNAme is profiled via bisulfite sequencing, a new method called cell heterogeneity–adjusted clonal methylation (CHLAM) can be applied to further correct the DNAme quantification. The method exploits the fact that each sequencing read approximately represents a single cell within the bulk sample profiled. The developer showed that CHALM significantly improves the correlation between the DNAme and gene expression level. As a result, such method will directly benefit a new integrative biomarker prioritizing strategy based on the concept of expression quantitative trait methylation, or eQTM [13].

As reasoned in Chapter 13, it is insufficient to consider only transcriptomics to study gene expression and regulation. Complicated by multitiered regulations and post-translational modifications, protein expression exhibits substantial spatial, temporal, and quantitative differences from mRNA expression profiles. Large-scale proteomic quantification methods based on mass spectrometry (MS) have enabled a systemic investigation of downstream signaling events and potential therapeutic targets. MS-derived proteomics data also creates considerable analytical challenges, including heterogeneous distributions, substantial noise, outliers, strong batch effects, and a high rate of missingness. The chapter by Song provided detailed analysis guidelines and protocols (based on R functions) for preprocessing, peak density normalization, outlier detection, batch effect correction, and missing data imputation. The chapter also presented state-of-the-art methods and example codes for performing network analysis, clustering analysis, integrative proteogenomic analysis, and multi-omics association analysis. Recent large-scale proteomic initiatives such as the Clinical Proteomic Tumor Analysis Consortium (CPTAC) and International Cancer Proteogenome Consortium (ICPC) have made great progress in providing high-quality curated multi-omics data, providing a foundation for further methodology development on integrative analysis. Metabolomics is one of the newest and fast-growing techniques in the omics family. MS-based metabolomics shares many similar analytical problems as proteomics data (such as peak calling, imputation, and batch effect correction). Chapter 12 discussed these steps and focused more on describing the underlying statistical models used in the downstream analysis. The chapter also suggested that machine learning methods, such as random forest-based imputation methods, provide a powerful alternative to conventional statistical methods.

The human microbiome, a key interface between the human genome and the environment, is a new omic discipline that is complementary to other "host"-level omic data types. The microbiome has been increasingly recognized as a pivotal regulator of human metabolome, immune response, and disease etiology. Chapter 10 provides a comprehensive technical review of microbiome analysis, in which both bioinformatics procedures for processing metagenomic sequencing and statistical models supporting the subsequent association analyses are covered. Chapter 11 further elaborates on statistical methods for identifying strains in metagenomic samples, which are based on deconvolving single nucleotide variant (SNV) profiles. Microbiome data is inherently compositional, and thus special considerations should be made when comparing species abundance, global diversity, and functional annotations. Regression-based association-testing models also need to account for zero-inflated and skewed distributions. There is a clear need to develop more effective models to facilitate integrative microbiome-host omics analysis. Recently, the neural network architecture has shown promising results in studying microbiome-metabolome dynamics [14].

Largely driven by large-scale GWAS and biobank resources, such as the UK Biobank, disease risk prediction using genetic variants from the entire genome or exome has become increasingly popular in recent years. The score derived, such as the polygenic risk score (PRS), has established its clinical utility for patient classification, risk surveillance, and preventive medicine. Chapter 15 provided an updated overview of statistical methods used in risk prediction, including PRS-based scores, linear mixed model (LMM)-based scores, and penalized regression-based methods. The authors conclude that, in the post-GWAS era, genetic risk modeling can be further improved by incorporating information from other large-scale functional and gene expression databases. Furthermore, the prediction will benefit tremendously from improved phenotype data obtained from electronic health records (EHRs), as these databases become more comprehensive and accessible. It is now more evident that a person's genetic makeup affects not only the predisposition to a particular disease but also how they respond to medications and other therapeutic interventions. Chapter 14 introduced basic concepts and statistical analysis approaches commonly used in pharmacogenomics (PGx), a fast-growing field that aims to discover genetic predictors underlying drug/treatment response and toxicity. Although many of the methods used in typical GWAS can still be applied, PGx data analysis presents unique challenges, as discussed in the chapter. First, the phenotype data in PGx is more complicated in its format, often with longitudinal outcomes, and limited to patients who are eligible for drug treatment. Second, both population-based and cell-line-based studies are leveraged in PGx analysis. Third, genetic data can be limited by specific genotyping platforms used in clinical labs, and the downstream genetic association analysis needs to consider additional strategies to maximize the statistical power. Last and most important, all the designs and analyses in a PGx study need to consider the information that can be acquired and then translated into clinical practice. A coherent conclusion is that, due to the real-world limitations in testing and validation, the integration of other high-throughput omics modalities (such as transcriptomic data) within the framework of PGx analysis will be a promising direction to further facilitate the personalization of treatment regimens.

There are several advanced analytical strategies currently available for multi-omics data integration, using both unsupervised and supervised learning schemes. One of the most widely used integrative algorithms is iCluster [15], which performs joint unsupervised clustering analysis of multi-omics data. iCluster is also able to effectively identify key features contributing to the joint sample clustering result. The algorithm has shown great success in the global molecular classification of more than 10,000 tumors in a recent large-scale pan-cancer study [16]. As elaborated in Chapter 5, the iCluster model can be viewed as a Gaussian latent variable model that decomposes input data matrices (X) into the product of modality-specific weight matrices (W) and a shared factor matrix (Z). The latent variables in Z are thus able to capture the correlative structure of multi-omics data. In solving for W, a lasso-like penalty term is added to the algorithm to induce sparsity and thus shrink the weights of non-informative features towards zero. For various molecular data types, different penalty forms can be incorporated—for example, the elastic net penalty for correlated features and the fused lasso penalty for DNA copy number variation data [17]. The iCluster algorithm has been further improved to model count data, binary data, and other noncontinuous data types in omics [18, 19]. The chapter also noted a few nonparametric methods for integrative clustering that do not rely on distribution assumptions, such as those using nonnegative matrix factorization (NMF). In contrast, the focus of Chapter 2 was more on supervised learning and discussed a series of high-dimensional statistical methods and machine learning approaches for discovering biomarkers and optimized models for prognostic prediction with multi-omics data, where the censored survival outcome is the main target. Modern machine learning techniques, such as gradient boosting and deep neural networks, offer many appealing advantages in handling missing data, high-dimensional features, large sample sizes, and nonlinear relationships. By specifying proper loss functions, many off-the-shelf machine learning software packages are available for accommodating different learning tasks in omics-based biomarker studies. However, certain disadvantages were also noted, namely the increased parameter tuning burden, the lack of transparency in feature selection, and the difficulties in interpreting the results. A converging finding from these discussions is that classic (or parametric) statistical methods and modern machine learning methods are complementary to each other, and thus both should be considered when building a more reproducible multi-omics analysis pipeline.

3 Concluding Remarks

In summary, multi-omics studies have been greatly aided by recent advances in statistical methods and computational algorithms for omics deconvolution and integration. Both supervised and unsupervised machine learning approaches have attractive potential in tackling the analytical challenges in various scenarios discussed above. Without a doubt, omic data analyses will continue to face challenges caused by batch effects, unidentified heterogeneity factors, rapidly growing data volumes, as well as continuously evolving sequencing platforms and the emergence of new data modalities. There is a huge need for new analytical methodology development to effectively and seamlessly integrate data generated from different modalities. Given the data and platform dynamics, building reproducible analytical pipelines is essential to prevent spurious findings caused by sample variability and to promote extensive collaborative research. Together, addressing both cellular and sample heterogeneities explicitly in the analysis will be a key step to improving the generalizability of study-specific findings, forming the basis for future translational applications.

References

1. Zhang Y, Parmigiani G, Johnson WE (2020) ComBat-seq: batch effect adjustment for RNA-seq count data. NAR Genom Bioinform 2(3):lqaa078
2. Johnson WE, Li C, Rabinovic A (2007) Adjusting batch effects in microarray expression data using empirical Bayes methods. Biostatistics 8(1):118–127
3. Newman AM, Liu CL, Green MR, Gentles AJ, Feng W, Xu Y, Hoang CD, Diehn M, Alizadeh AA (2015) Robust enumeration of cell subsets from tissue expression profiles. Nat Methods 12(5):453–457
4. Newman AM, Steen CB, Liu CL, Gentles AJ, Chaudhuri AA, Scherer F, Khodadoust MS, Esfahani MS, Luca BA, Steiner D (2019) Determining cell type abundance and expression from bulk tissues with digital cytometry. Nat Biotechnol 37(7):773–782
5. Yu X, Chen Y, Conejo-Garcia JR, Chung CH, Wang X (2019) Estimation of immune cell content in tumor using single-cell RNA-seq reference data. BMC Cancer 19(1):1–11

6. Luca BA, Steen CB, Matusiak M, Azizi A, Varma S, Zhu C, Przybyl J, Espin-Perez A, Diehn M, Alizadeh AA, van de Rijn M, Gentles AJ, Newman AM (2021) Atlas of clinically distinct cell states and ecosystems across human solid tumors. Cell 184(21):5482–5496. e5428. https://doi.org/10.1016/j.cell.2021.09.014
7. Steen CB, Luca BA, Esfahani MS, Azizi A, Sworder BJ, Nabet BY, Kurtz DM, Liu CL, Khameneh F, Advani RH, Natkunam Y, Myklebust JH, Diehn M, Gentles AJ, Newman AM, Alizadeh AA (2021) The landscape of tumor cell states and ecosystems in diffuse large B cell lymphoma. Cancer Cell 39(10):1422–1437. e1410. https://doi.org/10.1016/j.ccell.2021.08.011
8. Bacher R, Kendziorski C (2016) Design and computational analysis of single-cell RNA-sequencing experiments. Genome Biol 17(1): 1–14
9. Stuart T, Butler A, Hoffman P, Hafemeister C, Papalexi E, Mauck WM III, Hao Y, Stoeckius M, Smibert P, Satija R (2019) Comprehensive integration of single-cell data. Cell 177(7):1888–1902. e1821
10. Chakravarthy A, Furness A, Joshi K, Ghorani E, Ford K, Ward MJ, King EV, Lechner M, Marafioti T, Quezada SA (2018) Pan-cancer deconvolution of tumour composition using DNA methylation. Nat Commun 9(1):1–13
11. Teschendorff AE, Zhu T, Breeze CE, Beck S (2020) EPISCORE: cell type deconvolution of bulk tissue DNA methylomes from single-cell RNA-Seq data. Genome Biol 21(1):1–33
12. Zhu T, Liu J, Beck S, Pan S, Capper D, Lechner M, Thirlwell C, Breeze CE, Teschendorff AE (2022) A pan-tissue DNA methylation atlas enables in silico decomposition of human tissue methylomes at cell-type resolution. Nat Methods 19(3):296–306
13. Yu X, Cen L, Chen YA, Markowitz J, Shaw TI, Tsai KY, Conejo-Garcia JR, Wang X (2022) Tumor expression quantitative trait methylation screening reveals distinct CpG panels for deconvolving cancer immune signatures. Cancer Res 82(9):1724–1735
14. Reiman D, Layden BT, Dai Y (2021) MiMeNet: exploring microbiome-metabolome relationships using neural networks. PLoS Comput Biol 17(5):e1009021
15. Shen R, Mo Q, Schultz N, Seshan VE, Olshen AB, Huse J, Ladanyi M, Sander C (2012) Integrative subtype discovery in glioblastoma using iCluster. PLoS One 7(4):e35236
16. Hoadley KA, Yau C, Hinoue T, Wolf DM, Lazar AJ, Drill E, Shen R, Taylor AM, Cherniack AD, Thorsson V (2018) Cell-of-origin patterns dominate the molecular classification of 10,000 tumors from 33 types of cancer. Cell 173(2):291–304. e296
17. Shen R, Wang S, Mo Q (2013) Sparse integrative clustering of multiple omics data sets. Ann Appl Stat 7(1):269
18. Mo Q, Wang S, Seshan VE, Olshen AB, Schultz N, Sander C, Powers RS, Ladanyi M, Shen R (2013) Pattern discovery and cancer gene identification in integrated cancer genomic data. Proc Natl Acad Sci 110(11): 4245–4250
19. Mo Q, Shen R, Guo C, Vannucci M, Chan KS, Hilsenbeck SG (2018) A fully Bayesian latent variable model for integrative clustering analysis of multi-type omics data. Biostatistics 19(1): 71–86

Chapter 2

Statistical and Machine Learning Methods for Discovering Prognostic Biomarkers for Survival Outcomes

Sijie Yao and Xuefeng Wang

Abstract

Discovering molecular biomarkers for predicting patient survival outcomes is an essential step toward improving prognosis and therapeutic decision-making in the treatment of severe diseases such as cancer. Due to the high-dimensionality nature of omics datasets, statistical methods such as the least absolute shrinkage and selection operator (Lasso) have been widely applied for cancer biomarker discovery. Due to their scalability and demonstrated prediction performance, machine learning methods such as XGBoost and neural network models have also been gaining popularity in the community recently. However, compared to more traditional survival methods such as Kaplan-Meier and Cox regression methods, high-dimensional methods for survival outcomes are still less well known to biomedical researchers. In this chapter, we will discuss the key analytical procedures in employing these methods for identifying biomarkers associated with survival data. We will also identify important considerations that emerged from the analysis of actual omics data. Some typical instances of misapplication and misinterpretation of machine learning methods will also be discussed. Using lung cancer and head and neck cancer datasets as demonstrations, we provide step-by-step instructions and sample R codes for prioritizing prognostic biomarkers.

Key words Cox regression, Survival analysis, Lasso, Elastic net, Gradient boosting, Machine learning

1 Introduction

Survival is a key patient-level outcome for many vital diseases, such as cancer, and it is a well-established efficacy endpoint in oncology trials. Identifying reliable biomarkers that are associated with patient survival can not only inform prognostic assessment but also provide important insights into potential mechanisms underlying disease development and progression. In cancer, for example, many tumorigenesis-associated genes hold both diagnostic and prognostic impacts. Lower gene expression of tumor suppressor genes (such as *TP53* and *CDKN2A*) is often associated with worse patient survival, while lower expression of oncogenes (such as *EGFR* and *MET*) is often associated with better survival. The rapid advancement of high-throughput and cost-effective

Brooke L. Fridley and Xuefeng Wang (eds.), *Statistical Genomics*,
Methods in Molecular Biology, vol. 2629, https://doi.org/10.1007/978-1-0716-2986-4_2,

sequencing technologies allows a huge number of genomic biomarkers to be profiled simultaneously in a single experiment. However, identifying a subset of genes or biomarkers with the most significant prognostic values via a genome-wide scan remains a challenging task due to the high dimensionality and complicated correlations among the markers. In biomedical research, conventional survival analysis methods such as Kaplan-Meier and Cox regression methods have long been the mainstay for examining the prognostic significance of a targeted gene or signature. In general, two main strategies can be considered to circumvent the high feature dimension and low sample size situation (or the so-called "large p, small n" situation). The first strategy is based on hypothesis-driven approaches and selects biomarkers to be tested based on known mechanisms or discoveries derived from prior research. In many scenarios, the purpose of examining the prognostic implication of a targeted gene is to further validate the hypothesized functional role at the population level. The second strategy is to employ high-dimensional regularized regression methods, such as Lasso and elastic net, to simultaneously select features and build predictive models. The data-driven strategy has also been greatly augmented by modern machine learning methods that have been adapted to model censored survival outcomes, such as kernel methods [1, 2], gradient boosting [3], and neural networks [4, 5]. In cancer biomarker research, variations of hybrid approaches are often implemented to combine the above two strategies. For example, it is a common practice to narrow down candidate genes by focusing on genes that exhibit differential expression in tumor tissues as compared to normal samples.

This chapter endeavors to provide an overview and practical guidelines for applying survival regression analysis in prognostic biomarker discovery. Most guidelines are targeted at clinicians, basic scientists, and bioinformaticians who are less familiar with statistical modeling and the theories behind censored survival outcomes. In the following sections, we will first introduce standard survival analysis steps, followed by regularized regression models and newly developed machine learning methods. We will use a publicly available head and neck cancer dataset to demonstrate the discussed analytical steps, and all codes are based on existing R packages.

2 Methods

2.1 Examination of Survival Outcomes

In this chapter, we will only consider right-censored survival data, occurring when the true event (death or progress) time is greater than (to the right of) the censoring time. Therefore, each survival outcome data (also termed time-to-event data) contains a survival time (interval between disease onset to death/follow-up) and a status indicator (event observed or censored). The two types of survival outcomes most studied in cancer research are overall survival (OS) and progression-free survival (PFS). OS is defined as the time from the date of treatment or diagnosis to the date of death or the latest follow-up, and PFS is the time interval from the treatment/diagnosis to disease progression or death or the latest follow-up, whichever occurs first. In some studies, the concept of progression-free interval (PFI) is used instead of PFS to only include death events associated with tumors. Because date-type data is highly prone to human and excel-specific errors, it is highly recommended to apply an initial quality check on the survival data received. For example, one should "red flag" those data points when (1) the start time point occurs later than the second time point and (2) the second time point was applied inconsistently when calculating the time interval. Other key metrics for evaluating the quality of the survival data include (1) the total number of events, which reflects the effective number of sample size in a survival analysis; (2) the average follow-up time period of all patients; and (3) the median survival time, which is the time point at which 50% of the patients are alive or have an event. A Kaplan-Meier for all patients without stratification, as demonstrated in the example R code below, can be used to examine and visualize the metrics mentioned above.

```
##R example code, Section 2##

library(survival)
library(survminer)
KM.fit<- survfit(Surv(time=time, event=status) ~
1, data=lung)
ggsurvplot(KM.fit, data=lung, censor.shape= "|",
risk.table=T, surv.median.line= "hv")
```

2.2 Univariable and Multivariable Cox Regression Analysis Pipeline

In current biomedical literature, univariable survival analysis based on the Kaplan-Meier (KM) approach with log-rank testing (for comparing two patient groups) and multivariable Cox regression for adjusting additional covariates remain the mainstay methods for testing and validating the prognostic significance of a gene or biomarker. The Cox regression estimates (hazard ratios and their confidence intervals) can be visualized by forest plots by applying the `ggforest` function in the `survminer` package or the `forest_model0` function in the `forestmodel` package, as demonstrated in the example code below. Note that a hazard ratio is the exponential of the corresponding regression coefficient, and its magnitude is scale-dependent. Try to compare the hazard ratio estimates when the covariate for the ECOG score is treated as continuous and categorical variables in the Cox model. It is a common practice to split patients based on quartiles of a continuous measure and only compare the upper and lower quartile groups in the KM or Cox models, which often results in a larger effect size than the median-dichotomized approach. Although less popular in biomedical applications, accelerated failure time (AFT) regression can be considered as an alternative to the Cox regression in case the required proportional hazard (PH) assumption is not fulfilled. A graphical verification of the PH assumption can be based on the Schoenfeld residual as implemented in the `ggcoxzph` function in the `survminer` package.

```
##R example code, Section 3##

sex.fit<- survfit(Surv(time=time, event=status) ~ sex, data=lung)
ggsurvplot(sex.fit, data=lung, risk.table=T, surv.median.line= "hv",pval=TRUE)
#Multivariable Cox regression
Fit1<- coxph(Surv(time=time, event=status) ~ sex + ph.ecog, data=lung)
ggforest(Fit1) #or forestmodel::forest_model(Fit1)
lung$ecog.group<- cut(lung$ph.ecog,breaks=c(0,1,3),include.lowest=TRUE)
Fit2<- coxph(Surv(time=time, event=status) ~ sex + ecog.group, data=lung)
#Test PH assumption
test.ph <- cox.zph(Fit1); survminer::ggcoxzph(test.ph)
#AFT as alternative model
AFT.fit<- survreg(Surv(time=time, event=status) ~ sex+ ph.ecog, dist= 'exponential', data=lung); summary(AFT.fit)
```

2.3 Prognostic Biomarker Screening Based on the Lasso and Elastic Net Methods

Regularized regression methods, like Lasso and elastic net, are the de facto standard method for constructing a predictive model when there are more genes/biomarkers than the sample size (or $p > n$ case) [6]. `glmnet` is an R package that is most widely used to perform regularized regression, including regularized Cox models [7]. In order to optimize the model's prediction performance, two key hyperparameters need to be determined by cross-validation: ***lambda*** (λ) and ***alpha*** (α). The regularization parameter ***lambda***, also known as the shrinkage factor, determines how severe the model penalty is (on larger regression coefficients) or how much the coefficients will shrink. The model will become a standard multivariable Cox regression when alpha is set at 0. The function `cv.glmnet` function in `glmnet` is the main function to perform an automatic grid search on lambda and find the value that gives the lowest cross-validation error (*lambda.min*) or the largest value within one standard error of the best model (*lambda.1se*). When selecting the best model, *lambda.1se* is often preferred because it is slightly larger than *lambda.min* and thus controls overfitting better. However, in many real data applications, we found that it can be too stringent for Cox regression. The Cox regression coefficients can be visualized via the generic `plot` function and extracted via the `coef` function, as demonstrated below. The parameter ***alpha*** is the elastic net mixing parameter, controlling how much L1 (Lasso) regularization and L2 (ridge) regularization will be allocated. When *alpha* = 1, the model is equivalent to Lasso. Because the elastic net is capable of providing more robust results when correlation exists among the predictors, it is a highly recommended method for selecting gene expression biomarkers based on whole-transcriptome data. However, the parameter tuning on alpha is currently not supported by `glmnet`. One may install additional packages (e.g., `c060`) [8] which allow for the automatic tuning of both parameters in an efficient manner. Alternatively, the standard multi-hyperparameter tuning strategies based on grid search or random search can be applied.

```
##R example code, Section 4##
library(glmnet); library(survival)
data(CoxExample); x=CoxExample$x; y=CoxExample$y
#glmnet internal dataset
#y<-Surv(time=lung$time, event=lung$status); x<-
as.matrix(lung[,4:5]) #lung data
#fit <- glmnet(x,y, family= "cox", alpha=1) #basic
function
cv.fit <- cv.glmnet(x, y, family = "cox", alpha=1,
nfolds=5, type.measure= "C")
coef(cv.fit, s="lambda.1se") #extract coeffi-
cients at the optimal lambda
coef(cv.fit, s="lambda.min")
```

Since the Lasso and elastic net have been successfully implemented, many derivatives have been developed to further improve the high-dimensional feature selection under different settings. In this section, we introduce two ready-to-implement extended approaches that are more relevant to prognostic biomarker discovery in cancer research. The first approach aims to achieve more stable variable selection based on random subsampling or resampling approaches. Similar to the ideas of random forest and random Lasso [9, 10], only a random subset of samples and features is fit by the regularizing Cox regression in each run. The subsampling/bootstrapping steps will create desirable perturbations that help break highly correlated markers and have the ability to exclude outlier data points. The final importance score and coefficients can be obtained by averaging results from all sampling results. The prioritized features can be visualized using a volcano plot where the x-axis is the averaged regression coefficients and the y-axis is the total time a feature is selected by the regularizing model. A final model can be refit by only including top-prioritized markers in a multivariable Cox regression or by running another adaptive lasso (or elastic net) with feature weights determined by the associated importance scores. The computationally intensive approach provides a powerful strategy to compensate for the limited sample size seen in most cancer studies. In the second approach, we propose to apply network-constrained survival analysis for testing of prognostic significance of a gene set or pathway, by applying the function implemented in packages such as `Coxnet` (https://github.com/yuanjiawang/Coxnet). Briefly, a network constraint based on the Laplacian matrix can be incorporated into the regularized Cox model. The Laplacian matrix measures the connection or similarity between any pairs of genes in a pathway or gene set, which can be formed by normalized Pearson's correlation of gene expression values or external data sources such as the protein-protein interaction (PPI) information from the STRING database. It can be shown that when the Laplacian matrix is an identity matrix, the question reduces to an adaptive elastic network model. The main advantage of the network-constraint model is that it can handle correlated features better than an elastic net, as it encourages smoothness among correlated genes (avoiding "all-in" or "all-out" issues). The adaptive nature of the model also allows it to handle variables more efficiently when their effects have different signs. Finally, the model can integrate prior knowledge of gene relationships and thus will generalize well across different datasets.

2.4 Efficient Gradient Boosting Methods

Although the lasso and elastic net methods have shown success in high-dimensional biomarker screening settings, the search space of the model is limited to linear effects only. Other machine learning methods are needed to explore potential interactions and nonlinear effects. In this chapter, we will focus on two machine learning methods, e.g., gradient boosting and the deep neural network (to be discussed in the next section). A decision tree is one such method that can augment linear effect models, and it is easy to implement and visualize. The major disadvantage of the decision tree is that it is prone to overfitting, and the resulting models and feature cutoffs often do not generalize well to new validation datasets. The gradient boosting method or machine (GBM) [11] was developed to overcome this drawback, in which base learners (e.g., decision tree) are combined to construct an ensemble model that can greatly improve the prediction performance. One key benefit of GBM is that the model is flexible and can be easily extended to different base learners and different loss functions, including the Cox PH model-based loss function for survival outcomes. Different versions of the GBM algorithm have been implemented in multiple popular R packages, most notably "gbm" and "caret." Extreme gradient boosting (XGBoost [12] or XGB) is a more recent and improved version of GBM that has quickly gained popularity because of its enhanced computational efficiency and superior performance in many applications. The key differences between the XGB and the original GMB algorithm are that (1) XGB is optimized based on both the first and the second derivatives of the loss function; (2) XGB introduces extra penalty terms for the leaf weight and the number of terminal nodes, which significantly mitigate the overfitting issue; and (3) when using a regression tree as the base learner, XGB provides an extremely efficient framework to calculate and update the gain function for each split. Therefore, the XGB algorithm can be easily extended to the Cox model by plugging in the first derivative (gradient) and second derivative of the Cox partial likelihood function (as the loss function). LightGBM is a similar but more efficient implementation, which further optimizes the memory usage and computational speed by constructing the tree using strategies such as the left-wise growth strategy and the gradient-based one-sided sampling [13]. Both the XGB and LightGBM algorithms for the survival outcomes are implemented in the "Xsurv" package, recently developed by our group. The example R code is provided below.

```
##R example code, Section 5##
  # devtools::install_github("topycyao/Xsurv")
#install Xsurv package
  library(Xsurv);library(survival);library
(glmnet);library(rpart);
  data(CoxExample); x=CoxExample$x; y=CoxExample
$y; #glmnet internal dataset

  xgb.fit<-Xsurv(x,y,option='xgb',top_n=5);
#xgboost fitting
 model.xgb<-xgb.fit$mod; #extract xgboost model
  xgb.importance(colnames(as.data.frame(x)),mod-
el=model.xgb) #feature importance
 xgb.fit$SHAP #SHAP importance plot from xgboost

 #pipeline for Lightgbm
  lgb.fit<-Xsurv(x,y,option='lgb',top_n=5);
#lightgbm fitting
 model.lgb<-lgb.fit$mod; #extract lgb model
  lgb.importance(model.lgb);#feature importance
from lgb model
 lgb.fit$SHAP; #SHAP importance plot from lgb model
```

Similar to other machine learning methods, hyperparameter tuning is an essential step in model training and optimization. There are three main groups of hyperparameters to be tuned in XGB-like algorithms: (1) basic parameters that define the base learner (decision tree or linear models); (2) a series of booster parameters such as tree numbers, the maximum tree depth, learning rate, and other regularization terms involved; and finally (3) specific learning task parameters that define the object and evaluation functions. Since the parameters in the first and the last group are relatively fixed for a specific learning task, most tuning efforts tend to concentrate on the second group, i.e., booster parameters. Therefore, the parameter tuning in XGB is much more involved than in the elastic net, in which only two parameters need to be optimized. A grid search over all combinations of booster parameters will quickly become computationally intractable as the number of candidate values grows. The R package Xsurv provides a cross-validation function (Xsurv.cv) to automatically optimize a subset of key parameters based on either the grid search or the random search method. Nevertheless, we remark that there exists no uniformly optimal form of tuning that fits all data scenarios, and often a hybrid form of search strategy is applied. To achieve a more efficient tuning, we recommend the following general guidelines. First, we suggest a first run based on the default parameters set by a software or package and use the generated results as the benchmark. Second, try a quick grid search by expanding the key parameters around the default parameters and determining the model sensitivity in each parameter space. Third, it is recommended to tune the number of boosting rounds (i.e., the number of trees), followed by other booster parameters, and finally the regularization parameters. Finally, it is common practice to fix a reasonable learning rate in the beginning (for example, at 0.1) and then adjust based on the initial learning curve and prediction performance.

In addition to the Cox-based model, the Xsurv package also implements a boosting algorithm based directly on using the C-index as the objective function, where the first and second derivatives of the C-index loss functions are approximated by a smoothed concordance index. The C-index-based boosting will provide a robust alternative to the Cox model, especially when the proportional hazard assumption was not strictly met in the practical survival analysis. After the model is learned, the top predictive features can be visualized by the chart of the feature importance score, which is a measure of how much gain in prediction performance was achieved due to selecting this feature in the final boosted tree. Alternatively, a Shapley Additive Explanations (SHAP) value plot can also be used to explore the contribution of each predictor, which further facilitates model interpretation and feature prioritization.

2.5 Deep Learning Methods

Deep neural network (NN)-based algorithms have received considerable attention in recent years due to rapid advances in deep learning technologies and the development of easy-to-implement libraries such as Keras and PyTorch. In a typical NN, the complex nonlinear model between features and the final outcome is formed by multilayer perceptrons, which are basically mathematical functions connecting all input and output nodes. In each node, all the input signals are combined by a simple linear function followed by a nonlinear transformation based on the activation function like ReLU or, classically, the sigmoid function. The NN model can be easily extended to the survival outcome by incorporating the Cox PH loss into the final layer [5]. The determination of the NN architecture is complex and out of the scope of this chapter. Overall, there are two key hyperparameters that control the topology of a typical NN, i.e., the number of hidden layers and the number of nodes in each layer. The number of layers directly controls the model's complexity. The NN with more layers will cause a greater number of parameters to be estimated, more computational burden, and a higher risk of overfitting. While more layers are often used in imaging data analysis, we do not recommend including more than three layers for genomic data in the initial training. The geometric pyramid rule [14] is often used to determine the number of nodes, in which the number of nodes in a middle hidden layer is approximately the square root of the product of the number of nodes in the input and output layers. Additional dropout layers and L1/L2 regularization are often useful components in the architecture of NN to control overfitting. For example, the NN in the DeepSurv algorithm is constructed by stacking multiple fully connected layers and Dropout layers. It is also important to understand two concepts (or hyperparameters) that are unique to NN models: batch size and epoch. Batch size (greater than 1 and smaller than sample size n) is the number of samples to be included in one iteration of training before all model parameters are updated. A smaller batch size has a lower memory requirement and usually yields results that generalize better, but the trade-off is having slow convergency and unstable results. One epoch is one cycle of training across the entire training dataset, i.e., all samples are passed through the NN once. Without early stopping, the optimum epoch should be the one that generates the lowest validation loss. As an important monitoring tool, most API or deep learning software will output the updated loss values or performance gains after every epoch. The plot of loss over epochs basically forms the "learning curve" of the deep learning model. Based on the behavior of the learning curves, we can determine an optimal learning rate parameter, which is arguably the most critical NN-specific hyperparameter to tune when the NN architecture is fixed. We should reduce the learning rate if the loss over the first few training epochs is not decreasing (or oscillating). As a robust alternative, it is always helpful to consider different loss functions that are not based on the Cox PH model, such as the smoothed index function discussed in the previous section.

In general, we recommended deep learning methods as the primary choice in two scenarios: (1) The training sample size is sufficiently large. Given the model complexity and much-expanded space in the hyperparameter tuning, the NN models generally require more sample size to achieve a prediction performance superior or comparable to other machine learning methods. (2) The training data involves imaging or cross-modality data, where the convolutional neural network (CNN) and stacked NN methods are the most suitable modeling approaches.

```
##R example code, Section 6##
library(survivalmodels);library(survival);library(glmnet);
data(CoxExample); train_x=CoxExample$x[1:750,]; train_y=CoxExample$y[1:750,]; train_data<-as.data.frame(cbind(train_x,train_y));
test_x=CoxExample$x[751:1000,]; test_y=CoxExample$y[751:1000,];
test_data<-as.data.frame(cbind(test_x,test_y));
#install_pycox(pip = TRUE, install_torch = FALSE);# if pycox not installed
dp.fit<- deepsurv(data=train_data,frac=0.2,activation='relu',num_nodes=c(4,8,4,2),dropout=0.1, epochs=50);
predict(dp.fit,newdata=test_data,type='risk'); #predict relative risk
```

3 Conclusions

This chapter presented a brief overview of existing statistical and machine learning methods for predicting survival outcomes. We focus on practical applications (with an R code example) and implementation issues rather than theoretical aspects. Modern machine learning methods not only offer appealing advantages in handling high-dimensional and nonlinear relationships but also require more parameter tuning burden. It is therefore expected that machine learning methods with less carefully tuned hyperparameters may lead to even worse performance than the standard Cox regression model. Among the methods discussed above, the deep learning method requires the most parameter tuning efforts, followed by gradient boosting methods, while the lasso and elastic net regression involves the least tuning effort (*lambda* and *alpha*). One unique advantage of lasso-based regression models is that they can fit a model and perform feature selection simultaneously. The results generated from boosting and deep learning methods are less transparent but feature important scores that can be used to determine the most influential variables. Additional visualization tools can be considered to improve the interpretability of the fitted prognostic models, such as a survival decision tree fitted based on top-ranked features (as implemented in the Xsurv package). In summary, we conclude that the survival (especially Cox-based) machine learning methods and their implementations are mature and well established and could play a more prominent role in the future prognostic model discovery.

References

1. Van Belle V, Pelckmans K, Suykens JA, Van Huffel S, Survival SVM (2008) A practical scalable algorithm. In: ESANN, 2008. Citeseer, p 94
2. Wilson CM, Li K, Sun Q, Kuan PF, Wang X (2021) Fenchel duality of cox partial likelihood with an application in survival kernel learning. Artif Intell Med 116:102077
3. Li K, Yao S, Zhang Z, Cao B, Wilson CM, Kalos D, Kuan PF, Zhu R, Wang X (2022) Efficient gradient boosting for prognostic biomarker discovery. Bioinformatics 38(6): 1631–1638
4. Steingrimsson JA, Morrison S (2020) Deep learning for survival outcomes. Stat Med 39(17):2339–2349
5. Katzman JL, Shaham U, Cloninger A, Bates J, Jiang T, Kluger Y (2018) DeepSurv: personalized treatment recommender system using a Cox proportional hazards deep neural network. BMC Med Res Methodol 18(1):1–12
6. Zou H, Hastie T (2005) Regularization and variable selection via the elastic net. J R Stat Soc Ser B Stat Methodol 67(2):301–320
7. Simon N, Friedman J, Hastie T, Tibshirani R (2011) Regularization paths for Cox's proportional hazards model via coordinate descent. J Stat Softw 39(5):1
8. Sill M, Hielscher T, Becker N, Zucknick M (2015) c060: extended inference with lasso and elastic-net regularized Cox and generalized linear models. J Stat Softw 62:1–22
9. Wang S, Nan B, Rosset S, Zhu J (2011) Random lasso. Ann Appl Stat 5(1):468
10. Kim S, Baladandayuthapani V, Lee JJ (2017) Prediction-oriented marker selection (PROMISE): with application to high-dimensional regression. Stat Biosci 9(1):217–245
11. Friedman JH (2001) Greedy function approximation: a gradient boosting machine. Ann Stat 29:1189–1232
12. Chen T, Guestrin C (2016) XGBoost: a scalable tree boosting system. In: Proceedings of the 22nd ACM SIGKDD international conference on knowledge discovery and data mining (KDD'16), vol 785. San Francisco, CA, p 794
13. Ke G, Meng Q, Finley T, Wang T, Chen W, Ma W, Ye Q, Liu T-Y (2017) Lightgbm: a highly efficient gradient boosting decision tree. Adv Neural Inf Proces Syst 30:3146–3154
14. Masters T (1993) Practical neural network recipes in C++. Academic Press Professional, Inc., Boston

Chapter 3

Cell-Type Deconvolution of Bulk DNA Methylation Data with EpiSCORE

Tianyu Zhu and Andrew E. Teschendorff

Abstract

DNA methylation data generated from bulk tissue represents a mixture of many different cell types. Variation in the cell-type composition of tissues is thus a major confounder when inferring differential DNA methylation. Due to the high cost of single-cell methylome sequencing, computational methods that can dissect the cell-type heterogeneity of bulk DNA methylomes offer an efficient and cost-effective solution, especially in the context of large-scale EWAS. In this chapter, we present a step-by-step tutorial of Epigenetic cell-type deconvolution using Single-Cell Omic References (EpiSCORE), a reference-based method that leverages the high-resolution nature of single-cell RNA-Seq datasets to facilitate microdissection of bulk-tissue DNA methylomes.

Key words DNA methylation, EWAS, Cell-type heterogeneity, Single-cell RNA-Seq, Cell-type deconvolution

1 Introduction

Most DNA methylation (DNAm) data is generated from bulk tissue, which comprises a variety of different cell types in variable proportions. Because DNAm is highly cell-type-specific, variations in cell-type composition between samples (i.e., individuals) and phenotypes/exposures can therefore confound analysis. Often, it will lead to inflated signals (high false positive and discovery rate). For example, an EWAS in rheumatoid arthritis (RA) by Liu et al. showed that there is notable reduction of RA-associated differentially methylated cytosines (DMCs) after adjusting for cell-type heterogeneity [1]. Alternatively, variation in cell-type composition which is unrelated to the phenotype of interest can lead to deflation (low power or high false negative rate). For example, an epigenome study performed in cervical smears to detect DMCs associated with

Brooke L. Fridley and Xuefeng Wang (eds.), *Statistical Genomics*,
Methods in Molecular Biology, vol. 2629, https://doi.org/10.1007/978-1-0716-2986-4_3,

prospective cervical neoplasia is hampered by the large variations in the contaminating immune-cell fraction, thus requiring alternative statistical models to improve sensitivity [2, 3].

Cell-type heterogeneity can also hamper the identification of cell-type-specific signals. In relation to this, it is worth stressing that single-cell methylome sequencing technology has not been widely used, due to the sparsity of the data and the high prohibitive cost to generate such data in hundreds or thousands of samples [4]. Thus, to address the cell-type heterogeneity challenge in the EWAS context, a number of reference-based and reference-free cell-type deconvolution methods have been developed, with reference-based methods offering the greatest potential [5]. Reference-based methods such as constrained projection (CP) [6], CIBERSORT [7], and robust partial correlations (RPC) [8] require a priori-defined reference DNAm matrix containing representative DNAm profiles of different cell types within the tissue of interest. Typically, this DNAm reference matrix is only necessary for a group of informative CpGs displaying highly variable and discriminatory DNAm levels across the different cell types. These references have been generated for tissues such as whole blood [9], cord blood [10], prefrontal cortex [11], and breast [12, 13]. Given the references, the algorithm assumes that each bulk DNAm profile is a weighted linear sum of these reference profiles, with the weights being the relative cell-type fractions. The algorithms estimate the weights based on a multivariate linear regression framework, usually with a least-squares optimization procedure. Constraints on the weights can be imposed either during or after the estimation, since cell-type fractions have to be nonnegative and add to a number less than or equal to 1. The estimated cell-type fractions can then be used as covariates in supervised models, calibrating the model and correcting the bias caused by cell-type heterogeneity.

Most existing DNAm references have been generated for tissues like blood, where experimental purification of individual cell types is feasible. But for solid tissues like skin and liver, it is technically very challenging to generate DNAm profiles for purified samples representing all major cell types within a tissue. Epigenetic cell-type deconvolution using Single-Cell Omic References (EpiSCORE) is a publicly available software designed for high-resolution cell-type deconvolution of bulk DNA methylomes derived from complex solid tissues [14]. EpiSCORE generates DNAm reference profiles by translating them from scRNA-Seq data with a predictive semi-Bayesian model. In addition, EpiSCORE provides a DNAm-atlas of tissue-specific references defined for 13 solid tissue types and 40 cell types [15], publicly available from https://figshare.com/projects/EpiSCORE-atlas_version-1_/111473. In this chapter, we will introduce how EpiSCORE works and how to use it. This step-by-step tutorial will cover the construction of the DNAm reference matrix as well as applications to real datasets, using skin and brain as illustrative examples.

2 Materials

2.1 Software and Hardware Required

- A standard desktop or laptop computer with Windows, macOS, or Linux operating system. We recommend at least 16 GB of RAM.
- R statistical computing environment (version 4.0.0 or later) from The Comprehensive R Archive Network (https://cran.r-project.org/).

2.2 Installation of EpiSCORE R Package

1. The EpiSCORE package can be downloaded from GitHub (https://github.com/aet21/EpiSCORE). A number of packages need to be installed beforehand. Below is the R code for quick installation:

```
install.packages("MASS")
BiocManager::install(c("AnnotationDbi", "EpiDISH", "org.Hs.eg.db"))
remotes::install_github("immunogenomics/presto")
remotes::install_github("aet21/EpiSCORE")
```

2.3 Single-Cell RNA-Seq Datasets Used in Case Study

Skin scRNA-Seq data for reference construction: The skin scRNA-Seq data for reference construction can be downloaded from https://figshare.com/articles/dataset/Skin_sample_data_for_EpiSCORE_tutorial/19312007. The file is named *Skin_scRNA-seq_construction.rda*. This is the processed dataset from Solé-Boldo et al. [16]. It includes a count matrix (*skin.m*) and a vector of cell-type annotation (*skin.label*). The scRNA-Seq data has been normalized to log scale. The cells were classified into 2981 endothelial cells (EC), 5948 fibroblasts (Fib), 1339 differentiated keratinocytes (Kera_diff), 1187 undifferentiated keratinocytes (Kera_undiff), 2228 macrophages (Macro), 123 melanocytes (Mela) and 1281 T cells (Tcell).

Dataset overview:

```
dim(skin.m)
## [1] 21352 15147
table(skin.label)
## skin.label
##         EC        Fib   Kera_diff Kera_undiff      Macro       Mela
##       2981       5948        1399        1187       2228        123
##      Tcell
##       1281
```

Skin scRNA-Seq data for reference validation: The skin scRNA-Seq data for reference validation can be downloaded from the same link above. The file is named *Skin_scRNAseq_validation.rda*. The dataset is log-normalized. It includes a count matrix (*sampledata.m*) and a vector indicating cell types (*samct.v*). These cells (except melanocytes) are randomly selected from the original dataset from Deng et al. [17], which consist of 500 endothelial cells, 500 fibroblasts, 500 keratinocytes, and 500 immune cells. The 178 melanocytes are all the melanocytes from the original dataset.
Dataset overview:

```
1.dim(sampledata.m)
2.## [1] 23629  2178
3.table(samct.v)
4.## samct.v
5.##  Endothelial   Fibroblast  Immune_cell Keratinocyte  Melanocytes
6.##          500          500          500          500          178
```

2.4 DNAm Datasets Used in Case Study

Skin dermis and epidermis DNAm dataset: The processed DNAm data of skin dermis and epidermis samples [18] can be downloaded from the same link above. The file is named *Skin_DNAm_validation.rda*. This is a 450k dataset with 40 dermis and 38 epidermis samples. It includes a normalized beta matrix (*AF.bmiq*) and a data frame of phenotype information (*AF.pheno*).
Dataset overview:

```
1.dim(AF.bmiq)
2.## [1] 485509     78
```

Skin cutaneous melanoma (SKCM) TCGA DNAm dataset: The processed DNAm data of TCGA SKCM samples [19] can be downloaded from the same link above. The file is named *Skin_DNAm_SKCM.rda*. This is a 450k dataset with two normal skin, 104 primary SKCM, 369 metastatic SKCM samples. It includes a normalized beta matrix (*skcmbmiq.m*) and a data frame of phenotype information (*PhenoTypesTCGA_SKCM.df*).

Dataset overview:

```
dim(skcmbmiq.m)
## [1] 395692    475
```

Purified brain NeuN$^+$ and NeuN$^-$ DNAm samples: The processed DNAm data of purified dorsolateral prefrontal cortex samples [11] can be downloaded from the same link above. The file is named *Brain_DNAm_purified.rda*. This is a 450k dataset with 29 NeuN$^+$ and 29 NeuN$^-$ samples. It includes a normalized beta matrix (*brainfl.m*) and a vector of sample type (*brainfl_ct.v*), where "N" indicates neuron (NeuN$^+$) and "G" indicates glia (NeuN$^-$).
Dataset overview:

```
dim(brainfl.m)
## [1] 485165    58
table(brainfl_ct.v)
## brainfl_ct.v
##  G  N
## 29 29
```

Bulk brain DNAm samples from schizophrenia EWAS: The processed DNAm data of prefrontal cortex samples [20] can be downloaded from the same link above. The file is named *Brain_DNAm_SZ.rda*. This is a 450k dataset with 240 schizophrenia samples and 191 controls. It includes a normalized beta matrix (*jaffebeta.m*), a data frame of phenotype information (*phenotype.df*), and a cell-type fraction matrix (*estbrJ.m*) estimated by EpiSCORE.
Dataset overview:

```
dim(jaffebeta.m)
## [1] 473536    431
```

3 Methods

3.1 Overview of EpiSCORE

We display an overview of EpiSCORE algorithm (Fig. 1) and EpiSCORE DNAm atlas (Fig. 2) to help readers get basic ideas of EpiSCORE.

3.2 Construction of a Tissue-Specific scRNA-Seq Reference Matrix

Before constructing the reference, the scRNA-Seq count matrix should be processed with basic quality control steps and normalized to a log scale, with the cells annotated to known cell types. At this stage, we need to decide on the cell-type resolution that we desire, i.e., the number of distinct cell types for which reference profiles will be built. Note that there is always a compromise between cell-type resolution and robustness of the reference matrix, as it's more challenging to discriminate cell types originating from similar lineages (*see* **Note 1**).

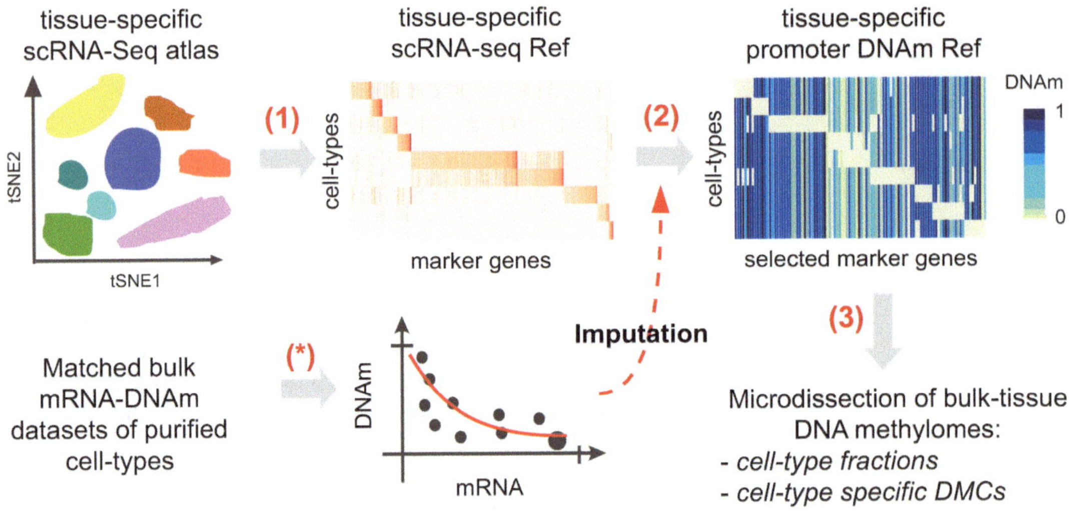

Fig. 1 EpiSCORE algorithm. (1) First step in EpiSCORE is the construction of a tissue-specific mRNA expression matrix from scRNA-Seq data. The expression reference matrix is defined over a set of marker genes that are differentially expressed between cell types. (*) A semi-Bayesian predictive model constructed with independent matched bulk RNA-Seq and DNAm data of purified samples from the Epigenomics Roadmap and SCM2, allows genes to be identified for which differential promoter DNAm across samples is predicted by corresponding gene expression variation. (2) Imputation of DNAm reference by applying the model learned in (*) to the scRNA-Seq reference matrix, weighting the marker genes according to how well the imputed DNAm values reflect gene expression. (3) Using the imputed DNAm reference, we can now estimate proportions for the corresponding cell types in a bulk DNAm profile and identify cell-type specific DNAm changes.
Note: Users can apply available EpiSCORE DNAm references to bulk-tissue DNAm datasets directly with step 3 or perform steps 1 and 2 first to generate their own reference. Steps 1, 2, and 3 are implemented as EpiSCORE functions (Figs. 1 and 2). The predictive model in step (*) is programmed into the EpiSCORE functions, and users do not need to do this step in practice

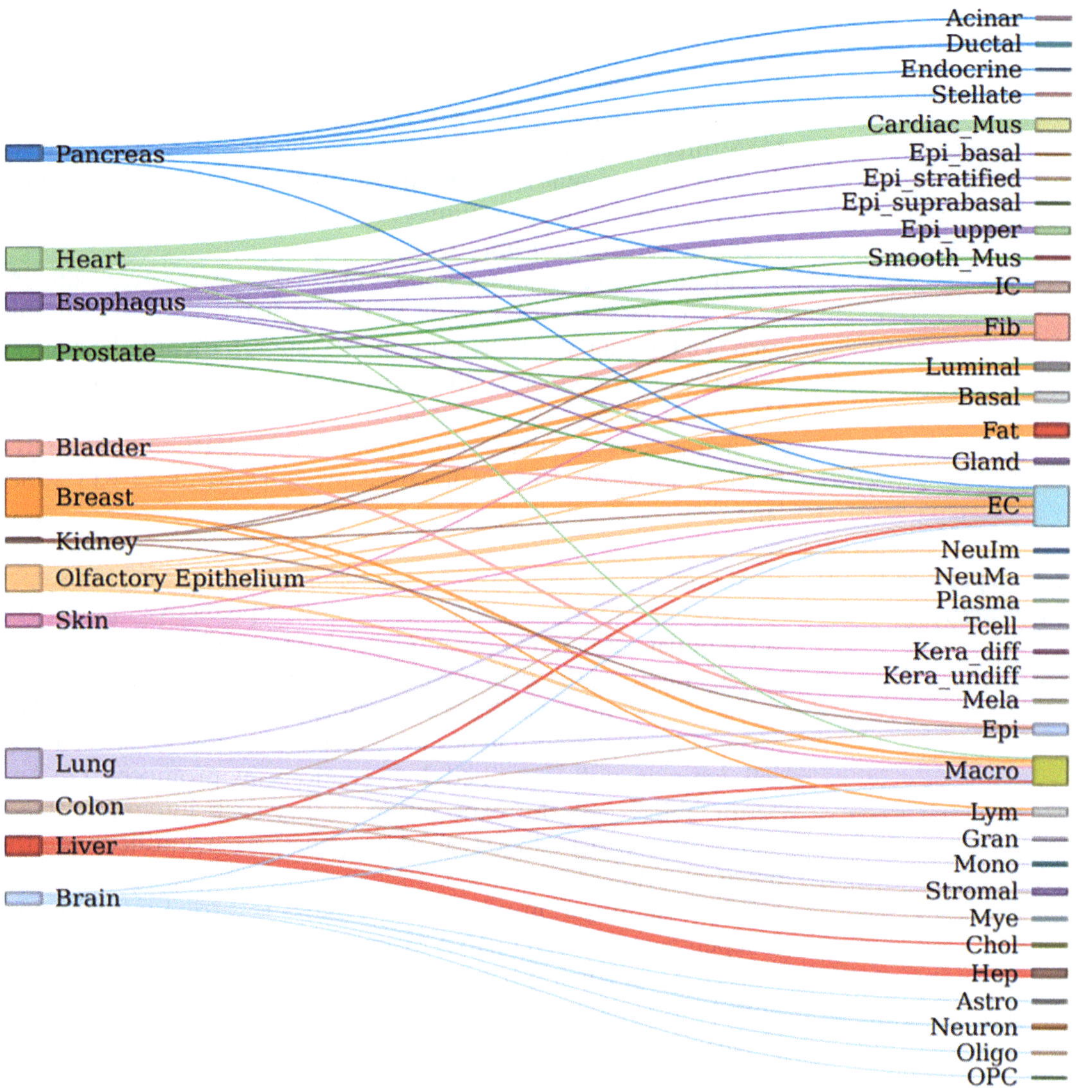

Fig. 2 EpiSCORE DNAm atlas. Sankey plot showing EpiSCORE DNAm atlas over 13 solid tissues. Each link represents a cell type (right) which exists in the corresponding tissue-specific reference (left). The width of the link indicates the number of DNAm markers of the given cell type in the given tissue-specific reference. The color of the link indicates the tissue type

EpiSCORE selects marker genes by performing differential expression (DE) between a given cell type and all others with the Wilcoxon Rank Sum Test. For each cell type, the candidate marker genes are those satisfying the following conditions: (i) a false discovery rate (FDR) < 0.05, (ii) the highest and nonzero median expression is in the given cell type, (iii) and a marker specificity score (MSS) of K-1. The MSS of a given marker gene in a given cell type is defined as the number of other cell types (maximum value is therefore K-1) for which the median expression in each of these cell types is zero. The above selection procedure guarantees that marker genes are statistically significantly overexpressed in the given cell type whilst maximizing their cell-type specificity. However, the MSS threshold could be relaxed to a value less than K − 1 if not enough marker genes are found (*see* **Note 2**). Indeed, it is important to ensure a sufficient number of marker genes in the expression reference for the next imputation steps.

Finally, we remark that EpiSCORE will construct two scRNA-Seq reference matrices, one using the median over all cells within each cell type and another using the average. In theory, the average should be preferred since the experimentally measured bulk tissue DNAm value is an experimental average over cells. However, the median-based reference is more robust, and importantly, it is statistically justified if cell-type fractions are inferred using RPCs (see later).

3.3 Imputation of DNAm Reference

Given the scRNA-Seq expression reference matrix, the next step is to impute a DNAm reference matrix with a predictive model. The key rationale underlying the model is that for a subset of the marker genes, variation in expression between cell types can be well explained by the corresponding variation in promoter DNAm. More generally, we call genes for which expression variation across cell types is well described by variations in promoter DNAm, "imputable," and the expectation is that such imputable genes would be enriched for marker genes as identified in the previous step.

The identification of imputable genes is carried out with two independent matched mRNA expression and DNAm datasets. One dataset is sequencing-based and is derived from the NIH Epigenomics Roadmap [21, 22], consisting of bulk RNA-Seq and WGBS/MeDIP-Seq DNAm data for a common set of 45 samples. The other matched dataset is array-based, is derived from the Stem-Cell Matrix Compendium-2 (SCM2) [23], and consists of Illumina BeadArray mRNA expression and 450k DNAm profiles for a common set of 34 samples. Both datasets represent a wide variety of cell and tissue types. The DNAm profiles were averaged at the promoter level and significant associations between mRNA expression and DNAm identified with Pearson correlation coefficients. The sets of significantly anticorrelated genes are precomputed and stored in the EpiSCORE package.

The imputation model works as follows: For the anticorrelated genes, high expression is associated with low or near-zero promoter DNAm levels, so for a sample with high expression of the given gene, EpiSCORE would impute a promoter DNAm value of zero. For a sample where the gene is not expressed, promoter DNAm is less informative (since the silencing could be caused by repressive histone marks), and additional work is required to predict the likely DNAm level. Prior to imputation, a two-state mixture of gamma distributions is fitted to all the expression values in the given matrix (SCM2 and RMAP) in order to learn the parameters for the unexpressed and expressed states. Parameters of the gamma distribution and the mixing weights were learned with an EM algorithm [24]. The posterior probabilities that a gene is expressed can be computed as:

$$p(E_{gs}=1|x_{gs}) = p\left(x_{gs}|\widehat{a}^{(E)},\widehat{s}^{(E)}\right)\widehat{\pi}^{(E)}/p(x_{gs})$$

where $p\left(x_{gs}|\widehat{a}^{(E)},\widehat{s}^{(E)}\right)$ is a gamma distribution with estimated parameters $\widehat{a}^{(E)},\widehat{s}^{(E)}$, $\widehat{\pi}^{(E)}$ is the estimated mixing weight for the expressed state, x_{gs} denotes the expression value of gene g in sample s, and where E_{gs} denotes a binary random variable with one indicating expressed state and zero the non-expressed state. EpiSCORE would then calculate the median DNAm level over all the samples where a gene is not expressed ($p(E_{gs} = 1 | x_{gs}) < 0.2$) and use that value as the imputed promoter DNAm level. If for a given marker gene, no sample passes the $p(E_{gs} = 1 | x_{gs}) < 0.2$ threshold, there are alternative ways to impute the DNAm-value [14], but they are not yet implemented in the current version of EpiSCORE.

In addition to the imputation, EpiSCORE also defines a weight for each marker gene as the average imputed DNAm value over the cell types where the marker gene is not expressed. Thus, for highly informative marker genes, where DNAm is close to one in the cell types where the gene is not expressed, the weight is also close to one, whereas for uninformative marker genes, the weight is close to zero. These weights are used as "quality scores" when estimating cell-type fractions.

3.4 Estimation of Cell-Type Fractions and Identification of Cell-Type-Specific Differential DNAm

Given the imputed DNAm reference matrix and a bulk DNAm profile summarized at the promoter level, we can run a weighted robust linear multivariate model with Huber's robust M estimator [25] by minimizing the following function:

$$\min_{\vec{f}}\left\| W^{\frac{1}{2}}\left(\vec{y} - M^{(R)}\vec{f}\right)\right\|$$

where $\vec{y}$ is the DNAm values of a bulk sample over the genes in the DNAm reference and $M^{(R)}$ and W denote the DNAm reference and quality weight matrices, respectively. We note that W is by definition a diagonal matrix. The above optimization can be viewed as performing robust partial correlations. The vector $\vec{f}$ which minimizes the above objective function represents the estimated cell-type fractions in the given bulk sample. Because the above optimization does not necessarily yield nonnegative regression coefficients, EpiSCORE imposes the nonnegativity constraint a posteriori, i.e., negative estimated regression coefficients are set to zero and the rest are scaled to ensure their sum adds up to one, as required for cell-type fractions. The imposition of the nonnegativity and normalization constraints after the inference of the regression weights has been shown to work extremely well also in the gene-expression context [7]. Importantly, this procedure also justifies the use of median DNAm values in the DNAm reference matrix.

Besides estimating cell-type fractions, in the context of EWAS we are often interested in identifying cell-type-specific differential methylation. We can achieve this using CellDMC algorithm [8, 26]. Briefly, CellDMC runs the following linear model:

$$\vec{y_c} = \sum_{k=1}^{K} \mu_{ck} \vec{\hat{f}_k} + \sum_{k=1}^{K} \beta_{ck}^{(I)} \vec{\hat{f}_k} * \vec{z} + \vec{\varepsilon}$$

where $\vec{y_c}$ denotes the vector of DNAm values of CpG c across all bulk samples, $\hat{f}_k$ denotes the corresponding vector of cell-type fraction estimates for cell type k across all bulk samples, $\vec{z}$ denotes the phenotype of interest, μ_{ck}, $\beta_{ck}^{(I)}$ are regression coefficients to be estimated, and * denotes the interaction term, and where we assume that errors are Gaussianly distributed with a variance that may depend on the specific CpG c. The regression coefficients $\beta_{ck}^{(I)}$ inform us whether there is a significant interaction between the phenotype and the corresponding fraction for cell type k. If significant, this can be seen as evidence that DNAm of this CpG may change in relation to the phenotype especially in the corresponding cell type k. We note that the interpretation of identified cell-type-specific DNAm changes can be complicated if the mean and variance of cell-type fractions vary a lot between cell types. We refer the reader to You et al. [27] for simulated estimates of sensitivity, precision, and specificity.

3.5 Case Study

Skin references construction and validation: Load *Skin_scRNA-seq_construction.rda* into R, and assign an index to each of the cell types. In this case, there are seven skin cell types (endothelial cells, fibroblasts, differentiated keratinocytes, undifferentiated keratinocytes, macrophages, melanocytes, and T cells), and the index would be one to seven, respectively.

Run EpiSCORE to construct the skin scRNA-Seq reference with MSS = (3,3,3,3,3,6,3). We choose this MSS threshold because it yields a sufficient number of marker genes for each cell type and at the same time guarantees the maximum specificity of the markers.

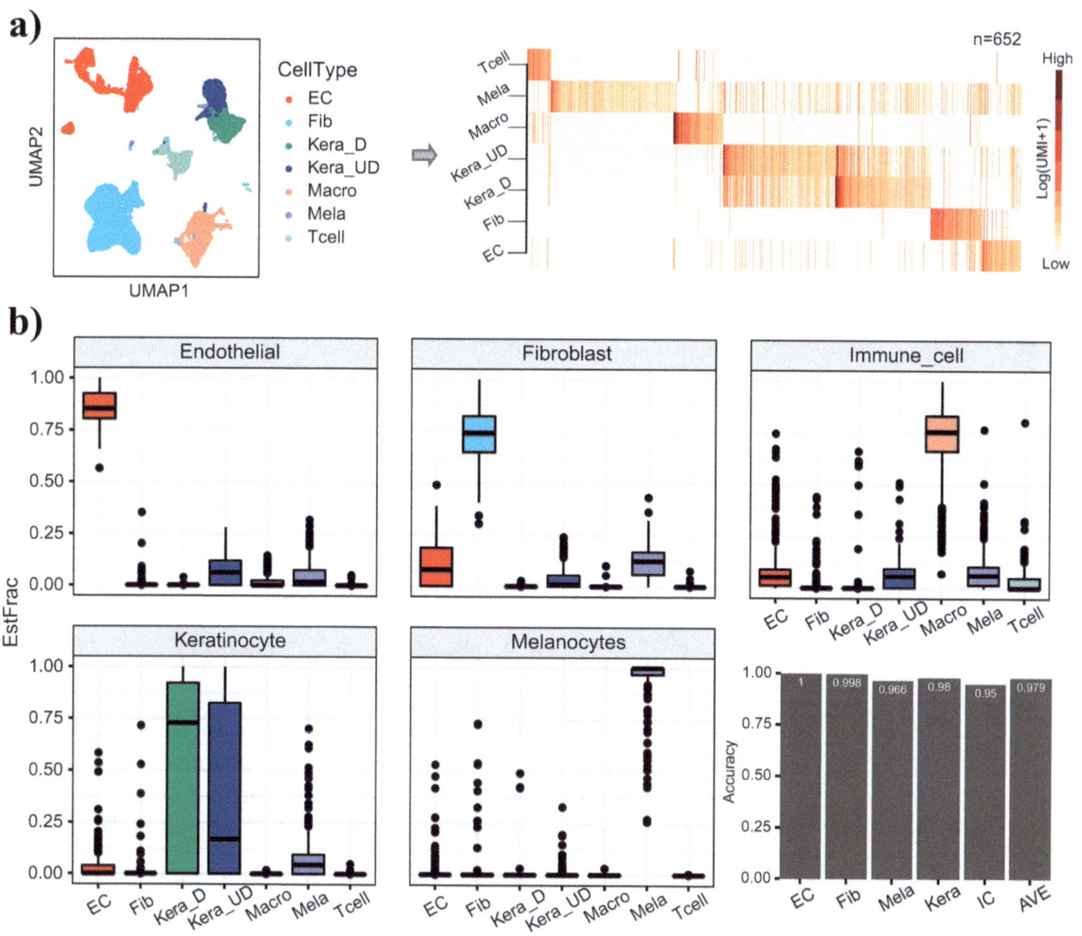

Fig. 3 Construction and validation of skin reference. (**a**) tSNE diagram and the scRNA-Seq expression reference matrix over 7 main skin cell subtypes and 652 marker genes. (**b**) Boxplots of estimated cell type fractions for the independent validation set. Cell-type fractions were estimated using RPC with the scRNA-Seq expression reference given in (**a**). Bottom right: Barplot displaying the overall accuracy of the classification. A cell is accurately predicted if the annotated cell type has the highest estimated fraction among all the cell types in the reference. Due to the difference in the annotation of datasets, the estimated keratinocyte fraction in the validation set is defined as the sum of differentiated keratinocytes (Kera_D) fraction and undifferentiated keratinocytes (Kera_UD) fraction; the estimated immune cell fraction in the validation set is defined as the sum of macrophages (Macro) fraction and T cells fraction

```
expref.o <-
    ConstExpRef(skin.m,skin.idx,alltypes,markspecTH=c(3,3,3,3,3,6,3))
```

This resulted in a 652-marker expression reference (Fig. 3a), which includes 53 endothelial, 67 fibroblast, 127 differentiated keratinocyte, 148 undifferentiated keratinocyte, 66 macrophage,

160 melanocyte, and 31 T cell-specific marker genes. The following shows the first ten rows of the median-based expression reference. These markers are specifically expressed in T cells.

```
expref.m <- expref.o$ref$med
head(expref.m)
##          EC Fib Kera_diff Kera_undiff     Macro Mela    Tcell
## CD52      0   0         0           0 0.0000000    0 2.722037
## IL32      0   0         0           0 0.0000000    0 2.776347
## CD3D      0   0         0           0 0.0000000    0 1.854410
## ARHGDIB   0   0         0           0 0.7630141    0 2.192969
## CXCR4     0   0         0           0 0.0000000    0 1.921457
## CD69      0   0         0           0 0.0000000    0 1.75998
```

We now validate the expression reference in the independent skin scRNA-Seq dataset (*Skin_scRNAseq_validation.rda*). There are many reference-based methods that can be used. Here we used robust partial correlation (RPC) [28]. The overall accuracy is 97.9% (Fig. 3b).

```
library(EpiDISH)
estFrac.m<-epidish(beta.m = data.m,ref.m = expref.m,method='RPC',maxit =
  200)$estF
```

When we construct the DNAm reference matrix, EpiSCORE first generates two separate DNAm reference matrices from the SCM2 and RMAP databases, respectively. Typically, we find a very good overlap of marker genes and agreement between the two DNAm reference matrices [14], which justifies merging them to maximize the number of available marker genes.

```
refMscm2.m <- ImputeDNAmRef(exprefSkin.m,db="SCM2",geneID="SYMBOL");
refMrmap.m <- ImputeDNAmRef(exprefSkin.m,db="RMAP",geneID="SYMBOL");
Mref.m <- ConstMergedDNAmRef(refMscm2.m,refMrmap.m)
```

After removing the marker genes which are not imputable (appear as NA in the imputed reference), this results in a skin DNAm reference with 114 markers (Fig. 4a). Below shows the first ten rows, which are the T-cell markers hypomethylated in the promoter region.

```
> head(Mref.m)
##             EC       Fib Kera_diff Kera_undiff     Macro      Mela Tcell
## 915  0.8230553 0.8230553 0.8230553   0.8230553 0.8230553 0.8230553     0
## 397  0.9400000 0.9400000 0.9400000   0.9400000 0.0000000 0.9400000     0
## 969  0.7106283 0.7106283 0.7106283   0.7106283 0.7106283 0.7106283     0
## 5790 0.8066197 0.8066197 0.8066197   0.8066197 0.8066197 0.8066197     0
## 5788 0.9416881 0.9416881 0.9416881   0.9416881 0.9416881 0.9416881     0
## 241  0.7428700 0.7428700 0.7428700   0.7428700 0.7428700 0.7428700     0
##         weight
## 915  0.8230553
## 397  0.9400000
## 969  0.7106283
## 5790 0.8066197
## 5788 0.9416881
## 241  0.7428700
```

Next, we can validate the DNAm reference in skin dermis and epidermis DNAm samples (*Skin_DNAm_validation.rda*). Firstly, we summarize the beta value at promoter DNAm level using the *constAvBetaTSS* function:

```
avbeta.m<-constAvBetaTSS(beta.m=AF.bmiq,type = '450k')
```

Then we run weighted robust partial correlations to estimate cell-type fractions:

```
estF.m <- wRPC(data=avbeta.m,ref=Mref.m,useW=TRUE,wth=0.4,maxit=200)$estF
```

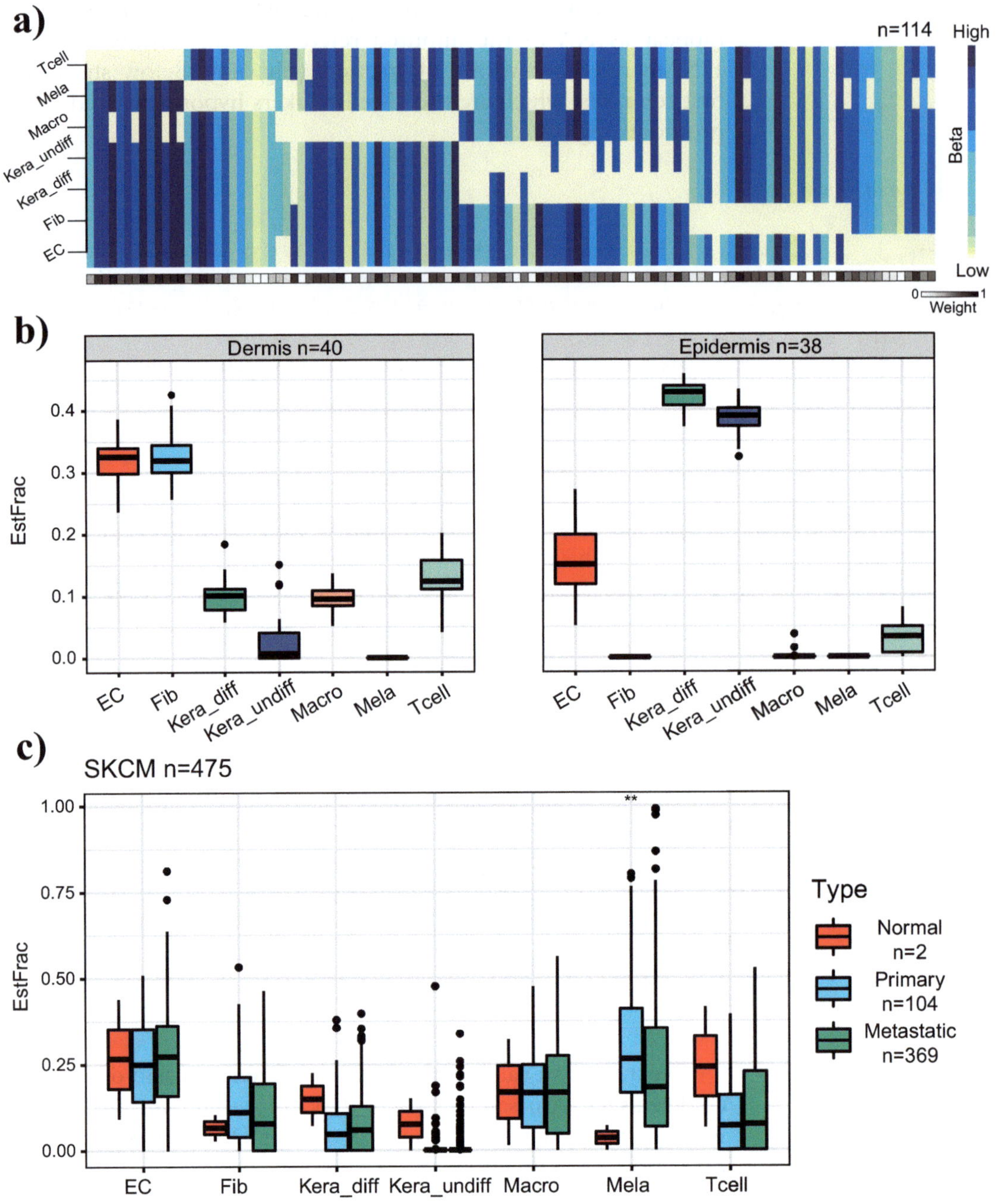

Fig. 4 Construction and validation of skin DNAm reference. (**a**) Heatmap displaying the imputed DNAm reference matrix for 114 marker genes. (**b**) Boxplots display the estimated cell-type fractions for each skin cell type with normal bulk skin data from the dermis and epidermis. (**c**) Boxplots display the estimated cell-type fractions for the TCGA SKCM 450k DNAm dataset

The result accurately predicts that endothelial cells and fibroblasts mainly exist in the dermis and keratinocytes mainly exist in the epidermis (Fig. 4b). Moreover, the melanocyte fraction is negligible in line with the fact that melanocytes represent a minority cell type. However, in skin cutaneous melanoma (SKCM), the melanocyte fraction is significantly higher in cancer samples (Fig. 4c). (The SKCM data can be loaded from *Skin_DNAm_SKCM.rda* and the cell-type fractions can be estimated similarly as above.)

Application of brain DNAm reference to a schizophrenia EWAS: The brain reference can be loaded from the EpiSCORE package. The brain DNAm reference contains 133 marker genes over six cell types (neurons, oligodendrocytes (Oligo), astrocytes (Astro), oligodendrocyte progenitor cells (OPC), endothelial cells (Endo), microglia).

```
data('BrainRef')
```

We first validate the brain DNAm reference with purified neuron and non-neuron samples. Load *Brain_DNAm_purified.rda* into R, and estimate the cell-type fractions with wRPC. The result accurately predicts that $NeuN^{+}$ samples are primarily composed of neurons and $NeuN^{-}$ samples are primarily composed of non-neurons (Fig. 5a).

```
avbrain.m<-constAvBetaTSS(beta.m = brainfl.m,type = '450k')
estf.m <-
  wRPC(data=avbrain.m,ref=mrefBrain.m,useW=TRUE,wth=0.4,maxit=200)$estF
```

Load *Brain_DNAm_SZ.rda* into R. Given the cell-type fractions, we can identify schizophrenia-associated differentially methylated cell types (SZ-DMCTs) with the CellDMC algorithm (which is implemented in the EpiDISH package). In the model, we adjust for age and plate, which are the major confounders in the analysis. The result shows that most SZ-DMCTs occurred in neurons, with lower numbers in oligodendrocytes and OPCs (Fig. 5b).

```
cdmc_br.ls<-CellDMC(jaffebeta.m,phenotype.df$Group,
                    estbrJ.m,
                    adjPMethod = 'fdr',
                    adjPThresh = 0.05,
                    mc.cores = 20,
                    cov.mod = model.matrix(~Age+Pplate,data=phenotype.df))
```

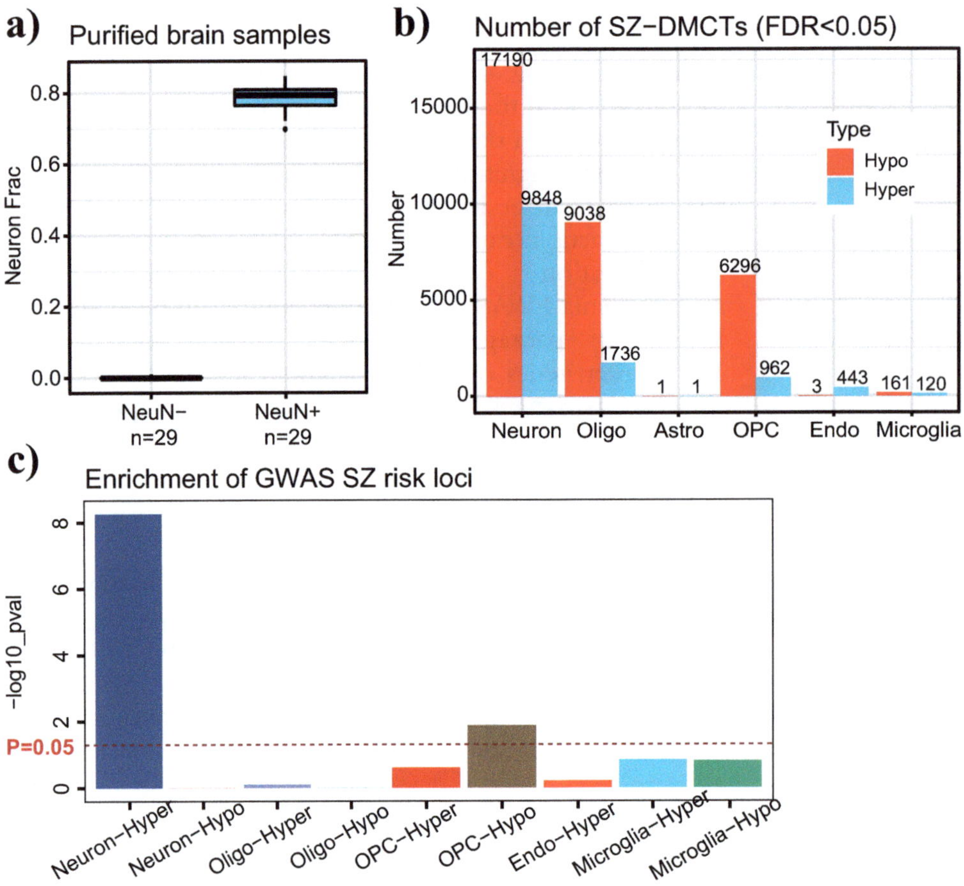

Fig. 5 Validation and application of brain DNAm reference. (**a**) Boxplots display the estimated neuron fractions in FACS-sorted NeuN$^+$ and NeuN$^-$ DNAm samples. (**b**) Bar plots display the number of cell-type-specific SZ-associated DMCTs, as inferred using CellDMC in an EWAS of SZ performed in the prefrontal cortex. (**c**) Enrichment of SZ-associated GWAS loci among the DMCTs displayed in (**b**)

Next, we can test for enrichment of GWAS SZ-risk loci in those DMCTs with Fisher's test. We can see that only hypermethylated neuron-DMCTs are strongly enriched for GWAS SZ-risk loci, indicating a neuronal origin for SZ (Fig. 5c). We don't show the code here because it is beyond the scope of this chapter. Please see [15] for further details. We also recommend readers check https://doi.org/10.24433/CO.8741744.v3 to find more details and illustrative examples of applying EpiSCORE.

3.6 Summary

To help users better understand how to use EpiSCORE in practice, we provide a workflow in Fig. 6, which can serve as a guideline for its implementation.

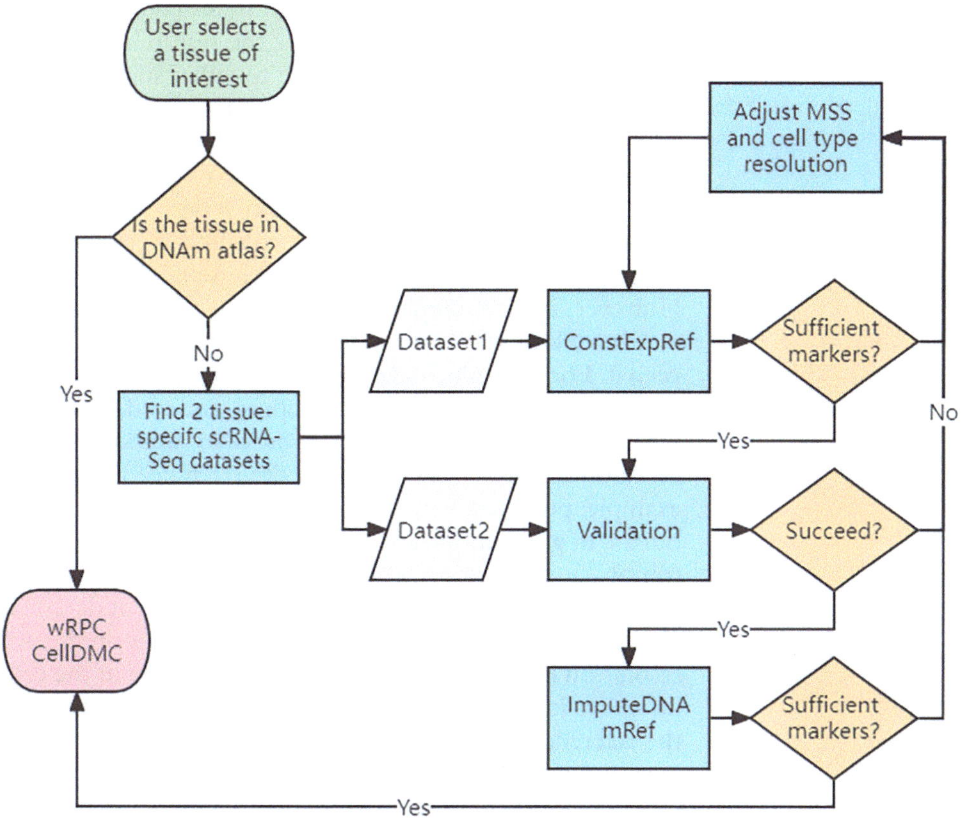

Fig. 6 EpiSCORE workflow. Once we select a tissue of interest, we may check if the tissue is in the EpiSCORE DNAm atlas. If yes, we can estimate cell-type fractions with wRPC and identify DMCTs with CellDMC. If not, we should find at least two scRNA-Seq datasets, one for reference construction and the other for validation. If the validation is successful, we can then impute the DNAm reference. If there are sufficient marker genes in the DNAm reference that we have constructed, we can then apply wRPC and CellDMC. If the validation of the mRNA expression reference matrix is not good or if there are insufficient numbers of markers in either of the two reference matrices, we can relax the MSS requirements and/or reduce the cell-type resolution, so as to reconstruct the mRNA expression reference matrix with more marker genes

4 Notes

1. *Choice of cell-type resolution*. Most scRNA-Seq atlases depict an almost complete landscape of existing cell types in a tissue, so there are plenty of cell types annotated or even newly discovered ones. As discussed before, there is a trade-off between resolution and robustness. There are a few issues we need to consider when deciding on the cell types to include, such as the biological importance of cell types, the abundance of cells per cell type, the accuracy in validation, and the number of resulting DNAm markers. Firstly, we must avoid including too many cell types; otherwise, we lose the power to identify sufficient numbers of

specific marker genes. Hence, if distinguishing similar cell types is not of biological importance for the research question being pursued, we recommend grouping together such cell types. For example, in lung tissue, we could group together CD4+ T cells, CD8+ T cells, natural killer cells, and B cells into a general lymphocyte group, assuming our prime interest is in the epithelial or fibroblast populations. If a cell type is very rare (e.g., ionocytes in the lung), we also recommend grouping it with other similar cell types or alternatively removing it completely. However, if there are reasons to believe that this rare cell type may be of biological importance in a disease context, we need to keep it. For example, melanocytes are very rare in skin tissue, but they are the cell of origin of melanoma, and so if we are building a skin DNAm reference matrix in order to subsequently apply to melanomas, we should have it in the reference matrix. Another example: pulmonary neuroendocrine cells are rare in lung tissue but are of great importance when studying lung neuroendocrine tumors. Furthermore, it is critically important to also validate the expression reference matrix in an independent dataset and to check how validation accuracy varies depending on the cell-type resolution of the built reference matrix. If some of the cell types do not validate well, we may consider changing the resolution of the reference matrix or relaxing the MSS. Finally, if we do not have enough markers per cell type (i.e., less than 10) in the final DNAm reference, we may need to lower the cell-type resolution or relax the MSS. Usually, merging similar cell types into one group or relaxing the MSS for the problematic cell types would result in more marker genes and increase the robustness of the reference. In practice, we recommend aiming for a DNAm reference matrix encompassing not more than eight or nine cell types and with at least ten markers per cell type.

2. *The choice of Marker Specificity Score (MSS)*. In cases where the number of marker genes for specific cell types in the expression reference matrix is too low (say less than 50 or so), the marker specificity score requirement for that cell type can be relaxed. Indeed, we note that it is difficult to know in advance how many of our marker genes will be selected for the DNAm imputation step, but typically this fraction is only around 20–30%. Thus, if for a given cell type we have 50 marker genes in the expression reference matrix, then only around ten of these will appear in the final DNAm reference. Ideally, we would also desire balanced numbers of marker genes for each cell type. Thus, it's typically recommended to use all markers derived in the expression reference matrix and to relax the MSS score in case there are too few markers for a given cell type, or to ensure better balance. For instance, in the case of skin tissue, the scRNA-Seq and DNAm references were constructed at the resolution of seven cell types

(endothelial cells, fibroblasts, differentiated keratinocytes, undifferentiated keratinocytes, macrophages, melanocytes, and T cells) using MSS thresholds 3, 3, 3, 3, 3, 6, and 3, respectively. Here, the maximal value is six because there are maximally only six other cell types where a given marker gene is not expressed. This resulted in a 652-marker expression reference, which includes 53 endothelial cell, 67 fibroblast, 127 differentiated keratinocyte, 148 undifferentiated keratinocyte, 66 macrophage, 160 melanocyte, and 31 T cell-specific marker genes. These choices were motivated so as to ensure reasonable numbers of markers for each cell type.

References

1. Liu Y, Aryee MJ, Padyukov L, Fallin MD, Hesselberg E, Runarsson A, Reinius L, Acevedo N, Taub M, Ronninger M (2013) Epigenome-wide association data implicate DNA methylation as an intermediary of genetic risk in rheumatoid arthritis. Nat Biotechnol 31(2):142
2. Teschendorff AE, Relton CL (2018) Statistical and integrative system-level analysis of DNA methylation data. Nat Rev Genet 19(3): 129–147. https://doi.org/10.1038/nrg.2017.86
3. Teschendorff AE, Jones A, Fiegl H, Sargent A, Zhuang JJ, Kitchener HC, Widschwendter M (2012) Epigenetic variability in cells of normal cytology is associated with the risk of future morphological transformation. Genome Med 4(3):24. https://doi.org/10.1186/gm323
4. Smallwood SA, Lee HJ, Angermueller C, Krueger F, Saadeh H, Peat J, Andrews SR, Stegle O, Reik W, Kelsey G (2014) Single-cell genome-wide bisulfite sequencing for assessing epigenetic heterogeneity. Nat Methods 11(8): 817–820. https://doi.org/10.1038/nmeth.3035
5. Zheng SC, Beck S, Jaffe AE, Koestler DC, Hansen KD, Houseman AE, Irizarry RA, Teschendorff AE (2017) Correcting for cell-type heterogeneity in epigenome-wide association studies: revisiting previous analyses. Nat Methods 14(3):216
6. Houseman EA, Accomando WP, Koestler DC, Christensen BC, Marsit CJ, Nelson HH, Wiencke JK, Kelsey KT (2012) DNA methylation arrays as surrogate measures of cell mixture distribution. BMC Bioinf 13(1):86
7. Newman AM, Liu CL, Green MR, Gentles AJ, Feng W, Xu Y, Hoang CD, Diehn M, Alizadeh AA (2015) Robust enumeration of cell subsets from tissue expression profiles. Nat Methods 12(5):453
8. Zheng SC, Breeze CE, Beck S, Teschendorff AE (2018) Identification of differentially methylated cell types in epigenome-wide association studies. Nat Methods 15(12): 1059–1066. https://doi.org/10.1038/s41592-018-0213-x
9. Reinius LE, Acevedo N, Joerink M, Pershagen G, Dahlén S-E, Greco D, Söderhäll C, Scheynius A, Kere J (2012) Differential DNA methylation in purified human blood cells: implications for cell lineage and studies on disease susceptibility. PLoS One 7(7):e41361
10. Gervin K, Page CM, Aass HCD, Jansen MA, Fjeldstad HE, Andreassen BK, Duijts L, van Meurs JB, van Zelm MC, Jaddoe VW (2016) Cell type specific DNA methylation in cord blood: a 450K-reference data set and cell count-based validation of estimated cell type composition. Epigenetics 11(9):690–698
11. Guintivano J, Aryee MJ, Kaminsky ZA (2013) A cell epigenotype specific model for the correction of brain cellular heterogeneity bias and its application to age, brain region and major depression. Epigenetics 8(3):290–302
12. Teschendorff AE, Gao Y, Jones A, Ruebner M, Beckmann MW, Wachter DL, Fasching PA, Widschwendter M (2016) DNA methylation outliers in normal breast tissue identify field defects that are enriched in cancer. Nat Commun 7:10478
13. Gao Y, Widschwendter M, Teschendorff AE (2018) DNA methylation patterns in normal tissue correlate more strongly with breast Cancer status than copy-number variants. EBioMedicine 31:243–252. https://doi.org/10.1016/j.ebiom.2018.04.025
14. Teschendorff AE, Zhu T, Breeze CE, Beck S (2020) EPISCORE: cell type deconvolution of bulk tissue DNA methylomes from single-cell RNA-Seq data. Genome Biol 21(1):221.

https://doi.org/10.1186/s13059-020-02126-9

15. Zhu T, Liu J, Beck S, Pan S, Capper D, Lechner M, Thirlwell C, Breeze CE, Teschendorff AE (2022) A pan-tissue DNA methylation atlas enables in silico decomposition of human tissue methylomes at cell-type resolution. Nat Methods 19(3):296–306. https://doi.org/10.1038/s41592-022-01412-7
16. Solé-Boldo L, Raddatz G, Schütz S, Mallm J-P, Rippe K, Lonsdorf AS, Rodríguez-Paredes M, Lyko F (2020) Single-cell transcriptomes of the human skin reveal age-related loss of fibroblast priming. Commun Biol 3(1):188. https://doi.org/10.1038/s42003-020-0922-4
17. Deng CC, Hu YF, Zhu DH, Cheng Q, Gu JJ, Feng QL, Zhang LX, Xu YP, Wang D, Rong Z, Yang B (2021) Single-cell RNA-seq reveals fibroblast heterogeneity and increased mesenchymal fibroblasts in human fibrotic skin diseases. Nat Commun 12(1):3709. https://doi.org/10.1038/s41467-021-24110-y
18. Vandiver AR, Irizarry RA, Hansen KD, Garza LA, Runarsson A, Li X, Chien AL, Wang TS, Leung SG, Kang S, Feinberg AP (2015) Age and sun exposure-related widespread genomic blocks of hypomethylation in nonmalignant skin. Genome Biol 16(1):80. https://doi.org/10.1186/s13059-015-0644-y
19. Cancer Genome Atlas N (2015) Genomic classification of cutaneous melanoma. Cell 161(7): 1681–1696. https://doi.org/10.1016/j.cell.2015.05.044
20. Jaffe AE, Gao Y, Deepsoboslay A, Tao R, Hyde TM, Weinberger DR, Kleinman JE (2016) Mapping DNA methylation across development, genotype and schizophrenia in the human frontal cortex. Nat Neurosci 19(1): 40–47
21. Kundaje A, Meuleman W, Ernst J, Bilenky M, Yen A, Heravi-Moussavi A, Kheradpour P, Zhang Z, Wang J, Ziller MJ (2015) Integrative analysis of 111 reference human epigenomes. Nature 518(7539):317
22. Stunnenberg HG, Abrignani S, Adams D, de Almeida M, Altucci L, Amin V, Amit I, Antonarakis SE, Aparicio S, Arima T (2016) The international human epigenome consortium: a blueprint for scientific collaboration and discovery. Cell 167(5):1145–1149
23. Nazor KL, Altun G, Lynch C, Tran H, Harness JV, Slavin I, Garitaonandia I, Müller FJ, Wang YC, Boscolo FS, Fakunle E, Dumevska B, Lee S, Park HS, Olee T, D'Lima DD, Semechkin R, Parast MM, Galat V, Laslett AL, Schmidt U, Keirstead HS, Loring JF, Laurent LC (2012) Recurrent variations in DNA methylation in human pluripotent stem cells and their differentiated derivatives. Cell Stem Cell 10(5):620–634. https://doi.org/10.1016/j.stem.2012.02.013
24. Dempster AP, Laird NM, Rubin DB (1976) Maximum likelihood from incomplete data via the EM algorithm. In: Paper presented at the royal statistical society, December 8
25. Huber PJ (1981) Robust statistics. Wiley, New York
26. Li Z, Wu Z, Jin P, Wu H (2019) Dissecting differential signals in high-throughput data from complex tissues. Bioinformatics 35(20): 3898–3905. https://doi.org/10.1093/bioinformatics/btz196
27. You C, Wu S, Zheng SC, Zhu T, Jing H, Flagg K, Wang G, Jin L, Wang S, Teschendorff AE (2020) A cell-type deconvolution meta-analysis of whole blood EWAS reveals lineage-specific smoking-associated DNA methylation changes. Nat Commun 11(1):4779. https://doi.org/10.1038/s41467-020-18618-y
28. Teschendorff AE, Breeze CE, Zheng SC, Beck S (2017) A comparison of reference-based algorithms for correcting cell-type heterogeneity in epigenome-wide association studies. BMC Bioinf 18(1):105

Chapter 4

Profiling Cellular Ecosystems at Single-Cell Resolution and at Scale with EcoTyper

Chloé B. Steen, Bogdan A. Luca, Ash A. Alizadeh, Andrew J. Gentles, and Aaron M. Newman

Abstract

Tissues are composed of diverse cell types and cellular states that organize into distinct ecosystems with specialized functions. EcoTyper is a collection of machine learning tools for the large-scale delineation of cellular ecosystems and their constituent cell states from bulk, single-cell, and spatially resolved gene expression data. In this chapter, we provide a primer on EcoTyper and demonstrate its use for the discovery and recovery of cell states and ecosystems from healthy and diseased tissue specimens.

Key words Tissue heterogeneity, Transcriptomics, Tumor microenvironment, Cell states, Ecosystems, Ecotypes, Single-cell RNA sequencing, Spatial transcriptomics

1 Introduction

Recent advances in multiplexed imaging, single-cell sequencing, and computational deconvolution have revealed new insights into the cellular diversity and modularity of complex tissues [1, 2]. For example, it is now well-established that individual cells can exhibit context-dependent phenotypic states that organize into functionally distinct cellular ecosystems [3]. Such ecosystems, commonly referred to as communities or ecotypes [4–6], are thought to play critical roles during tissue development, homeostasis, repair, and disease. In cancer, immune and stromal-rich ecotypes have been linked to heterogeneity in patient outcomes, including differential response to therapy [4, 6]. Understanding how ecotypes shape clinical trajectories in cancer and other diseases has significant potential for the discovery of more effective diagnostics and therapeutic targets.

EcoTyper is a suite of machine learning tools for characterizing transcriptionally defined cell states and their co-association patterns (i.e., ecotypes) from complex tissue specimens [4, 6]. It performs

Brooke L. Fridley and Xuefeng Wang (eds.), *Statistical Genomics*,
Methods in Molecular Biology, vol. 2629, https://doi.org/10.1007/978-1-0716-2986-4_4,

two key tasks: (i) the discovery of cell states and ecotypes from bulk or single-cell expression data, and (ii) the recovery of previously defined cell states and ecotypes from bulk, single-cell, or spatially resolved expression data (Fig. 1).

We recently applied EcoTyper to illuminate the cell state and ecotype landscape of human carcinoma – the most common human malignancy – and diffuse large B-cell lymphoma, the most prevalent blood cancer [4, 6]. A companion website, available at https://ecotyper.stanford.edu/, includes the following functions and features:

1. Recovery of carcinoma- and lymphoma-specific cell states and ecotypes in user-provided expression datasets.
2. Interactive exploration and visualization of gene expression signatures, survival associations, and predicted ligand-receptor pairs for carcinoma- and lymphoma-specific cell states and ecotypes.
3. Raw data associated with both publications.

Here we demonstrate how to run EcoTyper on a personal computer, server, or high-performance computing cluster. By illustrating key functions and commands using hands-on examples, we show how EcoTyper can be used for the discovery (*see* Subheadings 3.1 and 3.2) and recovery (*see* Subheadings 3.3, 3.4, and 3.5) of cell states and ecotypes from bulk and single-cell expression data. For those primarily interested in the recovery of predefined carcinoma- or lymphoma-specific cell states/ecotypes, *see* Subheadings 3.3, 3.4, and 3.5 (Fig. 1a).

2 Materials

EcoTyper is written in R. The latest source code can be found at https://github.com/digitalcytometry/ecotyper or https://ecotyper.stanford.edu/. All instructions necessary to install EcoTyper and its dependencies are available from GitHub (*see* **Notes 1–3**).

3 Methods

3.1 Discovery of Cell States and Ecotypes from Bulk Data

In this section, we describe how to apply EcoTyper to the discovery of cell states and ecotypes from bulk tissue expression data profiled by RNA-seq or microarrays (Fig. 1). An example bulk RNA-seq discovery dataset consisting of lung adenocarcinoma (LUAD) and lung squamous cell carcinoma (LUSC) tumor samples from The Cancer Genome Atlas (TCGA) [7] is used to illustrate each step of the workflow. Once EcoTyper has been downloaded and installed

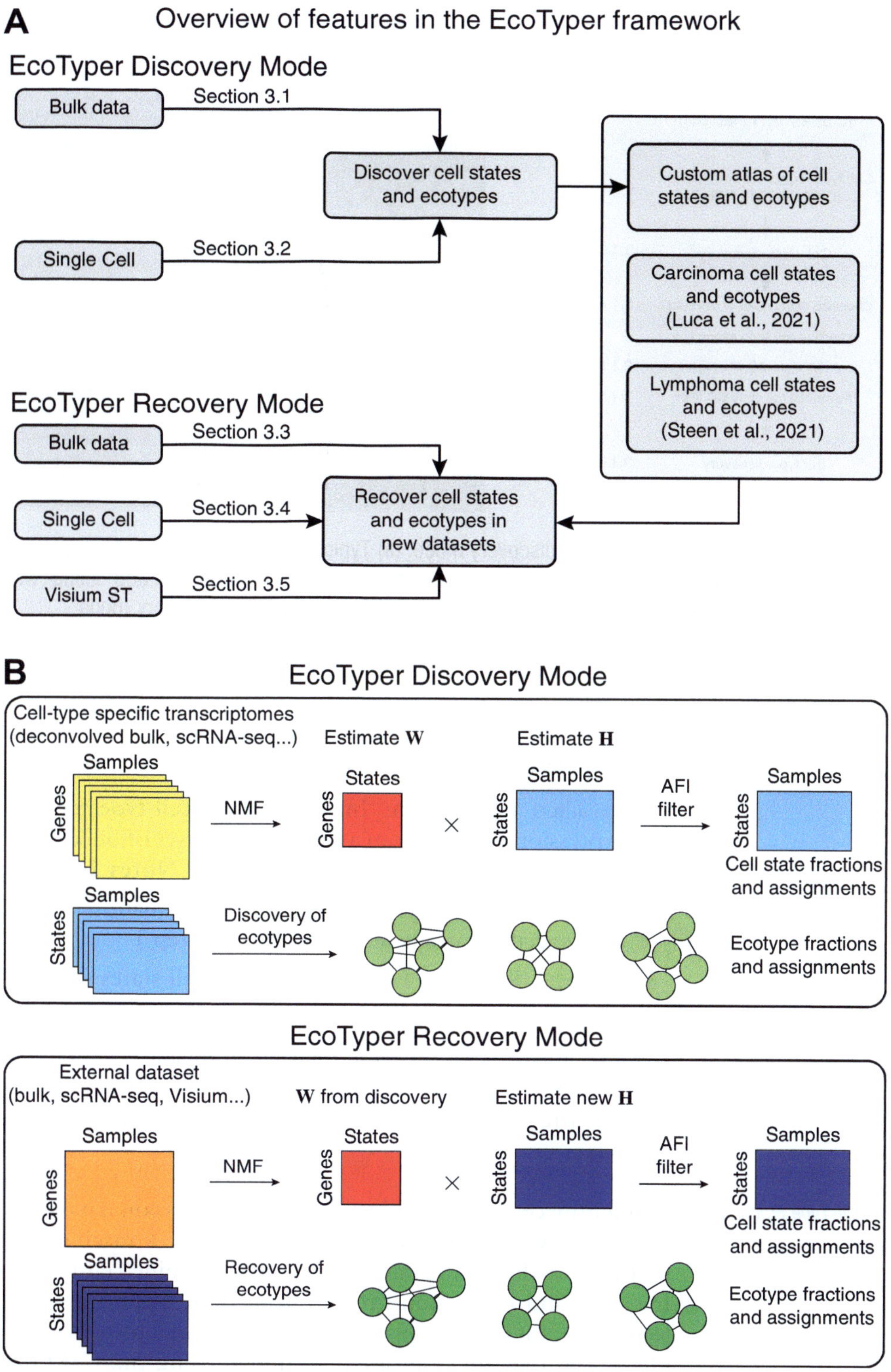

Fig. 1 The EcoTyper framework. (**a**) Key EcoTyper functions and their corresponding sections in this chapter. ST spatial transcriptomics. (**b**) Schematic depiction of the EcoTyper analytical pipeline, including the discovery and recovery of cell states and ecotypes from diverse transcriptomic platforms. For further information, see Luca et al. [4]

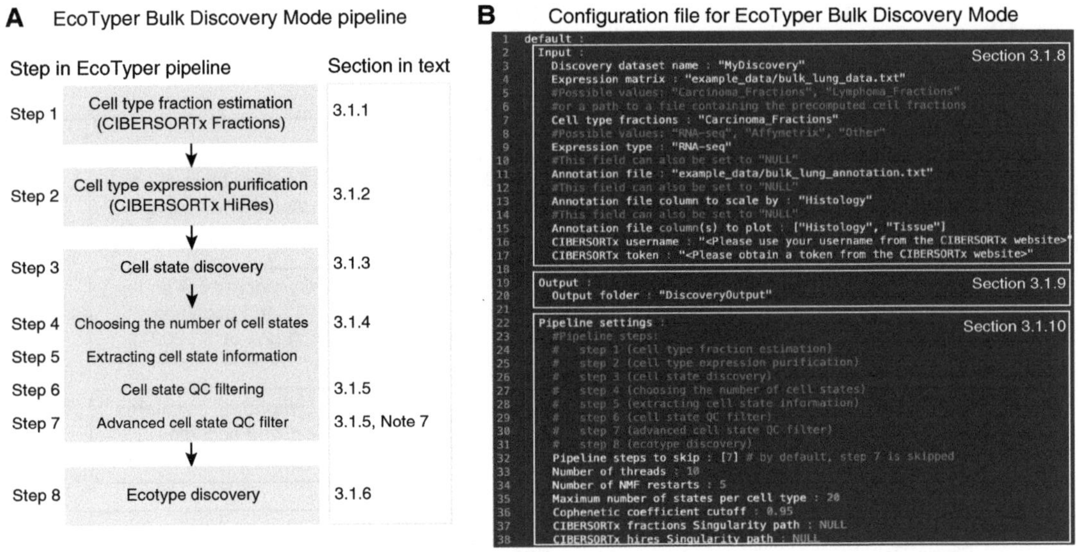

Fig. 2 Overview and configuration of bulk discovery mode. (**a**) Typical EcoTyper workflow for the discovery of cell states and ecotypes from bulk gene expression data. Key steps are indicated along with their corresponding sections in this chapter. (**b**) Example configuration file for bulk discovery mode

(*see* Subheading 2), the example dataset can be found in *example_-data/bulk_lung_data.txt* together with a sample annotation file, *example_data/bulk_lung_annotation.txt*.

This section involves the following key steps:

1. In silico purification: Imputation of cell-type-specific gene expression profiles from a discovery dataset of bulk tissue transcriptomes using CIBERSORTx [8] (*see* **Notes 2** and **4**).
2. Cell state discovery: Identification and quantitation of cell-type-specific states using the output of step 1.
3. Ecotype discovery: Co-assignment of cell states (from step 2) into ecotypes.

The above steps are described in Subheadings 3.1.1 through 3.1.6 below (see also Fig. 2a). The remainder of Subheading 3.1 reviews the relevant commands, input files, and configuration settings.

3.1.1 Cell-Type Fraction Estimation

To identify cell states within a discovery expression dataset, EcoTyper starts with CIBERSORTx (*see* **Note 2**), a machine learning method for digital cytometry that can determine cell-type-specific fractions and expression profiles from bulk tissue transcriptomes [8]. CIBERSORTx is applied in two sequential phases. In the first phase, cell-type fractions are determined using the *CIBERSORTx Fractions* module [8]. This step requires a signature matrix comprised of reference profiles that discriminate between major cell types within a tissue type of interest. For example, a signature matrix for a solid malignancy would ideally include key immune cell lineages (e.g., B cells, CD8 T cells, CD4 T cells, NK cells, monocytes/macrophages, dendritic cells, etc.), stromal subsets (fibroblasts and endothelial cells), and cancer cells. Such matrices are straightforward to derive from purified cell-type-specific expression profiles or scRNA-seq data, as described elsewhere [8, 9]. Instructions and vignettes for creating signature matrices are also available from the CIBERSORTx website (https://cibersortx.stanford.edu/).

In recent publications demonstrating the utility of EcoTyper [4, 6], we applied two well-validated signature matrices (LM22 and TR4, Table 1) covering the major immune, stromal, and cancer cell populations found in human carcinomas and lymphomas [8]. Both signature matrices are packaged with EcoTyper and can be applied as covered in Subheading 3.1.8, *Cell-type fractions* (*see* Table 1). To interrogate other cell types, users can apply CIBERSORTx (or an alternative deconvolution method with comparable functionality) to design their own signature matrix and then obtain cell-type fractions prior to running EcoTyper.

3.1.2 Cell-Type Expression Purification

In the second phase, once cell-type fractions are determined, CIBERSORTx is applied to impute cell-type-specific gene expression profiles from the discovery dataset. For this purpose, EcoTyper employs the *CIBERSORTx Impute High-Resolution Cell Expression* module ("CIBERSORTx HiRes").

As input, CIBERSORTx HiRes requires bulk expression data and cell-type fractions obtained from the previous step (*step 1*, *Cell-type fraction estimation* in Fig. 2a). The output consists of a purified gene expression matrix for each cell type, with genes as rows and samples as columns (*see* **Note 5**). These cell-type-specific gene expression matrices serve as input for cell state identification.

3.1.3 Cell State Discovery

Given a collection of cell-type-specific gene expression profiles (GEPs), EcoTyper performs cell state identification and quantification (Fig. 1b). Cell states are identified through simultaneous deconvolution and clustering of cell-type-specific GEPs via non-negative matrix factorization (NMF). NMF is combined with specialized preprocessing and postprocessing steps to automatically select the number of states per cell type while minimizing false positives using a novel adaptive filtering approach (*see* **Note 6**; Fig. 1b). Importantly, NMF defines cell states and determines their relative fractional abundance in each sample, allowing for downstream analyses of cell state composition, including ecotype detection (e.g., Subheadings 3.1.6 and 3.1.13).

Table 1
Built-in signature matrices and cell-type mapping schemes for carcinoma EcoTyper and Lymphoma EcoTyper

Signature matrix	Cell type	Carcinoma EcoTyper	Lymphoma EcoTyper
LM22	B cells naive	B cells	B cells
	B cells memory	B cells	B cells
	Plasma cells	PCs	PCs
	T cells CD8	T cells CD8	T cells CD8
	T cells CD4 naive	T cells CD4	T cells CD4
	T cells CD4 memory resting	T cells CD4	T cells CD4
	T cells CD4 memory activated	T cells CD4	T cells CD4
	T cells follicular helper	T cells CD4	T cells follicular helper
	T cells regulatory (Tregs)	T cells CD4	Tregs
	T cells gamma delta	T cells CD8	T cells CD8
	NK cells resting	NK cells	NK cells
	NK cells activated	NK cells	NK cells
	Monocytes	Monocytes and Macrophages	Monocytes and Macrophages
	Macrophages M0	Monocytes and Macrophages	Monocytes and Macrophages
	Macrophages M1	Monocytes and Macrophages	Monocytes and Macrophages
	Macrophages M2	Monocytes and Macrophages	Monocytes and Macrophages
	Dendritic cells resting	Dendritic cells	Dendritic cells
	Dendritic cells activated	Dendritic cells	Dendritic cells
	Mast cells resting	Mast cells	Mast cells
	Mast cells activated	Mast cells	Mast cells
	Neutrophils	Neutrophils	Neutrophils
	Eosinophils	Excluded	Excluded
TR4	Immune cells	Immune cells	Immune cells
	Endothelial cells	Endothelial cells	Endothelial cells
	Epithelial cells	Epithelial cells	Excluded
	Fibroblasts	Fibroblasts	Fibroblasts

For details about the derivation and application of both signature matrices, see Newman et al. [8], Luca et al. [4], and Steen et al. [6]

3.1.4 Determining the Number of Cell States

For each set of cell-type-specific GEPs, EcoTyper applies NMF across a range of ranks (2–20, by default) to determine the number of transcriptionally defined cell states. To select the most appropriate rank (i.e., the number of cell states), we developed a heuristic based on the cophenetic coefficient (CC) (*step 4, Choosing the number of cell states* in Fig. 2a). The rank at which the CC starts decreasing is typically selected manually; however, this rank may be suboptimal when the CC exhibits a multimodal shape [4]. Therefore, the number of cell states is automatically determined based on the first occurrence for which the CC drops below 0.95 (by default) and having been above this level for at least two consecutive ranks. As this threshold is predetermined and may not be ideal in every case, we recommend that users visually inspect the plots created by EcoTyper (*see* Subheading 3.1.12) and revise the threshold accordingly, with the goal of selecting a CC threshold that captures the first substantial drop in CC across the majority of cell types.

3.1.5 Cell State Quality Control

EcoTyper applies gene-level standardization (i.e., normalizing the mean of each gene to zero and the standard deviation to one) to improve the sensitivity of cell state discovery with NMF. However, as NMF requires a nonnegative input, EcoTyper applies a *posneg* transformation to enforce the nonnegativity requirement of NMF [4]. Since the posneg transformation can lead to the identification of spurious cell states driven by features with more negative values than positive ones, we devised an adaptive false positive index (AFI) (Fig. 1b) to eliminate likely false positives while maximizing sensitivity (Figure S1 in Luca et al. [4]). EcoTyper automatically filters states with AFI ≥ 1, and this is done in *step 6, Cell state QC filtering* of the EcoTyper pipeline (Fig. 2a). An additional filtering step is also performed in *step 7, Advanced cell state QC filter* (Fig. 2a), with further details provided in **Note 7**.

3.1.6 Ecotype (Cellular Community) Discovery

Following quality control, EcoTyper analyzes all remaining cell states to identify statistically robust patterns of co-occurrence across samples in the discovery dataset. These cellular communities constitute ecosystem subtypes or "ecotypes" (Fig. 1b). To identify ecotypes, EcoTyper calculates the degree of overlap between each pair of cell states across samples using the Jaccard index combined with statistical filtering. Ecotypes are subsequently identified by hierarchical clustering and silhouette width maximization. Ecotypes with less than three cell states are eliminated from further analysis by default.

3.1.7 Configuring the EcoTyper Run

The script for cell type and ecotype discovery from bulk data is *EcoTyper_discovery_bulk.R*.

This script takes as input a configuration file in YAML format, which has three parts: (i) input, (ii) output, and (iii) pipeline settings (Fig. 2b). In the following section, we describe the contents of this file and provide instructions for selecting the most appropriate settings for a given application. The configuration file used for the example discovery dataset is *config_discovery_bulk.yml.*

3.1.8 Input Section of Configuration File

The *Input* section of the configuration file is used for setting parameters tailored to the input data. Each field is described in detail below.

Discovery Dataset Name

The *Discovery dataset name* field is the identifier used by EcoTyper to internally save and retrieve information about the cell states/ ecotypes defined in the discovery dataset. Alphanumeric characters and '_' are accepted for this field. For the example dataset used in this section, this field is set to "MyDiscovery" (Fig. 2b); otherwise, it should be customized by the user.

Expression Matrix

The *Expression matrix* field refers to the path of a tab-delimited file where the discovery dataset is located. The expression matrix must be formatted with genes as rows and samples as columns and should be normalized to transcripts-per-million (TPM) for bulk RNA-seq and non-logarithmic (exponential) space for microarrays. It should have gene symbols in the first column and gene expression for each sample in the remaining columns. Column (sample) names should be unique (*see* **Note 8**).The example dataset used in this section is *example_data/bulk_lung_data.txt.*

Cell-Type Fractions

The *Cell-type fractions* field refers to the source data for cell-type fractions in the discovery dataset. If the major cell types expected in a given discovery dataset are the same as those analyzed with Carcinoma EcoTyper [4] (Table 1), this field should be set to *Carcinoma_Fractions* (case-sensitive) and EcoTyper will invoke CIBERSORTx to automatically estimate fractions for these populations in *step 1* of the workflow (*see* Fig. 2a; Subheading 3.1.1). Similarly, if the cell types analyzed with Lymphoma EcoTyper [6] are appropriate (Table 1), the user can set this field to *Lymphoma_-Fractions* and CIBERSORTx will be automatically applied to these cell types.

If neither of these cases applies, the user must provide a path to a tab-delimited file containing the fractions for the major cell types in their discovery dataset. This file can be obtained using CIBERSORTx, another deconvolution tool, or an independent source of cell-type fractions, outside of the EcoTyper pipeline. The first column should contain sample names that match the column names in the discovery dataset. The remaining columns should contain the fractions for each cell type, with no missing values. These fractions should sum to 1 for each row. For an example of a cell-type fraction file, see *example_data/bulk_fractions_example.txt.*

Expression Type

The *Expression type* field is only relevant if cell-type fractions are estimated automatically by EcoTyper (i.e., *Carcinoma_Fractions* or *Lymphoma_Fractions* are provided in the *Cell-type fractions* field of the configuration file, as described above). *Expression type* specifies the platform used to generate the data provided in the expression matrix. The accepted values are "RNA-seq" for bulk RNA-seq data, "Affymetrix" for data profiled using Affymetrix microarray platforms, and "Other" for data from non-Affymetrix microarray platforms. Once set, EcoTyper determines the appropriate parameters for the *CIBERSORTx Fractions* module.

Annotation File (Optional)

A path to an annotation file may be provided in the *Annotation file* field. This file must contain a first column called ID with the same names as the columns of the expression matrix. Additional columns can be used for defining sample batches (see section *Annotation file column to scale by* below) and for plotting color bars in the heat map output (see *Annotation file column(s) to plot* below). If not provided, this field must be set to "NULL." The example annotation file for the lung cancer discovery dataset is *example_data/bulk_lung_annotation.txt*.

Annotation File Column to Scale by (Optional)

To discover cell states and ecotypes across 16 types of human carcinoma while minimizing tumor-type-specific variation, we standardized genes to a mean of zero and a standard deviation of one within each type of carcinoma [4]. The field *Annotation file column to scale by* optionally allows users to specify a column name in the annotation file by which the samples will be grouped when performing standardization. The example discovery dataset has samples from LUAD and LUSC, as specified in the "Histology" column of the annotation file. To identify cell states preferentially shared by LUAD and LUSC, we will use the "Histology" column to perform standardization on the example dataset.

Of note, independently standardizing different sample types in the discovery dataset is a consideration that depends on the purpose of the analysis. For example, if users are instead interested in prioritizing cell states and ecotypes that are specific to each tumor type, this argument can be set to "NULL." In this case, standardization will be applied once across all samples in the discovery cohort (as was done for Lymphoma EcoTyper [6], in which a single cancer type was analyzed). The same will happen if the annotation file is not supplied.

Annotation File Column(s) to Plot

The *Annotation file column(s) to plot* field specifies which columns in the annotation file to use for custom color bar(s) in the heat map outputs.

CIBERSORTx Username and Token

The *CIBERSORTx username* and *CIBERSORTx token* fields should respectively contain the CIBERSORTx username and token necessary to run the CIBERSORTx executable. Once registered at https://cibersortx.stanford.edu, the token can be obtained by submitting a request from *Menu* > *Download*. For installation information, *see* **Note 2**.

3.1.9 Output Section of Configuration File

The *Output* section contains the *Output folder* field, which specifies the path where the final output will be saved. This folder will be created if it does not exist. For the lung cancer discovery dataset, the output folder is set to "DiscoveryOutput."

3.1.10 Pipeline Settings

This section determines how the EcoTyper bulk discovery workflow will be run.

Pipeline Steps to Skip

The *Pipeline steps to skip* option allows the user to omit key steps of the workflow (Fig. 2a). Please note that this option is only intended when the pipeline has already been run once and parameter adjustments are being made. For example, if the cophenetic coefficient threshold used in *step 4* (Fig. 2a) needs to be modified, the user should skip *steps 1–3* and rerun from *step 4*.

Number of Threads

This field specifies the number of threads (i.e., CPU cores) allocated to EcoTyper.

Number of NMF Restarts

The NMF approach used by EcoTyper can give slightly different results each time owing to stochastic initialization. To obtain a stable solution, EcoTyper runs NMF multiple times with different seeds, and the solution that best explains the discovery data is chosen. The parameter *Number of NMF restarts* specifies the number of NMF restarts for each rank and cell type. Since this is a very time-consuming process, we recommend 5 restarts for exploratory purposes and at least 50 restarts for publication-quality results.

Maximum Number of States per Cell Type

The maximum number of states per cell type is set to 20 by default.

Cophenetic Coefficient Cutoff

As described in Subheading 3.1.4, the cophenetic coefficient (CC) is used to select the number of cell states. This field indicates the CC cutoff (range of 0 to 1) used for determining the number of states in *step 4* (Fig. 2a). By default, the cutoff is set to 0.95, and in general, the lower the cutoff, the more clusters are identified. Users can select a more appropriate cutoff after inspecting the CC plots obtained after the first run. This is done by rerunning the pipeline and skipping *steps 1–3* (Fig. 2a).

CIBERSORTx Fractions/HiRes Singularity Path

These two fields specify the paths to the Singularity containers (.SIF files) for the CIBERSORTx Fractions and HiRes modules. These fields are relevant when running EcoTyper on servers or high-performance computing clusters that support Singularity instead of Docker as a container platform. If these paths are provided, cell-type fraction estimation (*step 1*, Fig. 2a) and expression purification (*step 2*, Fig. 2a) will be performed using Singularity. If set to "NULL," these steps will be performed using Docker.

3.1.11 Run Bulk Discovery Mode

After editing the configuration file (*config_discovery_bulk.yml*), the script can be run as indicated below. Please note that running this script on the example dataset might take ~2 h with 10 threads. Although EcoTyper can be run on the example dataset using a typical laptop with ~16GB of RAM and at least 10 CPU cores, larger datasets may require a server or cluster. In general, we recommend using a system with at least 10–20 cores and 32GB of RAM. For details on how to parallelize the code and run it on a cluster, *see* **Note 9**.

```
Rscript EcoTyper_discovery_bulk.R -c config_dis-
covery_bulk.yml
```

3.1.12 Overview of Output Files

When run in bulk discovery mode, EcoTyper outputs a directory containing the following key files and subdirectories:

1. *rank_plot.pdf*: This plot, shown in Fig. 3a for the example discovery dataset, consists of cophenetic coefficient (CC) plots calculated for each cell type (calculated for *step 4, Choosing the number of cell states*, Fig. 2a). The horizontal dotted line indicates the CC threshold specified in the configuration file (= 0.95 by default). The vertical dotted red line indicates the number of states automatically selected based on the CC cutoff.

   As indicated above (*see* Subheading 3.1.4), we recommend that users inspect this file to ensure that automatic CC selection provides sensible results. For example, a high cutoff may result in less than two states inferred for a given cell type after applying the AFI filter (*see* Subheading 3.1.5). This may indicate that the CC threshold is too high. In this case, users can lower it and rerun EcoTyper from *step 4* (Fig. 2a; *see* **Notes 10** and **11**).
2. The following output files are stored in subdirectories labeled by cell type:
   (a) *state_assignment_heatmap.pdf*: A heat map showing the expression of genes that have the highest $\log_2$ fold change in each cell state versus the remaining cell states (i.e., cell-state-specific marker genes). The columns represent samples in the discovery dataset, and the rows represent the marker genes for each cell state. Selected marker genes are shown on the left side of the heat map. The heat map is ordered according to cell state assignment, and the top color bar shows the dominant state for a given sample (see *state_assignment.txt* below). In addition, the heat map shown in Fig. 3b includes two color bars on top corresponding to *Tissue* and *Histology*. These are parameters that have been provided in the configuration file field *Annotation file column(s) to plot*.

   (b) *state_assignment.txt*: This file, consisting of samples as columns and cell states as rows, contains the assignment of samples in the discovery dataset to the cell state with the highest inferred abundance (i.e., the dominant cell state for each sample). If the cell state with the highest abundance in a given sample was filtered out in an earlier quality control step (*see* Subheading 3.1.5), the sample is considered unassigned and is not included in this output file.

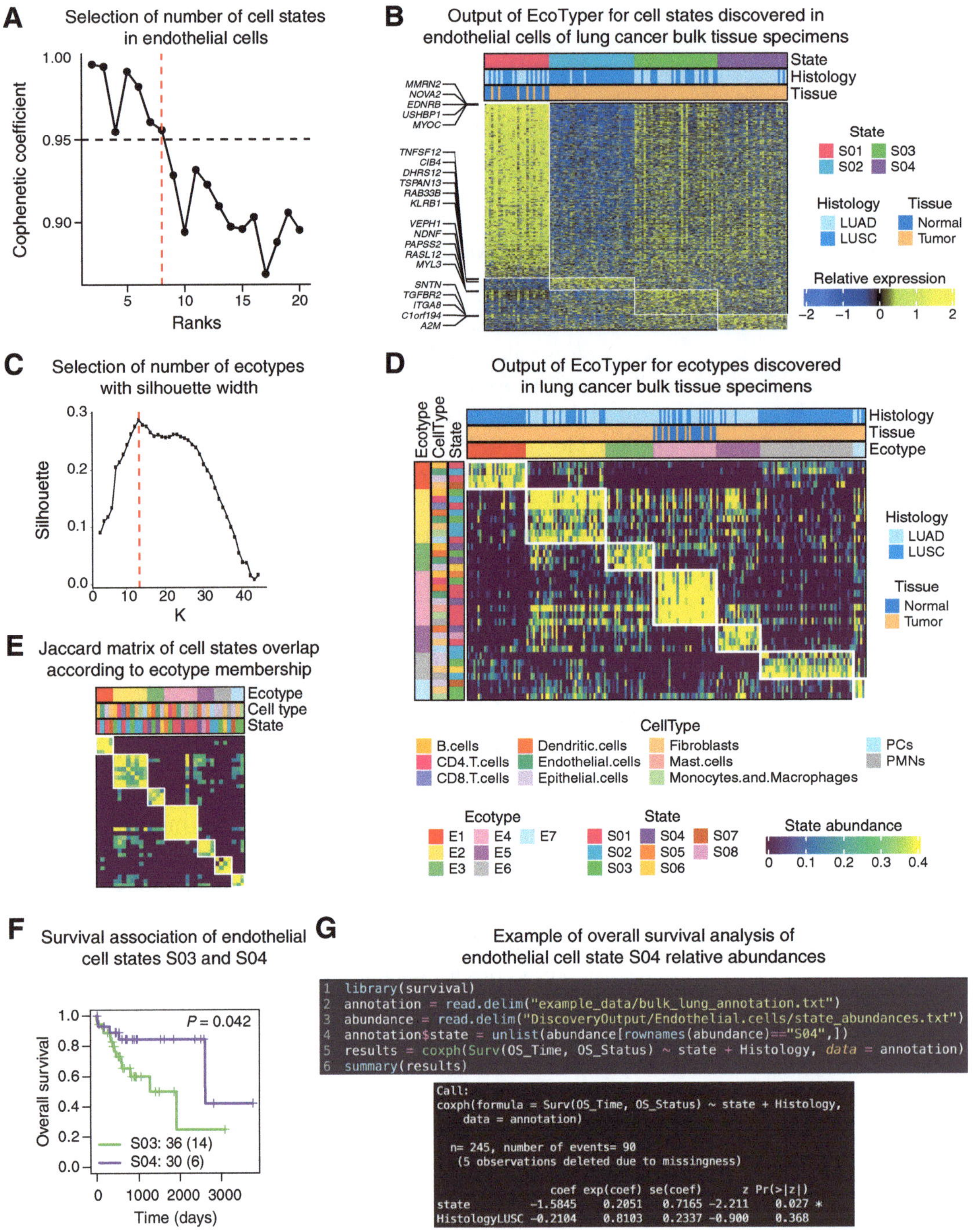

Fig. 3 Bulk discovery mode applied to non-small cell lung cancer. (**a**) Cophenetic coefficient plot used for the selection of cell state numbers, shown here for lung cancer epithelial cells deconvolved from the bulk discovery dataset. The black dashed line indicates the cophenetic coefficient threshold set by default, and the red dashed line denotes the corresponding matrix rank (i.e., number of cell states) selected. For further details, refer to Subheading 3.1.4. (**b**) Heat map showing endothelial cell states identified by EcoTyper from the bulk discovery dataset. Selected marker genes are indicated on the left side of the heat map. Color bars above the heat map denote the dominant cell state to which each sample is assigned and the corresponding tissue type. LUAD, lung adenocarcinoma; LUSC, lung squamous cell carcinoma; Normal, adjacent normal;

(c) *state_abundances.txt*: This file, consisting of samples as columns and cell states as rows, contains the relative abundance of each cell state in each sample. In contrast to the *state_assignment.txt* file, where only samples assigned to a cell state are included, all samples included in the discovery dataset are included here. This file can be used for downstream analyses where tissues are represented as mixtures of cell states (e.g., when evaluating continuous associations with overall survival; *see* Subheading 3.1.13).

(d) *gene_info.txt*: the $\log_2$ fold change of state-specific genes prior to the quality filtering steps performed in *steps 6* and *7* (Fig. 2a). Rows are genes, and columns are cell states. Because marker genes are identified prior to filtering steps, the columns are labeled *IS01–IS0X*, for *InitialState*. Each gene is assigned as a marker gene of a given state (shown in *InitialState*) based on the maximum fold change (shown in column *MaxFC*). The cell state labels after filtering (*S01–S0X*) are provided in the column *State*.

(e) *heatmap_data.txt*: The normalized gene expression data underlying the output heat map shown in *state_assignment_heatmap*.

(f) *heatmap_top_ann.txt*: The annotation file provided by the user after being (i) restricted to the samples assigned to cell states and (ii) populated with cell state labels, both pre- (column *initialState*) and post-quality control filtering (column *State*). Only samples assigned to a dominant cell state are included in this file.

Of note, only samples assigned to cell states remaining after the quality control in *steps 6* and *7* (Fig. 2a) are included in the output files *state_assignment.txt* and *state_assignment_heatmap.pdf*. The samples not assigned to one of these states are considered unassigned and are not included in these output files. In contrast, the file *state_abundances.txt* includes all samples in the discovery dataset.

Fig. 3 (continued) Tumor, LUAD or LUSC. (**c**) Silhouette plot showing the optimal number of ecotypes (red line) identified in the bulk discovery dataset. (**d**) Heat map of inferred ecotype abundances in the bulk discovery dataset. Rows are cell states ($n = 35$), columns are samples ($n = 190$), and both axes are ordered by cell state co-association patterns (i.e., ecotypes). The color bars above the heat map denote the ecotype label and tissue (as in panel **b**). (**e**) Jaccard matrix of cell state overlap patterns according to ecotype membership. (**f**) Kaplan-Meier plot showing differences in overall survival for lung cancer patients in the bulk discovery dataset assigned to endothelial cell states S03 ($n = 36$) and S04 ($n = 30$). Statistical significance was calculated by a two-sided log-rank test. (**g**) Example R script for analyzing the association between the abundance of endothelial state S04 and overall survival in the bulk discovery dataset. Cox proportional hazards regression was used to calculate statistical significance and adjust for histological subtype

3. EcoTyper also provides an output directory called *Ecotypes*, consisting of the following files:
   (a) *nclusters_jaccard.pdf*: A plot depicting the number of clusters (or "ecotypes") obtained by clustering the Jaccard index matrix and using the average silhouette width maximization method for cluster number selection (Fig. 3c). This is the number of ecotypes prior to filtering those with less than three cell states.
   (b) *ecotypes.txt*: The cell state composition of each ecotype (the set of cell states comprising each ecotype).
   (c) *ecotype_abundance.txt*: The relative abundance inferred for each ecotype across all samples in the discovery dataset.
   (d) *ecotype_assignment.txt*: The assignment of samples in the discovery dataset to ecotypes. A *Q*-value is calculated to assess the probability of ecotype assignment for each sample by comparing the cell state abundances of a sample in a given ecotype versus the cell state abundances of the same sample in all other ecotypes (a two-sided unpaired t-test with unequal variance corrected by the Benjamini-Hochberg method). Each sample is then assigned to the ecotype with the highest ecotype abundance if (i) its corresponding *Q*-value (provided in column *AssignmentQ*) is less than 0.25 and (ii) the sample is assigned to at least one cell state contributing to the ecotype. Otherwise, the sample is unassigned. The samples not assigned to any ecotype are filtered out from this file. Additional information from the user-provided annotation file is also included.
   (e) *heatmap_assigned_samples_viridis.pdf*: A heat map of cell state fractions inferred by EcoTyper across the samples assigned to ecotypes (Fig. 3d). If an annotation file has been provided and the field *Annotation file column(s) to plot* has been set, annotation variables will be shown as color bars over the heat map. For example, in the example dataset analyzed here, ecotype 4 (E4) is highly enriched in samples obtained from adjacent normal tissue (Fig. 3d). Similarly, ecotypes 1 and 6 are enriched in squamous cell carcinoma samples, while other ecotypes are enriched in adenocarcinoma samples.
   (f) *jaccard_matrix.pdf*: A heat map of the Jaccard index matrix, after filtering ecotypes with less than three cell states (Fig. 3e).

3.1.13 Example Downstream Analyses

Output files produced by bulk discovery mode may be used for a variety of downstream analyses.

For example, the *state_assignment.txt* files can be used for enrichment analyses and comparison with other sample labels that are available for the discovery dataset. In Fig. 3b, the samples assigned to endothelial cell state S01 are largely from adjacent normal tissue (shown in blue in the "Tissue" color bar above the heat map), while the samples assigned to endothelial cell states S02–S04 are exclusively derived from tumor tissue (shown in tan in the "Tissue" color bar). Similarly, endothelial cell state S02 is highly enriched for samples of lung squamous cell carcinoma, while endothelial cell state S04 is highly enriched for lung adenocarcinoma samples.

Another example is to interrogate the cell states for associations with clinical outcomes. Figure 3f shows a Kaplan-Meier plot and log-rank test comparing the overall survival associations between lung cancer patients in the discovery dataset assigned to endothelial cell states S03 and S04 (*state_assignment.txt* file). Patients assigned to endothelial cell state S04 show significantly longer survival time compared to patients assigned to endothelial cell state S03 (log rank $P = 0.042$). The survival follow-up data necessary to conduct this analysis is provided in the annotation file of the example dataset.

For downstream analyses including all samples in the input dataset, one can use the output files *state_abundances.txt* or *ecotype_abundance.txt*. Such analyses can model cell state or ecotype abundance as a continuous variable using Cox proportional hazards regression (*coxph* from the R package *survival* [10]). To calculate continuous survival associations for EcoTyper-derived states and ecotypes, the relative abundance estimates can be provided as explanatory variables to the model. In Fig. 3g, we show an example script where endothelial cell state S04 is provided as a continuous explanatory variable to *coxph()*, adjusting for the categorical variable "Histology," which specifies whether a sample is LUAD or LUSC. This analysis shows that endothelial cell state S04 is significantly associated with longer overall survival time after adjustment for histological subtype ($P = 0.027$).

Other examples of relevant downstream analyses, such as gene set enrichment analyses, ligand-receptor interactions, and associations with immunotherapies, can be found in Luca et al. [4] and Steen et al. [6].

3.2 Discovery of Cell States and Ecotypes from scRNA-Seq Data

In the previous section, we reviewed how to perform cell state and ecotype discovery from bulk tissue specimens. Here we describe how to discover cell states and ecotypes directly from scRNA-seq data (Fig. 1). Prior to analysis with EcoTyper, users must perform preprocessing of scRNA-seq data to remove low-quality cells, to annotate major cell types, and to integrate across technical batches (e.g., using Seurat or Scanpy). To illustrate the concepts, commands, and files covered in this section, we have included an example discovery dataset consisting of a down-sampled colorectal cancer scRNA-seq atlas [11] (*example_data/scRNA_CRC_data.txt*) along with a sample annotation file (*example_data/scRNA_CRC_annotation.txt*). Key steps are depicted schematically in Fig. 4a. As several steps are the same as those covered in Subheading 3.1, here we emphasize steps that are specific to scRNA-seq discovery mode.

3.2.1 Extract Cell-Type-Specific Genes from scRNA-seq Input Matrix (Optional)

To limit cell state discovery to states that are cell-type-specific, rather than ubiquitous, and to improve the performance of cell state and ecotype recovery in bulk data (*see* Subheading 3.3), EcoTyper starts (by default) with the removal of genes in scRNA-seq data that are not over-expressed in a given cell type, relative to other cell types. Given considerations of computational efficiency and balanced representation, up to 500 cells per cell type are randomly selected and used for this step. Differential expression is then calculated between each cell type and the remaining cell types. Genes with a Q-value >0.05 (two-sided Wilcoxon test with Benjamini-Hochberg correction) are omitted from further analysis. This operation is performed by default in *step 1*, *Extract cell-type-specific genes* of scRNA-seq discovery mode (Fig. 4a).

3.2.2 Cell State Discovery

Round One

Similar to bulk data (Subheading 3.1.3), EcoTyper applies NMF to identify cell states from scRNA-seq data. However, to combat the dropout and sparsity inherent to scRNA-seq data, EcoTyper first constructs a cross-correlation matrix for each cell type and then applies NMF to each cross-correlation matrix in order to identify cell states. For efficiency, EcoTyper randomly samples up to 2500 single-cell transcriptomes per cell type to build each cross-correlation matrix. As described in Subheading 3.1.3, NMF should be run ~5 times for exploratory purposes and up to 50 times for publication-quality results. This procedure is performed in *step 2*, *Cell state discovery on correlation matrices* of EcoTyper Bulk Discovery Mode (Fig. 4a; *see* Subheading 3.2.7).

Number of Cell States per Cell Type

The number of states per cell type is determined as described in Subheading 3.1.4.

Round Two

Although the use of cross-correlation matrices can overcome sparsity and dropout in scRNA-seq data, the resulting NMF models are dependent on the number of input cells and cannot be used to recover cell states in new data. To address this issue, EcoTyper performs a second round of NMF in which (i) the creation of cross-correlation matrices is bypassed and (ii) NMF is only applied to cell-state-specific marker genes in the scRNA-seq data. In this way, NMF will be conditioned to discover the same cell states as identified in round one while also generating models that enable the recovery of cell states (and ecotypes) in new datasets.

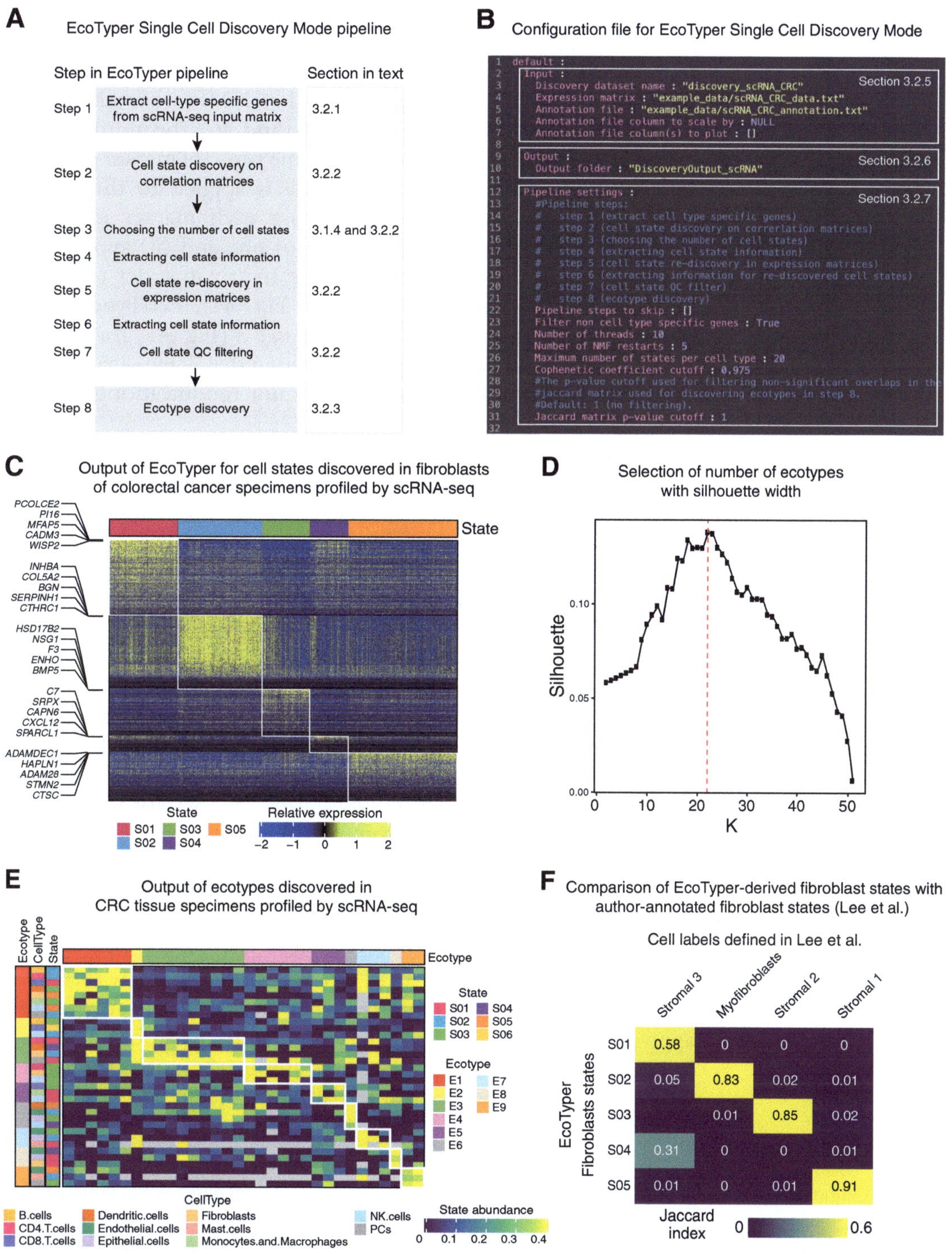

Fig. 4 Single-cell RNA-seq discovery mode and its application to colorectal cancer. (**a**) Typical EcoTyper workflow for the discovery of cell states and ecotypes from scRNA-seq data. Key steps are indicated along with their corresponding sections in this chapter. (**b**) Configuration file for scRNA-seq discovery mode using an example discovery dataset ($n = 13{,}780$ single-cell transcriptomes from 23 tumor and 10 normal tissue specimens obtained from 23 colorectal cancer patients; Lee et al. [11]). (**c**) Heat map of fibroblast states identified by EcoTyper in the scRNA-seq discovery dataset. Selected cell state marker genes are indicated on

To accomplish this, EcoTyper first identifies marker genes for each cell state by $\log_2$ fold change. Marker genes are ranked by $\log_2$ fold change within each cell state, and the top K/N markers per cell state are selected for the second round of NMF, where $K = 1000$ (or the total number of genes if <1000) and $N =$ the number of cell states (i.e., if 5 cell states are detected, the top 200 marker genes per state are selected). In this case, NMF is applied to the original scRNA-seq data after (i) standardizing each marker gene to a mean of zero and standard deviation of one and (ii) applying the posneg transformation to the resulting matrix in order to enforce nonnegativity. In this step (*step 5* in Fig. 4a), EcoTyper applies NMF once using the rank selected in *step 3* (Fig. 4a) for each cell type. Quality control filtering performed in *step 7* (Fig. 4a) proceeds as described for bulk data with one minor modification (*see* **Note 12**).

3.2.3 Ecotype Discovery

Ecotype detection from scRNA-seq data is the same as described in Subheading 3.1.6. As this step involves the identification of cell state co-association patterns across tissue samples, we recommend analyzing scRNA-seq data from at least 10 distinct samples to obtain the most reliable results.

3.2.4 Configuring the EcoTyper Run

The script for cell type and ecotype discovery from scRNA-seq data is *EcoTyper_discovery_scRNA.R*. It has three parts: (i) input, (ii) output, and (iii) pipeline settings (Fig. 4b). Given similarities with the bulk data configuration file (*see* Subheadings 3.1.7, 3.1.8, 3.1.9, and 3.1.10), we focus on aspects unique to scRNA-seq discovery mode below. The configuration file used for the example discovery dataset is *config_discovery_scRNA.yml*.

3.2.5 Input Section of Configuration File

Expression Matrix

The *Expression matrix* field specifies the path to a tab-delimited file containing the scRNA-seq discovery dataset, with genes as rows and cells as columns. We recommend TPM, counts-per-million (CPM), or SCTransform [12] normalization space without log adjustment. Users should perform their own quality control before applying EcoTyper (e.g., to filter low-quality cells, doublets, etc.); however, we do not recommend pre-filtering the matrix for variable genes as this will be done internally by default (*see* Subheading

Fig. 4 (continued) the left side of the heat map. The color bar above the heat map denotes the dominant cell state to which each single-cell transcriptome was assigned. (**d**) Silhouette plot showing the optimal number of ecotypes identified in scRNA-seq data (dashed red line). (**e**) Heat map showing ecotype abundances inferred by EcoTyper from the scRNA-seq discovery dataset. Rows are cell states ($n = 34$) columns are patient samples assigned to ecotypes ($n = 32$), and both axes are ordered by ecotypes. The color bar above the heat map denotes the dominant ecotype to which each sample is assigned. (**f**) Jaccard matrix comparing EcoTyper-derived fibroblast states with author-annotated fibroblast states (Lee et al. [11]) in the scRNA-seq discovery dataset

3.2.1). The matrix should have gene symbols in the first column and gene expression for each cell in the remaining columns. Column (cell) names should be unique. In this section, we use *example_data/scRNA_CRC_data.txt*.

Annotation File

A path to an annotation file, such as the example available in *example_data/scRNA_CRC_annotation.txt*, should be provided in the field *Annotation file*. This file must contain at least three columns: ID, CellType, and Sample. The "ID" column must have the same names (e.g., cell barcodes) as the columns of the expression matrix. The "CellType" column denotes the cell-type annotation of each cell. The "Sample" column indicates the sample identifier for each cell (e.g., "Tumor 1") and is used for ecotype discovery. We recommend at least 10 tissue samples for the de novo discovery of cell states and ecotypes.

The annotation file can contain any number of additional columns for use as meta-data. These columns can be used for defining sample batches and for plotting color bars in the heat map output (*see* Subheading 3.1.8).

3.2.6 Output Section of Configuration File

The Output section contains a single field, *Output folder*, which specifies the path where the final output will be saved. This folder will be created if it does not exist. For the example dataset in this section, set this field to "DiscoveryOutput_scRNA."

3.2.7 Pipeline Settings

Filter Non-Cell-Type-Specific Genes

This parameter controls whether to apply the filter for cell-type-specific genes as outlined in Subheading 3.2.1. We recommend setting this field to "TRUE" to limit cell state discovery to cell-type-specific states.

Jaccard Matrix p-Value Cutoff

Ecotype identification is performed by clustering a Jaccard index matrix that quantifies the sample overlap between each pair of states. Prior to ecotype identification, the Jaccard matrix values corresponding to pairs of states for which the sample overlap is not significant are set to 0. The value provided in the *Jaccard matrix p-value cutoff* field specifies the p-value cutoff above which cell state overlaps are considered nonsignificant. When the number of samples in the scRNA-seq dataset is modest (i.e., <50), such as in the current example, we recommend disabling this filter by setting the p-value cutoff to 1. However, if the dataset has at least 50 samples, we encourage users to set this cutoff to an alpha value that enforces nominal statistical significance (e.g., 0.05 or lower).

The remaining pipeline settings are the same as for bulk discovery (*see* Subheading 3.1.10).

3.2.8 Run Single-Cell Discovery Mode

After editing the configuration file (*config_discovery_scRNA.yml*), the scRNA-seq discovery script can be executed by running the line of code below. This script may take 24–48 h to run on 10 threads. Of note, EcoTyper may not be runnable on the example data from this tutorial using a typical laptop with 16GB of memory or less. Therefore, we recommend executing the script on a server or high-performance computing cluster with >50–100GB of RAM (*see* **Note 9**).

```
Rscript EcoTyper_discovery_scRNA.R -c config_dis-
covery_scRNA.yml
```

3.2.9 Overview of Output Files

The output files generated by scRNA-seq discovery mode are the same as those described for bulk discovery mode (*see* Subheading 3.1.12), with the following exception:

In contrast to cell-type-specific sample-level gene expression profiles obtained from bulk samples, which may represent mixtures of cell states, each single cell is only assigned to one specific cell state. Therefore, the main output file for each cell type is the file *state_assignment.txt*, which contains the cell state assignments of single cells assigned to clusters that passed quality control filters. Cells are columns and cell states are rows. Output files from the example dataset are shown in Fig. 4c–e.

3.2.10 Example Downstream Analysis

The scRNA-seq dataset analyzed in this section includes author-supplied cell-type annotations. In Fig. 4f, we show the overlap between fibroblast cell states identified by EcoTyper (Fig. 4c) and fibroblast states defined by Lee et al. for the same cells. The overlap is calculated using the Jaccard index. While EcoTyper states S01, S02, S03, and S05 show substantial overlap with author-defined states, state S04 represents a novel state. Despite similarities in cell state definitions, one advantage of using EcoTyper for cell state discovery is the ability to recover cell states and ecotypes in external single-cell and bulk expression datasets. Examples of cell state and ecotype recovery are described below.

3.3 Recovery of Cell States and Ecotypes from Bulk Samples

EcoTyper comes preloaded with the resources necessary for reference-guided recovery of previously defined carcinoma- and lymphoma-specific cell states and ecotypes [4, 6] in user-provided bulk expression data (*see* **Note 1**). Please note that the recovery procedure described here can also be applied to user-defined cell states and ecotypes (*see* Subheadings 3.1 and 3.2).

This section includes two example datasets in the *example_data* directory:

1. A subset of bulk LUAD transcriptomes from TCGA, available in *bulk_lung_data.txt* with a sample annotation file *bulk_lung_annotation.txt*. Of note, this is the same example dataset used in Subheading 3.1.
2. Bulk DLBCL tumor transcriptomes [13] available in *bulk_lymphoma_data.txt* along with the annotation file *bulk_lymphoma_annotation.txt*.

3.3.1 Run Recovery from Bulk Data

The script used to perform recovery in bulk data is *EcoTyper_recovery_bulk.R* and takes the following arguments:

-d/–discovery The name of the discovery dataset used for defining cell states. Users can specify *Carcinoma* or *Lymphoma* (case sensitive) to recover cell states and ecotypes that were previously defined across carcinomas and lymphomas, respectively [4, 6]. Alternatively, for user-defined cell states and ecotypes (*see* Subheadings 3.1 and 3.2), the name of the discovery dataset is the value provided in the *Discovery dataset name* field of the configuration file used for running cell state discovery (*see* Subheading 3.1.8).

-m/–matrix Path to the input expression matrix. The expression matrix should be in the TPM or FPKM space for bulk RNA-seq data and non-logarithmic (exponential) space for microarrays. It should have gene symbols in the first column and gene expression for each sample in the remaining columns. Column (sample) names should be unique (*see* **Note 8**).

-a/–annotation Path to a tab-delimited annotation file. This argument is optional. If provided, the annotation file should contain a column called ID with the same values as the columns of the expression matrix. Additionally, this file can contain any number of meta-data columns that can be used for plotting color bars in the heat map outputs (see argument *-c/--columns*).

-c/–columns A comma-separated list of column names from the annotation file (see argument *-a/--annotation*) to be plotted as color bars in the heat map outputs. By default, heat maps contain a color bar denoting the cell state label each cell is assigned to. The column names indicated by this argument will be added to this color bar.

-t/–threads Number of threads to be used for running the analysis; the default is 10.

-o/–output Output folder. The output folder will be created if it does not already exist.

3.3.2 Overview of Output Files

For each cell type analyzed, EcoTyper creates a directory containing the following files: *state_assignment.txt*, *state_abundances.txt*, *heatmap_data.txt*, *heatmap_top_ann.txt*, *ecotype_abundance.txt*, and *heatmap_assigned_samples_viridis.pdf*, which are identical to the files described in Subheading 3.1.12, with the following exception: the *state_assignment_heatmap.pdf* file displays heat maps of the discovery dataset (left) and the user-provided bulk dataset (right).

These output files can then be used for downstream analyses, such as the ones described in Subheading 3.1.13.

3.4 Recovery of Cell States and Ecotypes from Single-Cell RNA-seq Data

In this section, we show how cell states and ecotypes can be recovered from scRNA-seq data. Importantly, this type of analysis can validate a cell state and ecotype atlas at the single-cell level, though technical caveats, including dropout, sparsity, and dissociation-induced artifacts, may influence results [4, 8].

This section can be run with two example datasets provided in the EcoTyper *example_data* directory:

1. A down-sampled colorectal cancer scRNA-seq atlas [11] (*scRNA_CRC_data.txt*) and sample annotation file (*scRNA_CRC_annotation.txt*). Of note, this is the same example dataset used in Subheading 3.2.
2. A down-sampled B-cell lymphoma scRNA-seq atlas [6] (*scRNA_lymphoma_data.txt*) and sample annotation file (*scRNA_lymphoma_annotation.txt*).

3.4.1 Run Recovery from scRNA-seq Data

The script used to perform recovery in scRNA-seq data is *EcoTyper_recovery_scRNA.R*. It takes the same input arguments as the script described in Subheading 3.3.1, with the following additions:

-z/–z-score Boolean flag indicating whether the significance of cell state recovery should be calculated (default is FALSE). This optional procedure allows users to determine whether cell states are significantly recovered in a given dataset. This procedure is detailed in Luca et al. [4] (*Significance of cell state recovery* in STAR Methods). Of note, the significance calculation can be very slow, as the NMF model is applied 30 times per cell type to generate null distributions.

-s/–subsample An integer specifying the maximum number of cells per cell type. Cell types exceeding this number will be randomly down-sampled. For values <50, no down-sampling will be performed. Default = −1 (no down-sampling).

3.4.2 Overview of Output Files

For each cell type analyzed, EcoTyper outputs a directory containing text files with cell state details (*state_assignment.txt*), along with output plots (*state_assignment_heatmaps.pdf*). These files are identical to those described in Subheading 3.2.9, with the exception that the output plot displays the expression of cell state marker genes in the discovery dataset (left) and the scRNA-seq dataset (right), with the latter smoothed to mitigate the impact of scRNA-seq dropout.

3.5 Recovery of Cell States and Ecotypes from Spatial Transcriptomics Data

As described in Subheadings 3.3 and 3.4, EcoTyper provides a framework to interrogate cell states and ecotypes in external datasets, whether they are obtained from bulk or single-cell cell expression profiling platforms. This also extends to spatial transcriptomics data, for example, as obtained with the 10× Visium platform (10× Genomics). In this section, we show how to perform cell state and ecotype recovery from Visium data. A publicly available Visium profile of a breast cancer specimen is used as an example (https://www.10xgenomics.com/resources/datasets/human-breast-cancer-block-a-section-1-1-standard-1-0-0).

3.5.1 Overview of Input Files

The Visium input files required for EcoTyper are the filtered feature-barcode matrices *barcodes.tsv.gz*, *features.tsv.gz*, and *matrix.mtx.gz* in the format provided by 10× Genomics and *the tissue_positions_list.csv* file produced by the 10× Genomics Space Ranger pipeline, the latter of which provides the spatial position of each spot barcode. The example files for this section are available in *example_data/VisiumBreast.*

3.5.2 Configuration File

The script that performs cell type and ecotype discovery from bulk Visium data is *EcoTyper_recovery_visium.R*. This script takes as input a configuration file in YAML format. The example configuration file used here is *config_discovery_scRNA.yml.* It consists of three parts: (i) input, (ii) output, and (iii) pipeline settings (Fig. 5a). Here we focus on aspects that differ from Subheadings 3.1.7 and 3.2.4.

Input Visium Directory

The four input files enumerated above must be stored in one directory. The path to that directory is specified in this field. For the example dataset used in this section, the path is *example_data/VisiumBreast.*

Recovery Cell-Type Fractions

This field specifies the path to a file containing the cell-type fraction estimations for each spot on the Visium array. This field is ignored when the *discovery dataset name* field has been set to *Carcinoma* or *Lymphoma*, or when the discovery has been performed as described in Subheading 3.1.8 *Cell-type fractions* using *Carcinoma_Fractions* or *Lymphoma_Fractions.* However, if users have determined custom cell-type fractions in Subheading 3.1.8, this field must point to a tab-delimited text file containing cell-type fractions for each spot in the Visium array. The cell-type labels must match those used for cell state/ecotype discovery, and the cell-type fractions must sum to 1 for each spot. Users are free to select any suitable deconvolution method for this purpose, including cell2location [14], RCTD [15], or CIBERSORTx [8]. An example file is "*example_data/visium_-fractions_example.txt.*" However, for the example Visium array used in this section, this field should be set to "NULL."

Background Cell Type

If the tissue specimen analyzed is a tumor specimen, this field specifies the label of the malignant cell type being analyzed. This field is used for plotting a gray background in the resulting output plot, with the intensity of gray depicting the abundance of the malignant cell type. The field is ignored when the *discovery dataset name* field has been set to *Carcinoma* or Lymphoma, or when the discovery has been performed as described in Subheading 3.1.8 *Cell-type fractions*, using *Carcinoma_Fractions* or *Lymphoma_Fractions.* In these cases, the background cells are automatically considered to be originating from *Epithelial.cells* or *B.cells*, the malignant cell populations in carcinomas and B-cell lymphomas, respectively. Otherwise, this field may specify a column name in the file provided to the *Recovery cell-type fractions* field. If no column name is specified, all spots in the Visium dataset will be shown in gray.

3.5.3 Run Recovery from Spatial Transcriptomics Data

After editing the configuration file, the command line for recovering cell states and ecotypes in Visium data can now be run. Please note that if spot-level deconvolution results are not provided by the user, this script will invoke CIBERSORTx for cell-type fraction estimation, which can take up to 2 h to run.

```
Rscript EcoTyper_recovery_visium.R -c config_re-
covery_visium.yml
```

3.5.4 Output Files

The following output files are generated:

- "*state_abundances.txt*": The columns are spot IDs, X/Y coordinates for each spot in the Visium array, and the inferred cell state abundances across all cell types.
- "*CellTypeLabel_spatial_heatmaps.pdf*," where CellTypeLabel is the annotation given to each cell type: Plots showing cell state abundances for each cell type. The intensity of charcoal represents cell state abundance. The intensity of gray represents the fraction of the background population (*see* Subheading 3.5.2; Fig. 5b).

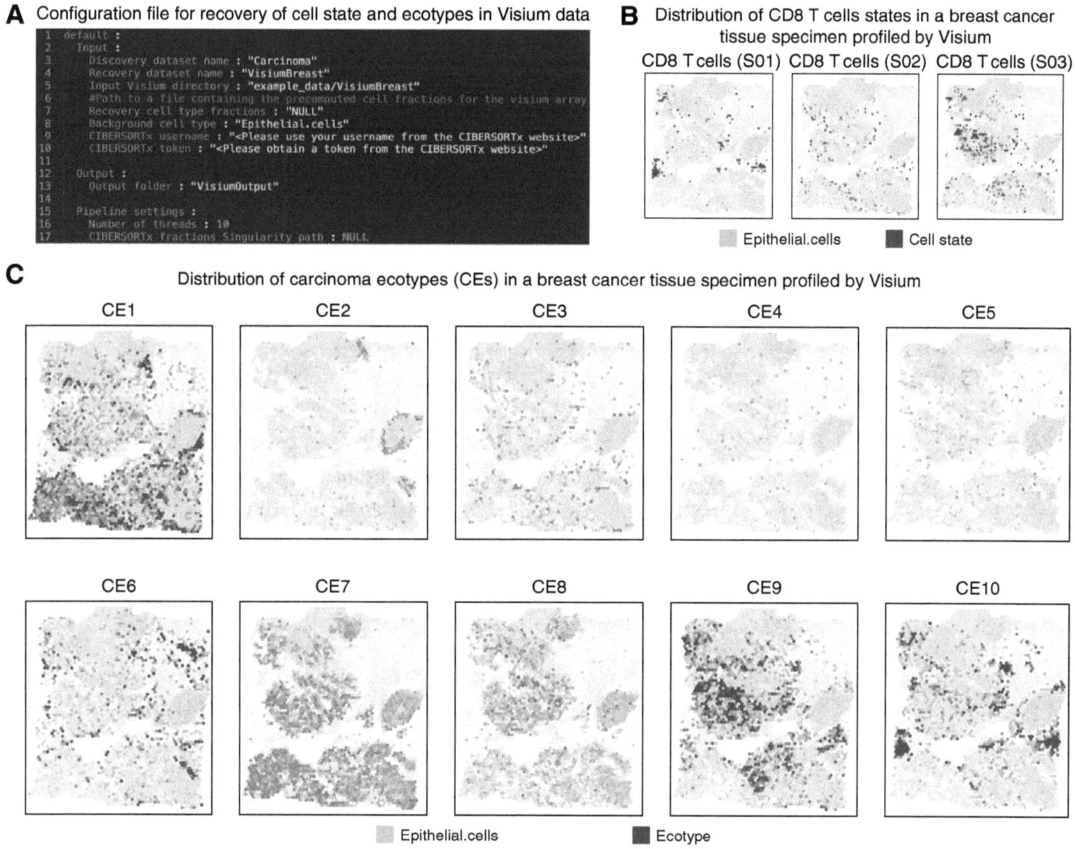

Fig. 5 Application of EcoTyper to spatial transcriptomics data. (**a**) Example configuration file for running EcoTyper on Visium data. (**b**, **c**) Heat maps showing the relative abundances of (**b**) CD8 T cell states and (**c**) carcinoma ecotypes (CEs) recovered by the Carcinoma EcoTyper [4] in a Visium breast cancer profile

- "*ecotype_abundances.txt*": The relative abundance inferred for each ecotype across all spots in the Visium array.
- "*ecotype_abundances.pdf*": Plots depicting inferred ecotype abundances. The intensity of charcoal represents ecotype abundance. The intensity of gray represents the fraction of the background population (*see* Subheading 3.5.2; Fig. 5c).

3.5.5 Downstream Analysis

Among various downstream analyses, users can determine (i) the distance of each cell state or ecotype to predefined geographic regions (e.g., tumor or stromal regions), (ii) patterns of spatial organization or aggregation, and (iii) potential cell state/ecotype interaction patterns, including heterotypic ligand-receptor pairings, as demonstrated in Luca et al. [4] and Steen et al. [6].

3.6 Conclusion

EcoTyper is the first dedicated machine learning framework for cell state and ecotype discovery from bulk and single-cell gene expression data. Once defined, cell states and ecotypes can be recovered in external datasets and across platforms, including bulk, single-cell, and spatially resolved gene expression data. As such, we anticipate that EcoTyper will prove useful for understanding the cellular organization of complex tissues in health and disease, with implications for the discovery of new candidate biomarkers and therapeutic targets.

4 Notes

1. The EcoTyper website enables the exploration and recovery of cell states and ecotypes previously identified in carcinoma and lymphoma [4, 6]. With the exception of melanoma, where carcinoma cell states have been previously observed and validated [4], users studying other tissue types or disease states should either (i) carefully validate their results obtained with Carcinoma or Lymphoma EcoTyper (*see* Subheadings 3.3 and 3.4), or (ii) perform a custom discovery of cell states and ecotypes (*see* Subheadings 3.1 and 3.2). Importantly, the EcoTyper website does not provide the latter functionality; discovery modules are only available via the EcoTyper source code, as described in this chapter.
2. In some instances, such as discovery from bulk expression data (*see* Subheading 3.1) or recovery from Visium spatial transcriptomics data (*see* Subheading 3.5), EcoTyper uses CIBERSORTx by default [8]. To run CIBERSORTx, the following resources are needed: (i) Docker or Singularity and (ii) the CIBERSORTx executables available from https://cibersortx.stanford.edu/ (*Menu* > *Download*). Users must follow the instructions on the Download section of the CIBERSORTx website to download the Docker images and obtain the token necessary for running them. If Singularity is used, the Docker images need to be converted to Singularity Image Files (SIF). A link to the instructions can be found in the EcoTyper GitHub repository, under the "Frequently Asked Questions" section.
3. Some EcoTyper functions are computationally intensive, especially the cell state discovery modules described in Subheadings 3.1 and 3.2. Therefore, EcoTyper is designed as a collection of modular command-line R scripts that can be run in parallel on a multiprocessor server or a high-performance computing cluster (also, *see* **Note 9**).

4. EcoTyper Bulk Discovery Mode may also be run starting from cell-type-specific expression matrices, obtained either through FACS-purifying cell populations of interest and then performing bulk tissue expression profiling of each cell population, or by performing in silico purification using a deconvolution method comparable to CIBERSORTx.
5. CIBERSORTx High Resolution mode employs an adaptive noise filter that eliminates cell-type-specific genes that cannot be reliably estimated [8]. The more abundant a cell type is, the more genes are imputed.
6. In *step 3*, *Cell State Discovery* (Fig. 2a), EcoTyper identifies cell states with NMF. If, for a given cell type, more than 1000 genes are available (e.g., >1000 genes imputed by CIBERSORTx High Resolution mode; Subheading 3.1.2), the top 1000 genes with the highest relative dispersion across samples are used for NMF (*Cell state discovery* in STAR Methods in Luca et al. [4]). Otherwise, all genes are used as input. If <50 genes are imputed for a given cell type, that cell type is not used for cell state identification [4].
7. When run in bulk discovery mode, EcoTyper calculates a "dropout score" to identify poor quality cell states based on marker genes that preferentially show a pattern of low variance and high expression across samples. The dropout score is used for cell state filtering in *step 7*, *Advanced cell state QC filter* (Fig. 2a). Calculation of the dropout score is described in Luca et al. (*Cell state quality control* in STAR Methods) [4].
8. Column names should not contain special characters that are modified by the R function *make.names*. For example, names should not contain digits at the beginning or special characters, including space, tab, or dash.
9. EcoTyper can be modified to run on a high-performance cluster by overriding the *pipeline/lib/multithreading.R* library. Currently the library provides two functions, *PushToJobQueue*, which adds a command line call to the job queue, and *RunJobQueue*, which waits for all the jobs in the queue to finish. The default implementation of these functions uses the R function *mclapply* to perform computations on multiple threads. Users can rewrite these two functions according to the requirements of their cluster infrastructure. A primitive example of how this can be achieved on a high-performance computing cluster built on the SLURM infrastructure is provided in the EcoTyper GitHub repository, under the "Frequently Asked Questions" section.

10. The plots displaying the cophenetic coefficient indicate the number of cell states obtained before applying the filters for low-quality states in *steps 6* and *7* (Fig. 2a). Therefore, the results will likely contain fewer cell states once filtering is complete.
11. The cophenetic coefficient plot shown in Fig. 3a might look slightly different than the one obtained by a user. This is because some EcoTyper steps, including the NMF algorithm initialization step, depend on stochastic initialization, and may show modest variation across runs. For stable results, we recommend at least 5 NMF restarts for exploratory analyses and 50 restarts when running EcoTyper for publication-quality results.
12. The "dropout score" (*see* **Note 7**) is not applied when the input is scRNA-seq data.

Acknowledgments

We would like to thank E. Brown for providing critical feedback on the manuscript. This work was supported by grants from the American Association for Cancer Research (C.B.S., 19-40-12-STEE), the National Cancer Institute (A.M.N., R01CA255450 and R00CA187192; A.J.G., U54CA209971 and U24CA224309; A.A.A., R01CA233975), the Stanford Bio-X Interdisciplinary Initiatives Seed Grants Program (IIP) (A.M.N.), the Donald E. and Delia B. Baxter Foundation (A.M.N.)., the Virginia and D.K. Ludwig Fund for Cancer Research (A.A.A., A.M.N.)., the Stinehart-Reed Foundation (A.A.A., A.M.N.), the Bakewell Foundation (A.A.A.), the SDW/DT and Shanahan Family Foundations (A.A.A.) and the Fund for Cancer Informatics (A.J.G.).

References

1. Armingol E, Officer A, Harismendy O, Lewis NE (2021) Deciphering cell-cell interactions and communication from gene expression. Nat Rev Genet 22(2):71–88. https://doi.org/10.1038/s41576-020-00292-x
2. Elmentaite R, Dominguez Conde C, Yang L, Teichmann SA (2022) Single-cell atlases: shared and tissue-specific cell types across human organs. Nat Rev Genet 23:395. https://doi.org/10.1038/s41576-022-00449-w
3. Keidar Haran T, Keren L (2022) From genes to modules, from cells to ecosystems. Mol Syst Biol 18(5):e10726. https://doi.org/10.15252/msb.202110726
4. Luca BA, Steen CB, Matusiak M, Azizi A, Varma S, Zhu C, Przybyl J, Espin-Perez A, Diehn M, Alizadeh AA, van de Rijn M, Gentles AJ, Newman AM (2021) Atlas of clinically distinct cell states and ecosystems across human solid tumors. Cell 184(21):5482–5496 e5428. https://doi.org/10.1016/j.cell.2021.09.014
5. Nirmal AJ, Maliga Z, Vallius T, Quattrochi B, Chen AA, Jacobson CA, Pelletier RJ, Yapp C, Arias-Camison R, Chen YA, Lian CG, Murphy GF, Santagata S, Sorger PK (2022) The spatial landscape of progression and immunoediting in primary melanoma at single-cell resolution. Cancer Discov 12(6):1518–1541. https://doi.org/10.1158/2159-8290.CD-21-1357

6. Steen CB, Luca BA, Esfahani MS, Azizi A, Sworder BJ, Nabet BY, Kurtz DM, Liu CL, Khameneh F, Advani RH, Natkunam Y, Myklebust JH, Diehn M, Gentles AJ, Newman AM, Alizadeh AA (2021) The landscape of tumor cell states and ecosystems in diffuse large B cell lymphoma. Cancer Cell 39(10):1422–1437.e1410. https://doi.org/10.1016/j.ccell.2021.08.011
7. Tatlow PJ, Piccolo SR (2016) A cloud-based workflow to quantify transcript-expression levels in public cancer compendia. Sci Rep 6: 39259. https://doi.org/10.1038/srep39259
8. Newman AM, Steen CB, Liu CL, Gentles AJ, Chaudhuri AA, Scherer F, Khodadoust MS, Esfahani MS, Luca BA, Steiner D, Diehn M, Alizadeh AA (2019) Determining cell type abundance and expression from bulk tissues with digital cytometry. Nat Biotechnol 37(7): 773–782. https://doi.org/10.1038/s41587-019-0114-2
9. Steen CB, Liu CL, Alizadeh AA, Newman AM (2020) Profiling cell type abundance and expression in bulk tissues with CIBERSORTx. Methods Mol Biol 2117:135–157. https://doi.org/10.1007/978-1-0716-0301-7_7
10. Therneau TM, Grambsch PM (2000) The cox model. In: Modeling survival data: extending the Cox model. Springer, pp. 39–77
11. Lee HO, Hong Y, Etlioglu HE, Cho YB, Pomella V, Van den Bosch B, Vanhecke J, Verbandt S, Hong H, Min JW, Kim N, Eum HH, Qian J, Boeckx B, Lambrechts D, Tsantoulis P, De Hertogh G, Chung W, Lee T, An M, Shin HT, Joung JG, Jung MH, Ko G, Wirapati P, Kim SH, Kim HC, Yun SH, Tan IBH, Ranjan B, Lee WY, Kim TY, Choi JK, Kim YJ, Prabhakar S, Tejpar S, Park WY (2020) Lineage-dependent gene expression programs influence the immune landscape of colorectal cancer. Nat Genet 52(6):594–603. https://doi.org/10.1038/s41588-020-0636-z
12. Hafemeister C, Satija R (2019) Normalization and variance stabilization of single-cell RNA-seq data using regularized negative binomial regression. Genome Biol 20(1):296. https://doi.org/10.1186/s13059-019-1874-1
13. Chapuy B, Stewart C, Dunford AJ, Kim J, Kamburov A, Redd RA, Lawrence MS, Roemer MGM, Li AJ, Ziepert M, Staiger AM, Wala JA, Ducar MD, Leshchiner I, Rheinbay E, Taylor-Weiner A, Coughlin CA, Hess JM, Pedamallu CS, Livitz D, Rosebrock D, Rosenberg M, Tracy AA, Horn H, van Hummelen P, Feldman AL, Link BK, Novak AJ, Cerhan JR, Habermann TM, Siebert R, Rosenwald A, Thorner AR, Meyerson ML, Golub TR, Beroukhim R, Wulf GG, Ott G, Rodig SJ, Monti S, Neuberg DS, Loeffler M, Pfreundschuh M, Trumper L, Getz G, Shipp MA (2018) Molecular subtypes of diffuse large B cell lymphoma are associated with distinct pathogenic mechanisms and outcomes. Nat Med 24(5):679–690. https://doi.org/10.1038/s41591-018-0016-8
14. Kleshchevnikov V, Shmatko A, Dann E, Aivazidis A, King HW, Li T, Elmentaite R, Lomakin A, Kedlian V, Gayoso A, Jain MS, Park JS, Ramona L, Tuck E, Arutyunyan A, Vento-Tormo R, Gerstung M, James L, Stegle O, Bayraktar OA (2022) Cell2location maps fine-grained cell types in spatial transcriptomics. Nat Biotechnol 40(5):661–671. https://doi.org/10.1038/s41587-021-01139-4
15. Cable DM, Murray E, Zou LS, Goeva A, Macosko EZ, Chen F, Irizarry RA (2022) Robust decomposition of cell type mixtures in spatial transcriptomics. Nat Biotechnol 40(4): 517–526. https://doi.org/10.1038/s41587-021-00830-w

Chapter 5

Statistical Methods for Integrative Clustering of Multi-omics Data

Prabhakar Chalise, Deukwoo Kwon, Brooke L. Fridley, and Qianxing Mo

Abstract

Cancers are heterogeneous diseases caused by accumulated mutations or abnormal alterations at multi-levels of biological processes including genomics, epigenomics, transcriptomics, and proteomics. There is a great clinical interest in identifying cancer molecular subtypes for disease prognosis and personalized medicine. Integrative clustering is a powerful unsupervised learning method that has been increasingly used to identify cancer molecular subtypes using multi-omics data including somatic mutations, DNA copy numbers, DNA methylation, and gene expression. Integrative clustering methods are generally classified into model-based or nonparametric approaches. In this chapter, we will give an overview of the frequently used model-based methods, including iCluster, iClusterPlus, and iClusterBayes, and the nonparametric method, integrative nonnegative matrix factorization (intNMF). We will use the integrative analyses of uveal melanoma and lower-grade glioma to illustrate these representative methods. Finally, we will discuss the strengths and limitations of these representative methods and give suggestions for performing integrative analyses of cancer multi-omics data in practice.

Key words Integrative clustering, iCluster, iClusterPlus, iClusterBayes, NMF, intNMF, TCGA, Uveal melanoma, Lower-grade gliomas

1 Introduction

Identification of molecular subtypes of cancers using multi-omics data such as gene expression, DNA methylation, DNA copy number, and protein expression has played a very important role in understanding the molecular basis of disease prognosis and treatment response. Similar-looking cells at the morphological level can have very different molecular profiles. A few examples of molecular heterogeneity within a cell include point mutations, deletions or insertions in tumor suppressor genes, differential promoter methylation, chromosomal translocations, and so on. Studies have shown that such differing molecular heterogeneity can result in multiple cancer subtypes [1–3]. People having the same cancer diagnosed with similar clinicopathological, morphological, and

Brooke L. Fridley and Xuefeng Wang (eds.), *Statistical Genomics*,
Methods in Molecular Biology, vol. 2629, https://doi.org/10.1007/978-1-0716-2986-4_5,

functional characteristics can exhibit very different clinical outcomes with respect to the severity of the disease, response trajectory to drugs, etc. Therefore, it is clinically important to identify the subtypes of cancers at the molecular level to understand their etiology and to develop personalized medicine. One such method to identify cancer subtypes is to perform clustering analysis, which subgroups the samples based on the similarity pattern of their molecular profiles. Within the subgroups, it is thought that the tumors will be more homogeneous and thus may have similar clinical response to a given therapy regimen. There have been many successful examples of the application of clustering method to identify cancer subtypes [4–6].

Clustering can be carried out using a single type of data at a time or using multiple types of omics (multi-omics) datasets comprehensively together [7]. Multiple layers of biological data have been increasingly generated due to the advent of high throughput technologies such as microarrays and next-generation sequencing technologies. For example, the multi-institution collaborative project, The Cancer Genome Atlas (TCGA), has generated multiple layers of omics data including genome, transcriptome, epigenome, and proteome information for a large number of subjects for many cancers [6, 8–13]. The different types of datasets are assayed at different layers of biological processes. The biological signal (a systematic pattern of molecular profiles) may or may not be present in all those datasets. For example, a similarity (or dissimilarity) pattern may be observed with gene expression data for a few subjects, while such signals may be clearer with methylation data for others. Also, there might be weak but consistent signals present across several datasets. Since multiple layers of data are assayed on the same subjects, they are interrelated. Clustering analysis using a single data cannot capture such signals. Integrative clustering is a powerful approach to identify latent subtype structures inherent in the datasets accounting for both between and within data correlations. The goal of the integrative clustering analysis is to identify the subgroups of samples into a distinct class (clusters), considering the biological phenomena at several levels including gene expression, DNA methylation, DNA copy number variation (CNV), protein expression, etc. [4, 14, 15].

Different molecular data have their own characteristics in terms of their measurement methods, scales of measurement, and distribution. For example, gene mutation is usually summarized as the presence/absence of somatic mutations, methylation is measured by the ratio of methylated and unmethylated signal intensities, and gene expression is measured by the amount of production of mRNAs using microarray or sequencing technologies. The data differ widely by their quantitative scales and statistical distributions. Such disparity of the datasets poses challenge for integrative analyses. The simplest approach to integration is by concatenating multiple datasets after appropriate normalization (i.e., rescaling of the data). Then well-known methods designed for single omics data clustering are applied to the combined data matrix. Such approaches are simple and computationally efficient but tend to dilute the small signal-to-noise ratio in multiple datasets [16]. Another approach is to carry out clustering analysis using one data at a time and manually integrate the results. Such approach is time-consuming, suffers from subjective bias, and is non-reproducible. Third, clustering analyses are carried out for each data separately, the clustering assignments are combined, and another clustering method is used on the combined clustering assignment data (cluster of clusters) [17]. However, defining the similarity metric or distance is not easy with such methods.

A few parametric (model-based) [15, 18–22] and nonparametric methods [14, 16, 23, 24] have been developed for integrative clustering analysis. In this chapter, we will focus on introducing the frequently used model-based methods including iCluster, iClusterPlus and iClusterBayes [15, 20–22] and the nonparametric method integrative nonnegative matrix factorization (intNMF) [14]. To illustrate these methods, we will present the analyses of the uveal melanoma and lower-grade glioma multi-omics data from TCGA. Finally, we will discuss the strengths and limitations of the model-based methods and the nonparametric methods.

2 Model-Based Methods of Integrative Clustering

2.1 The iCluster Method

The integrative clustering (iCluster) method is a model-based approach for unsupervised joint clustering analysis of multi-omics data [15]. The major goals of the iCluster are to obtain joint sample clusters and simultaneously identify the multi-omics features contributing to the joint sample clustering (Fig. 1). The iCluster method is designed to decompose multiple continuous data matrices into the product of omics-specific weight matrices and a shared factor matrix, which can be written as

$$\boldsymbol{X}_t = \boldsymbol{W}_t \boldsymbol{Z} + \boldsymbol{E}_t,$$

where $\boldsymbol{X}_t$ is the tth data matrix with dimension of $p_t \times n$ (p_t rows of features and n columns of samples), $\boldsymbol{W}_t$ is the weight (coefficient) matrix of dimension $p_t \times k$, $\boldsymbol{Z}$ is a matrix of dimension $k \times n$ that connects all the data matrices, and $\boldsymbol{E}_t$ is a random error matrix of dimension $p_t \times n$ (Fig. 1). The iCluster model can be viewed as a Gaussian latent variable model in which the ith column vector $\boldsymbol{z}_i = (z_{i1}, z_{i2}, \ldots, z_{ik})'$ of matrix $\boldsymbol{Z}$ is associated with sample $i\,(i = 1, 2, \ldots, n)$, which is used to capture the correlative structure of multi-omics data. To obtain a likelihood-based solution of the joint matrix decomposition, the components of $\boldsymbol{Z}$ and $\boldsymbol{E}_t$ are assumed to be normally distributed. A penalized expectation-

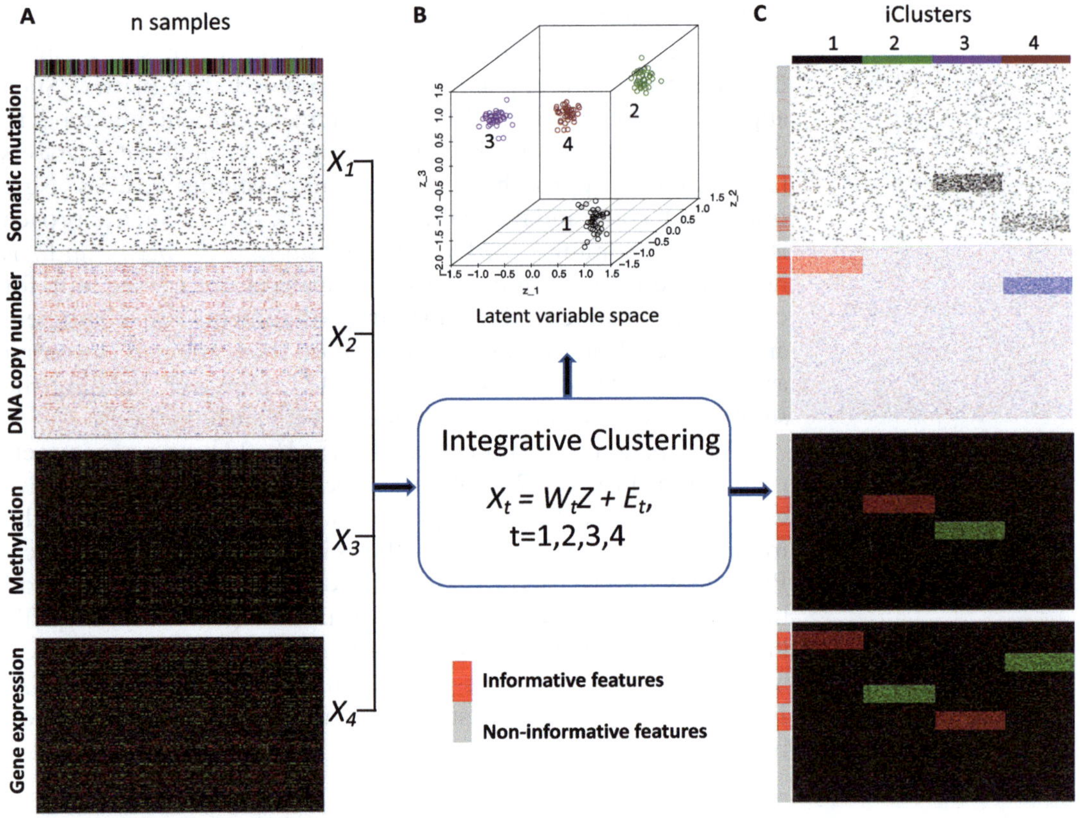

Fig. 1 Integrative clustering (iCluster) framework. (**a**) n cancer samples undergone somatic mutation, DNA copy number, promoter methylation, and mRNA gene expression analyses. Matrices X_t ($t = 1, 2, 3, 4$) represent the multi-omics datasets with n common samples (columns) and p_t ($t = 1, 2, 3, 4$) features (rows), respectively. (**b**) iCluster models are based on matrix factorization and can be considered a latent variable model in which Z is a $k \times n$ latent variable matrix and W_t is a $p_t \times k$ coefficient matrix for $t = 1, 2, 3, 4$. When $k = 3$, n samples can be divided into 4 clusters in the 3-dimensional latent variable space. (**c**) Besides performing sample clustering, iCluster models identify the informative (driver) features that drive the sample clustering. The simplified correlated patterns between DNA copy number and gene expression are shown in clusters 1 and 4, and the simplified correlated patterns between methylation and gene expression are shown in clusters 2 and 3

maximization algorithm is used to obtain optimized $\boldsymbol{W_t}$ and $\boldsymbol{Z}$ in which a lasso-like penalty term is added on $\boldsymbol{W_t}$ to generate a sparse $\boldsymbol{W_t}$ matrix. As a result, informative features that contribute to the sample clustering will have nonzero coefficients (rows) in $\boldsymbol{W_t}$ matrix, and non-informative features will have zero coefficients. Integrative sample cluster assignments are obtained by performing k-means clustering on the n column vectors of $\boldsymbol{Z}$. Therefore, the iCluster method can perform sample clustering and distinguish informative features from non-informatic features, which are important for integrative clustering analysis of multi-omics data. Different data types have different data structures that call for applying different penalty functions on the coefficient matrix $\boldsymbol{W_t}$.

For example, DNA copy number data tend to be spatially correlated along contiguous chromosomal regions, which makes the fused lasso penalty an appealing method for inducing sparsity on $\boldsymbol{W}_t$. A set of genes in a certain pathway or functional group may be co-regulated. The elastic net penalty shrinks coefficients of correlated features towards each other, resulting in a group effect by removing or selecting highly correlated omics features together. In order to accommodate different omics data structures, Shen et al. further extended the iCluster method by applying lasso, elastic net, and fused lasso penalty functions on $\boldsymbol{W}_t$ to facilitate multi-omics feature selection [22].

2.2 The iClusterPlus Method

The iCluster method is limited to modeling continuous multi-omics data, which are usually assumed to be normally distributed. In practice, multi-omics data are made up of many data types. For example, in a typical RNA-seq experiment, the expression level of a gene is usually represented by sequence count (the number of reads falling in the gene); for DNA copy number data, a chromosomal region can be classified as gain, loss, or normal; for somatic mutation data, the mutation status of a gene can be classified as mutation or normal. The iClusterPlus method can integrate four different data types including continuous, count, binary, and multi-categorical data types, which is a significant enhancement of the iCluster method [21]. Suppose n samples are analyzed by m high-throughput techniques generating m omics datasets and the tth dataset has p_t features. Let x_{ijt} denote an omics variable for the jth ($j = 1, 2, \ldots, p_t$) omics feature of the ith ($i = 1, 2, \ldots, n$) sample in the tth ($t = 1, 2, \ldots, m$) dataset. It is assumed that sample $\boldsymbol{i}$ is associated with latent variable $\boldsymbol{z_i} = (z_{i1}, z_{i2}, \ldots, z_{ik})'$, a column vector consisting of k unobserved latent variables that can be used for sample clustering. In the iClusterPlus framework, when x_{ijt} is a continuous variable, x_{ijt} and z_i are related through a standard linear regression:

$$x_{ijt} = \alpha_{jt} + \boldsymbol{\beta}_{jt}\boldsymbol{z_i} + \varepsilon_{ijt}, \varepsilon_{ijt} \sim N\left(0, \sigma_{jt}^2\right),$$

where α_{jt} is the intercept, $\boldsymbol{\beta_{jt}} = (\beta_{1jt}, \ldots, \beta_{kjt})$ are the slope coefficients, and ε_{ijt} is the random error that is assumed to be independent and normally distributed with mean 0 and variance σ_{jt}^2. When x_{ijt} is a count variable, the relationship between x_{ijt} and $\boldsymbol{z_i}$ is modeled by a Poisson regression:

$$\log\ (\lambda(x_{ijt}|\ \boldsymbol{z_i})) = \alpha_{jt} + \boldsymbol{\beta}_{jt}\boldsymbol{z_i},$$

where $\lambda(x_{ijt}|\ \boldsymbol{z_i})$ is the conditional mean of the count given $\boldsymbol{z_i}$. When x_{ijt} is a binary variable with a realized value of 0 or 1, the following logistic regression is used.

$$\log\left(\frac{P(x_{ijt}=1|\boldsymbol{z}_i)}{1-P(x_{ijt}=1|\boldsymbol{z}_i)}\right)=\alpha_{jt}+\boldsymbol{\beta}_{jt}\boldsymbol{z}_i,$$

where $P(x_{ijt} = 1|\boldsymbol{z}_i)$ is the probability of $x_{ijt} = 1$ given $\boldsymbol{z}_i$. Finally, when x_{ijt} is a multi-categorical variable, the following multinomial logistic regression is used.

$$P(x_{ijt}=c|\boldsymbol{z}_i)=\frac{\exp\left(\alpha_{jct}+\boldsymbol{\beta}_{jct}\boldsymbol{z}_i\right)}{\sum_{l=1}^{C}\exp\left(\alpha_{jlt}+\boldsymbol{\beta}_{jlt}\boldsymbol{z}_i\right),},c=1,2,\ldots,\mathrm{C},$$

where $P(x_{ijt}=c|\boldsymbol{z}_i)$ denotes the probability of x_{ijt} being the c category. To identify the omics features that make an important contribution to sample clustering, the lasso penalty is applied for the estimation of the model parameters. As a result, the coefficients $\boldsymbol{\beta}_{jt}$ are shrunk to $\boldsymbol{0}$ for the non-informative features, while the coefficients for the informative features contributing to sample clustering have nonzero components. Sample clusters are determined by the values of the latent variables $\boldsymbol{z}_i$, a length-k vector. K-means clustering is used to separate the n samples into $k + 1$ clusters. In order to achieve an optimal solution for the iClusterPlus model, a small number of k (e.g., $k = 1, 2, 3, 4, 5$) are tested, and a grid search of the lasso shrinkage parameters is performed. The optimal k is selected based on the plot of the deviance ratio or the Bayesian information criterion (BIC) versus k, where the deviance ratio or BIC becomes stable (see the example below).

2.3 The iClusterBayes Method

The iClusterBayes model is a fully Bayesian latent variable model built on the iClusterPlus framework [20]. This method applies the Bayesian variable selection algorithm to identify informatic features that contribute to sample clustering. Briefly, the iClusterBayes model for the continuous, count, and binary variables can be described as follows:

$$x_{ijt}=\boldsymbol{\beta}_{jt}\,\boldsymbol{\Gamma}_{jt}\,\boldsymbol{z}_i+\varepsilon_{ijt},\varepsilon_{ijt}\sim N\left(0,\sigma_{jt}^2\right),$$

$$\log\,\left(\lambda\left(x_{ijt}|\,\boldsymbol{z}_i\right)\right)=\boldsymbol{\beta}_{jt}\,\boldsymbol{\Gamma}_{jt}\,\boldsymbol{z}_i,$$

$$\log\left(\frac{P(x_{ijt}=1|\boldsymbol{z}_i)}{1-P(x_{ijt}=1|\boldsymbol{z}_i)}\right)=\boldsymbol{\beta}_{jt}\boldsymbol{\Gamma}_{jt}\,\boldsymbol{z}_i,$$

where $\boldsymbol{\beta}_{jt} = (\beta_{0jt},\beta_{1jt},\ldots,\beta_{kjt})$ are the coefficients for the jth feature in the tth dataset, $\boldsymbol{z}_i = (1, z_{i1}, z_{i2},\ldots, z_{ik})'$, and $\Gamma_{jt} = \mathrm{diag}(1, \gamma_{jt},\ldots, \gamma_{jt})$ is a diagonal matrix with $k + 1$ diagonal elements. The constant 1 in $\boldsymbol{z}_i$ and Γ_{jt} is designed to let the models have intercepts β_{0jt} (equivalent to α_{jt} in the iClusterPlus model). In the iClusterBayes model, an indicator variable γ_{jt} with value 1 or 0 is used for Bayesian variable selection. When $\gamma_{jt} = 1$, it indicates that the jth feature in the tth dataset contributes to the sample clustering.

When $\gamma_{jt} = 0$, it indicates that the contribution of the corresponding feature to sample clustering is negligible. To perform Bayesian analysis, the model parameters are assumed to have the following prior distributions:

$$\beta_{jt} \sim N\left(\mathbf{0}, \sum\nolimits_t\right), \sigma_{jt}^2 \sim \text{Inverse} - \text{Gamma}\left(\frac{v}{2}, \frac{v\sigma^2}{2}\right), \gamma_{jt} \sim \text{Bernoulli}\ (q_t).$$

In words, β_{jt} follows multivariate normal distribution with mean vector $\mathbf{0}$ and covariance Σ_t, σ_{jt}^2 follows inverse gamma distribution with shape parameter $v/2$ and scale parameter $v\sigma^2/2$, and γ_{jt} follows Bernoulli distribution with probability of $q_t$being a driver for sample clustering. With these assumptions, the model parameters and z_i can be drawn from their posterior distributions. The drivers of the sample clustering are these features with high posterior probability of $\gamma_{jt} = 1$. Like the iClusterPlus method, the sample cluster assignments are obtained by performing K-means clustering on the latent variable z_i. In addition, to achieve an optimal k, a small number of k are tested, and the optimal k is selected at the transition point on the plots of the deviance ratio and/or the Bayesian information criterion (BIC) versus k, where the deviance ratio or BIC become relatively stable (see the example below). To analyze the same multi-omics dataset, the iClusterBayes needs much less computing time than the iClusterPlus since it does not need to perform a grid search of the lasso parameters. Another advantage of the iClusterBayes method is that it calculates the posterior probabilities of being drivers for the omics features, while the iClusterPlus method does not calculate statistical significance for omics features.

The iCluster methods have been used by TCGA and other research groups to characterize a variety of cancers, including breast cancer, lung squamous cell carcinoma (LUSC), lung adenocarcinoma (LUAD), uterine corpus endometrial carcinoma (UCEC), stomach adenocarcinoma (STAD), skin cutaneous melanoma (SKCM), prostate adenocarcinoma (PRAD), oesophageal carcinoma (ESCA), liver hepatocellular carcinoma (LIHC), bladder urothelial carcinoma (BLCA), mesothelioma (MESO), sarcoma (SARC), and uveal melanoma (UM) [8–13, 25–31]. Recently, Hoadley et al. performed an unsupervised pan-cancer analysis of the 33 cancer types using the iCluster method and found that about one-third of the iClusters were almost homogeneous for a single cancer type, and the remaining two-thirds had various degrees of heterogeneity [32]. These findings demonstrated the dominant role of cell-of-origin in the classification of some cancer types, while, for the other cancer types, the cell-of-origin had less effect on cancer classification.

3 Nonparametric Methods of Integrative Clustering

In contrast to the model-based methods, the nonparametric methods do not rely on any distribution assumptions of the data. A few nonparametric methods have been proposed in recent years including the integrative nonnegative matrix factorization method (intNMF) [14], similarity network fusion (SNF) [16], perturbation clustering (PINS) [24], and neighborhood-based multi-omics clustering (NEMO) [23], etc. The nonparametric methods utilize the similarity among the subjects as a criterion to identify the subtypes. Different methods utilize different types of similarity metrics, such as distance, correlation, nearest neighborhood, or consensus. The integrative clustering intNMF utilizes state-of-the-art consensus clustering in integrating multiple types of molecular data. We briefly introduce a few nonparametric methods for integrative clustering and present more details on intNMF clustering method.

3.1 Similarity Network Fusion (SNF) [16]

For each omics data, SNF creates sample similarity matrices based on pairwise distance measures, termed as patient networks for each data. The samples represent the nodes, and similarity measures represent the edges between the nodes. Then, the networks are integrated to create a single common network matrix using a nonlinear combination method. SNF method utilizes the kernel method to create and fuse the network structures. Then spectral clustering is used on the final network structure to identify the disease subtypes.

3.2 Perturbation Clustering (PINS) [24]

PINS uses the perturbation technic, i.e., adding Gaussian noise to the data repeatedly and applying a classical clustering algorithm to the resulting data. For each type of the omics data, a connectivity matrix is constructed, which will be merged into a single combined similarity matrix. Then, classical clustering methods, such as k-means, hierarchical clustering, partitioning around medoid, or dynamic tree cut, are used to identify the subtypes and determine the cluster membership assignment to the subjects.

3.3 Neighborhood-Based Multi-omics Clustering (NEMO) [23]

NEMO constructs similarity networks for the samples as in SNF for each data and then modifies the similarity to relative similarity. The relative similarity networks are then averaged to create a common relative similarity network structure. Then, as in SNF, spectral clustering is used on the common similarity matrix to carry out the clustering of the subjects.

In this chapter, we describe the intNMF method and its application in a real-life data on lower-grade gliomas. The method is based on powerful nonnegative matrix factorization technic [33] and consensus clustering-based approaches [34]. Nonnegative matrix factorization method has been shown to be highly effective in pattern recognition studies [33]. Consensus clustering technic is considered to be the state-of-the-art approach for clustering algorithms [24, 34]. The integrative clustering intNMF utilizes nonnegative matrix factorization and consensus clustering in integrating multiple types of molecular data. The method has been found to be highly effective in identifying disease subtypes [14, 35].

4 Integrative NMF Clustering (intNMF)

Let us begin with a brief introduction of nonnegative matrix factorization (NMF) framework. By definition, NMF imposes a nonnegativity constraint on its estimation process. In other words, the linear combination of the estimated matrices has only the additive effect if any effect is present; otherwise, it is zero, i.e., there is no negative effect. NMF follows an intuitive approach in which the parts can be combined to form a whole object without having any cancellation effects during the process. NMF approach was first proposed by Paatero and Tapper [36] in 1994. In 1999, Lee and Seung [33] proposed an algorithm for matrix factorization and demonstrated the successful application of NMF in pattern recognition problem in human image analysis. Prior to NMF, the standard method in image analyses was singular value decomposition (SVD) or principal component-based method. Lee and Seung showed that although SVD-based methods are powerful and have unique stable decomposition, in pattern recognition problems, they often result in noisy patterns induced by the cancellation effects of negative signs. They concluded that NMF is more efficient than SVD for pattern recognition research.

Brunet et al. [37] incorporated the algorithm proposed by Lee and Seung and the consensus clustering technic [34] to determine subtypes of cancer. Suppose $\boldsymbol{X}_{p\times n}\in\mathcal{R}^{p\times n}$ is a matrix with p features and n subjects containing all the nonnegative entries. Then NMF factorizes the matrix $\boldsymbol{X}_{p\times n}$ into two nonnegative matrices $\boldsymbol{W}_{p\times k}$ and $\boldsymbol{Z}_{k\times n}$ approximately, ($\boldsymbol{X}_{p\times n}\approx\boldsymbol{W}_{p\times k}\boldsymbol{Z}_{k\times n}$), where k is user-specified latent factors, e.g., the number of groups or classes. The resulting matrices $\boldsymbol{Z}$ and $\boldsymbol{W}$ are called the matrix of basis vectors and the matrix of coefficient vectors, respectively. Using the algorithm, optimum basis matrix $\boldsymbol{Z}$ is identified and used to classify the subjects based on their rank in the columns.

In order to estimate the optimum $\boldsymbol{Z}$ an objective function is defined, such as the Frobenius norm $Q=\min_{W,Z}\|\boldsymbol{X}-W\boldsymbol{Z}\|_2$. Although divergence-based functions can also be used instead of the Frobenius norm, the results are very similar [14, 33]. The objective function is convex in $\boldsymbol{Z}$ when $\boldsymbol{W}$ is given, or convex in $\boldsymbol{W}$ when $\boldsymbol{Z}$ is given. But, since both $\boldsymbol{Z}$ and $\boldsymbol{W}$ are unknown, the objective function is non-convex in general. As a result, there is no global minimum of the optimization problem [36, 38]. In pattern recognition or clustering analysis applications, achieving the "best" local minimum out of several random initializations of $\boldsymbol{Z}$ and $\boldsymbol{W}$ in the algorithm is enough and has been successfully implemented and interpreted [37]. The clustering structure is determined based on the relative ranks of the subjects within the columns of $\boldsymbol{Z}$ matrix rather than the numerical values. Therefore, as long as the relative rank (i.e., cluster pattern) remains stable in a local optimum, we will be able to extract the latent structure successfully.

intNMF extends the NMF framework for more than one type of data (e.g., mRNA, DNA methylation, protein expression) collected on the same set of subjects. Let $\boldsymbol{X}_t$, $t = 1, 2, \ldots, m$ be matrices representing m data types assayed on n samples with p_t, features. Integrative clustering is carried out by estimating the common latent structure (basis matrix) $\mathbf{Z}$ and data-specific coefficient matrices $\boldsymbol{W}_t$ such that

$$\boldsymbol{X}_t \approx \boldsymbol{W}_t \boldsymbol{Z}, \; t = 1, 2, \ldots m,$$

where all entries of $\mathbf{Z}$ and $\boldsymbol{W}_t$ are nonnegative. Figure 2 shows the overall framework of the intNMF method. The objective function is then defined as the weighted Frobenius norm

$$Q = \min_{W,Z} \sum_{t=1}^{m} \theta^t \| \boldsymbol{X}_t - \boldsymbol{W}_t \boldsymbol{Z} \|_2$$

where $\theta^i > 0$ is the user-specified weight if such weights are available. The default weight in the current implementation of intNMF is 1, i.e., equal weight for all data. However, users can provide their own estimated weights as appropriate. One example of such weight estimation for each data can be based on the mean sums of squares given by $\left(\theta^t = \frac{\text{Max}\left\{ \text{mean}\left\{ \| \boldsymbol{X}_t \|_2 \right\}, t = 1, \ldots, m \right\}}{\text{mean}\left\{ \| \boldsymbol{X}_t \|_2 \right\}}, t = 1, \ldots, m \right)$.

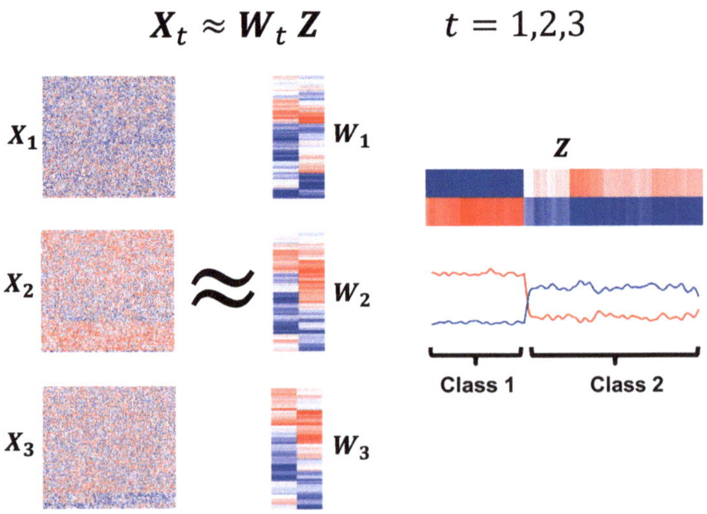

Fig. 2 The framework of integrative nonnegative matrix factorization (intNMF) method

Numerical optimization methods are used to find the local optimum by minimizing the non-convex objective function. intNMF utilizes nonnegative-constrained alternating least squares (NNALS) algorithm [39] to solve the optimization problem. The strategy is to initialize only one $\boldsymbol{Z}$ matrix regardless of how many data are being integrated. In the first part of each iteration, non-negative least squares method is used to estimate $\boldsymbol{W}_t$ matrices. Then, in the second part of the same iteration, the $\boldsymbol{Z}$ matrix is estimated by feeding those $\boldsymbol{W}_t$ matrices into the least squares method. In this way, the non-convex problem is converted into a convex problem piecewise for each of the alternating iterative steps. There are several advantages to this algorithm over the multiplicative update rule as proposed earlier [14, 40]. After optimum $\boldsymbol{Z}$ matrix is identified, the cluster memberships for the subjects are determined based on the relative rank of the values in each column. The convergence criterion is based on the stability of the clustering assignment using the idea of consensus clustering [14, 34].

In any clustering method, there are two major steps involved: (i) identification of optimum number of clusters k and (ii) assignment of the subjects to those clusters, i.e., cluster memberships. The meaningful cluster k needs to be small enough to reduce the noise but large enough to retain important patterns present in data. intNMF utilizes a resampling-based cross-validation technic to estimate the optimum number of clusters by calculating the Cluster Prediction Index (CPI). After the cluster numbers are identified, cluster memberships for the subjects are assigned by running the intNMF algorithm and by specifying the optimum number of clusters. The details of the algorithm and the clustering assignments can be found in Chalise and Fridley [14]. The method is implemented in the R package *intNMF* and is freely available to download.

5 Preparation of Multi-omics Data for Integrative Clustering Analysis

TCGA has generated integrative multi-omics data including genomic, epigenomic, transcriptomic, and proteomic data for 36 cancer types, which have become a great resource for the discovery of integrative cancer subtypes. The level 3 multi-omics data are available at http://firebrowse.org/ and can be used for integrative clustering analysis. Usually, the multi-omics data are processed to form multiple data matrices with columns corresponding to the samples and rows corresponding to omics features for each dataset. For example, somatic mutation data can be summarized using a binary matrix with values 1 or 0, which indicates a gene that contains mutations or no mutation, respectively. A gene is classified as "mutated" if it contains in-frame deletion/insertion, frame shift deletion/insertion, missense/nonsense/nonstop mutation, RNA splice site, and/or translation start site mutation. Genes with a low mutation rate (e.g., ≤2%) in the samples usually do not significantly contribute to sample clustering, and it is challenging to fit the iCluster models due to the sparsity of the data. Therefore, we suggest only including genes with mutation rates >2% in iCluster analysis. For the copy number data, the log2 ratio segment means can be used for iCluster analysis. Since neighboring genomic regions tend to have similar values, they can be merged into condensed regions using the methods described by Mo et al. [21]. Gene expression and methylation data usually contain tens of thousands of genes. Typically, the cluster driver genes are those with relatively large variances. In practice, we found that using the most variable genes (e.g., top 25%) was sufficient for iCluster analysis [30, 31]. To make the gene expression and methylation data fit the model better, for the iCluster method, we suggest performing log2 transformation of the gene expression values (e.g., normalized mRNA-seq counts) and logit transformation of the methylation beta values. Similar data preparation steps can be followed for the intNMF method as well. But since intNMF does not utilize any statistical distribution assumption, no such transformation is required. The specific requirement of intNMF is that each of the datasets must have only nonnegative entries. While a few types of data are naturally nonnegative, e.g., methylation, mRNA out of RNAseq, many others contain both negative and positive entries, e.g., microarray datasets. We recommend calculating the absolute value of the smallest negative number present in the data and adding that to the whole data matrix, thus shifting the data in the positive direction. In doing so, the variance of the data will remain unchanged and will not have any effect on clustering performance. Also, it is important to keep the scales of the multiple datasets similar in order to avoid the results being biased towards the data with a wider range. The data can be rescaled such that they fall within a certain range. One simple approach to do that is by diving the data by the maximum value present in the data.

6 Integrative Clustering Analyses of Uveal Melanoma

Uveal melanoma (UM) is a rare cancer that originates from melanocytes within the uveal tract of the eye [41]. UM is a highly aggressive disease that preferentially metastasizes to the liver within a few years after diagnosis, regardless of the successful local treatment of the disease. The three-year survival rate was less than 5% for patients with metastatic UM [42–44]. UM characterized by loss of a copy of chromosome 3 (monosomy 3), gain of chromosome 8q, and a high-frequency mutation of BRCA-associated protein 1 (BAP1) is associated with a high risk of metastasis [45–47]. UM characterized by the normal copy number of chromosome

3 (disomy 3) and gain of chromosome 6p is associated with a low risk of metastasis [47, 48]. TCGA generated multi-omics data for 80 UM samples and reported DNA copy number, methylation, and gene expression subtypes [49]. However, TCGA did not report integrative subtypes based on integrative clustering analysis of the multi-omics data. Therefore, we use the analysis of UM multi-omics data as an example to illustrate the iCluster methods.

We performed iCluster analysis of TCGA uveal melanoma (UM) multi-omics data including somatic mutation, DNA copy number, methylation, and gene expression [30]. The multi-omics data were processed as described in the previous section and analyzed using the iClusterPlus method. In order to identify an optimal number of clusters, the cluster number parameter $k = 1, 2, 3, 4, 5$ were tested, respectively, and 701 lasso parameters were searched for each k. Figure 3a shows the BIC value and deviance ratio at each k for the model such that the deviance ratio of the model is the maximum among the 701 possible models. It can be seen that the BIC values are the second smallest and the smallest when $k = 3$ and $k = 4$, respectively, suggesting that four and five clusters are two optimal solutions (Fig. 3a). By examining the two options, we found that the major difference between the four- and five-cluster solutions was that one cluster of the four-cluster solution was further divided into two smaller clusters, generating five clusters in total [30]. To avoid overfitting, the four-cluster solution was considered optimal. Figure 3b shows the distribution of the UM samples in the three-dimensional latent variable space in which the samples form two major clusters and each of the two major clusters is divided into two smaller clusters. This observation is consistent with the multi-omics patterns that drive the formation of the four iClusters (Fig. 3c). It is well known that loss of one copy of chr3 (monosomy 3) is the most common aberration in UM [46, 49]. Strikingly, the samples with chr3 loss form a cluster, and the samples with chr3 normal (disomy) form another cluster (Fig. 3c, copy number). Following the tradition, the two major clusters are named M3 and D3, respectively. Primarily driven by the chromosome 6q status, the M3 is further divided into M3.1 and M3.2 clusters, and the D3 is further divided into D3.1 and D3.2 clusters, where M3.2 and D3.1 are characterized by 6q loss (Fig. 3c, copy number). Notably, M3 is correlated with 8q gain, D3 is correlated with 6p gain, and M3 and 6p gain appears to be mutually exclusive (Fig. 3c, copy number). The top genes driving the sample clustering form two gene clusters (Me1, Me2) in the methylation data and two gene clusters (Ex1, Ex2) in the gene expression data, respectively (Fig. 3c, methylation, mRNA). The methylation and gene expression patterns are negatively correlated. Overall, the methylation patterns of Me1 and the expression patterns of Ex2 are positively correlated with the chr3 status. In contrast, the methylation patterns of Me2 and the expression

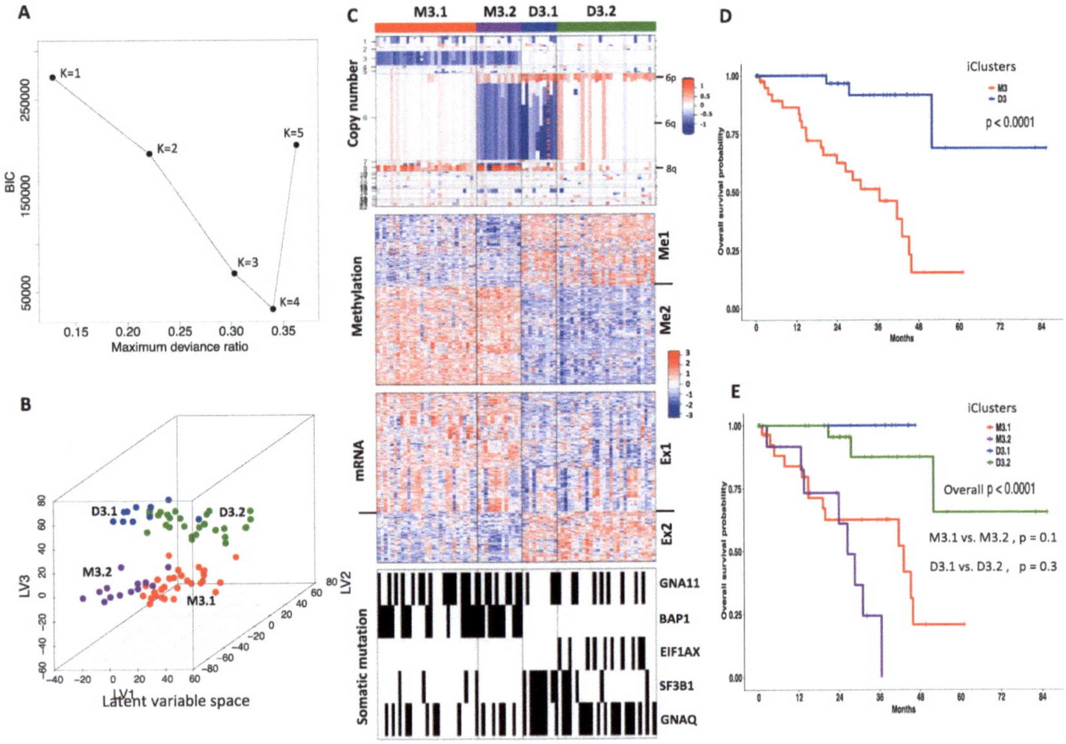

Fig. 3 UM integrative clusters (adapted from Mo et al. [30]). (**a**) Bayesian information criterion (BIC) and the maximum deviance ratio at the cluster parameter $k = 1, 2, 3, 4, 5$. Based on the plot, $k = 3$ (4 clusters) and $k = 4$ (5 clusters) are two optimal solutions. (**b**) the UM samples distributed in the latent variable space of the iClusterPlus model. (**c**) heatmaps of the multi-omics features driving the UM sample clustering. Copy number: copy number loss, normal, and gain are indicated by blue, white, and red, respectively. Methylation: hypomethylation and hypermethylation are indicated by blue and red, respectively; mRNA: high and low gene expression are indicated by blue and red, respectively. Somatic mutation: mutated genes are indicated by black bars. (**d**) Patient overall survival of the two major M3 and D3 iClusters. (**e**) Patient overall survival of the four iClusters

patterns of Ex1 are negatively correlated with the chr3 status. In addition, genes including *GNA11*, *BAP1*, *EIF1AX*, *SF3B1*, and *GNAQ* are identified as the driver genes with different mutation rates between the M3 and D3 iClusters (Fig. 3c, somatic mutation). The M3 iCluster is associated with a worse overall survival, compared to the D3 iCluster (Fig. 3d). However, the survival functions are not significantly different between M3.1 and M3.2 and between D3.1 and D3.2 (Fig. 3e). These observations suggest that it may be sufficient to classify UM into two major subtypes for personalized management in practice.

7 Integrative Clustering Analyses of Lower-Grade Glioma (LGG)

We illustrate the application of intNMF using data from TCGA and Ceccarelli et al. studies on lower-grade glioma (LGG) [5, 50]. TCGA studies consisted of LGG subjects only, while Ceccarelli et al. combined those subjects with glioblastoma subjects for their analysis. The data used in this example consists of only the LGG subjects, and we refer to the TCGA for the data we used hereafter.

The lower-grade gliomas (grades II and III) have extremely variable clinical behavior and disease progression which cannot be adequately predicted from the histological class diagnosis alone. Although at histological examinations the LGGs look similar, a few of them stay indolent while others progress rapidly and develop glioblastoma [5]. This shows that the disease is dissimilar at the molecular level, indicating the presence of subtypes characterized by molecular activities. The LGGs data was downloaded from the Genomics Data Commons (GDC) website, https://portal.gdc.cancer.gov/, which consists of mRNA (20,330 genes), DNA methylation (25,978 probes), and DNA copy number (24,776 genes) assayed on 511 subjects. Many genomic features are similar across the subjects (noise features), and those features do not contribute to the clustering analysis. Therefore, the most varying features as measured by standard deviations from each type of data were selected for the clustering analysis. Such selection of the features also helps in optimizing computational efficiency by reducing the dimension of the data. After selecting the top three percentile features ranked by standard deviation, the new dimensions of the datasets become 584 mRNAs, 553 methylation features, and 493 CNVs for 511 subjects. In addition, a few different percentile cutoffs of the top varying features were also used in the intNMF method, and we observed that the results were stable. For simplicity, we picked the results from the top three percentile features for the illustrative example here. The intNMF is applied to the datasets, and the resulting subtypes are cross compared with the subtypes identified by TCGA studies. The three subtypes identified by the TCGA study have been characterized by IDH mutation and 1p/19q co-deletion status; IDHmut-codel, IDHmut-non-codel, and IDHwt [5]. In our implementation of intNMF, we specified the search range of numbers of clusters k from 2 to 7. In order to identify the optimum number of clusters, we carried out cross-validation to estimate the cluster prediction index (CPI). We found that the CPI value was highest for $k = 2$ and only slightly lower for $k = 3$ (Fig. 4a). Beyond $k = 3$, the CPI values were much smaller. Therefore, dividing the LGG samples into three clusters was the optimum solution (Fig. 4d). The three sample clusters can be visualized on the tSNE (t-distributed stochastic neighbor

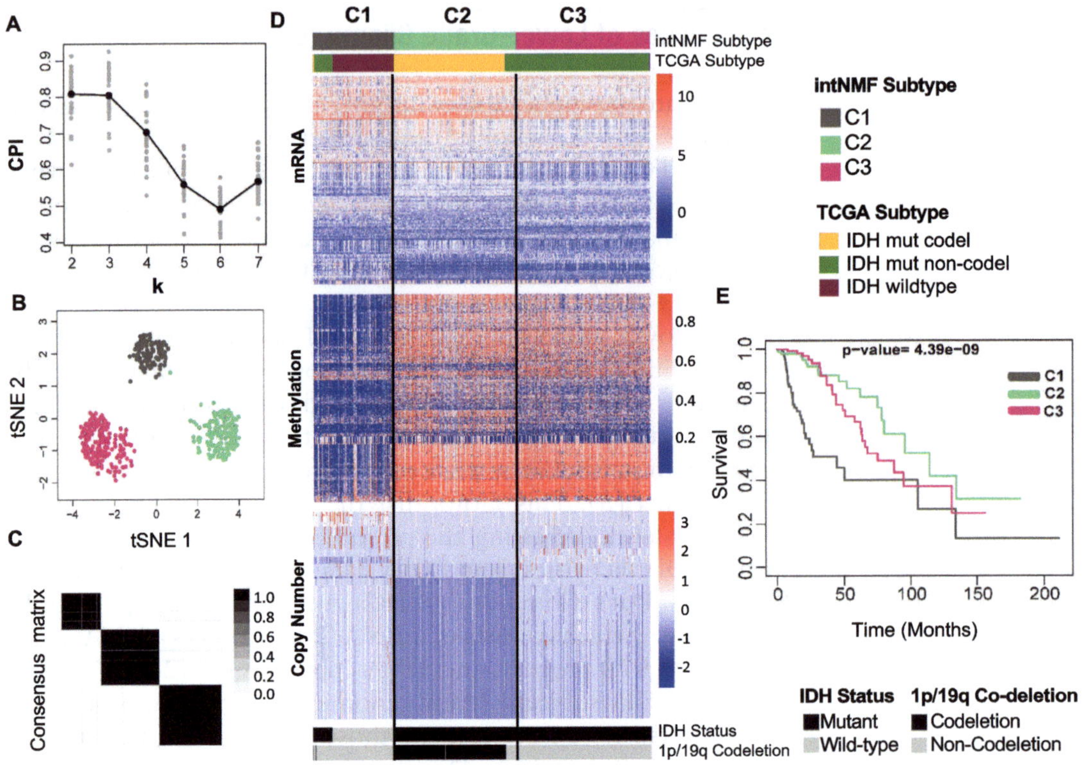

Fig. 4 Integrative clustering analysis of lower-grade glioma using three datasets: DNA methylation, mRNA expression, and copy number variation. (**a**) Plot of Cluster Prediction Index (CPI) against search range of a number of clusters (*k*) showing the optimum number of clusters. (**b**) Visualization of the subgroups using tSNE. (**c**) Sample block structure of the consensus matrix. Stability of this matrix is the stopping criteria of the algorithm. (**d**) Heatmaps of mRNA, methylation, and copy number showing the identified cluster groups on the top. The bottom panel shows the biologically important IDH mutation status and 1p/19q codeletion status. (**e**) Survival probability differences using Kaplan Meier method followed by log-rank test

embedding) plot (Fig. 4b). The three subtypes identified by intNMF highly agree with the subtypes identified by TCGA studies [5]. The subtypes identified by TCGA were based on a cluster-of-clusters analysis approach. The subtypes were characterized by IDH mutation and *1p/19q* co-deletion status. The intNMF-C1 was enriched with IDH wild type and non-codeletion, intNMF-C2 was made up of IDH mutation and highly enriched with *1p/19q* codeletion, and intNMF-C3 was entirely made up of IDH mutation with non-codeletion. Notably, the intNMF-C1 was widely separated from the rest two clusters and had a worse overall survival than the other two clusters (Fig. 4e). The survival differences among the subtypes were assessed by Kaplan-Meier analysis method followed by a log-rank test (p-value < 0.001). The subtypes identified by intNMF were biologically relevant and had differing survival probabilities.

8 Conclusion

A fundamental approach in many omics data analyses is to find an appropriate lower-dimensional representation of the data. One such approach is identifying the latent structure of the datasets so that subgroups of the subjects (and/or features) can be discovered. Clustering analysis is a strong method to identify such latent structures. Traditionally, clustering analyses are carried out by considering one omics data at a time. Although often independently analyzed, the molecular datasets are highly correlated since they are assayed at several layers of biological processes on the same set of patient samples. The true strong latent structure may not be present in all data types. Also, there might be weak but consistent signals across the multiple data types. Integrative analysis can leverage such strength across the omics data types.

There are a few model-based and nonparametric clustering methods that have been developed. In this chapter, we focus on one representative model-based (iCluster) and one nonparametric clustering method (intNMF) with illustrative examples. Model-based methods assume omics data follows certain distributions. For example, the iCluster methods assume microarray gene expression data following normal distribution and RNA-seq gene expression data (sequence counts) following Poisson distribution. In most cases, the raw omics data may not exactly follow the assumed distributions. However, we found that the iCluster methods performed reasonably well when the data were approximately distributed according to the assumed distributions, which usually can be achieved via data transformation. For example, methylation beta values (range: 0–1) do not follow normal distribution. However, after performing logit transformation, the transformed data can be modeled using linear regression in the iCluster models. In addition, after log transformation of sequence count data, the transformed data can also be modeled using linear regression. In contrast, nonparametric methods such as intNMF do not make distribution assumptions about the data. It only requires that the data be positive. Both the methods have pros and cons. For example, the iClusterPlus and iClusterBayes methods are computationally intensive methods, which need more computational resources than the intNMF method. However, the iCluster methods can perform feature selection to identify the driver features for sample clustering, while there is no corresponding function in the intNMF method. In addition, the iClusters methods can model different types of data (e.g., binary, categorical), while the intNMF method is limited to continuous data. Regardless, if there exists a decent cluster pattern in the omics datasets, both methods are able to identify the biologically relevant subtypes, and they mostly agree with each other. The iCluster methods are implemented in the iClusterPlus package (https://www.bioconductor.org/packages/release/bioc/html/iClusterPlus.html), and the intNMF method is implemented in the intNMF package (https://cran.r-project.org/web/packages/IntNMF/index.html), both of which are publicly available for downloading.

Acknowledgments

We thank the Cancer Genome Atlas project for the use of data from uveal melanoma and lower-grade gliomas. Mo and Fridley are supported in part by the National Cancer Institute Center Core Grant P30 CA076292.

References

1. Sørlie T, Perou CM, Tibshirani R, Aas T, Geisler S, Johnsen H, Hastie T, Eisen MB, van de Rijn M, Jeffrey SS, Thorsen T, Quist H, Matese JC, Brown PO, Botstein D, Lønning PE, Børresen-Dale A-L (2001) Gene expression patterns of breast carcinomas distinguish tumor subclasses with clinical implications. Proc Natl Acad Sci 98(19): 10869–10874. https://doi.org/10.1073/pnas.191367098
2. Sotiriou C, Neo S-Y, McShane LM, Korn EL, Long PM, Jazaeri A, Martiat P, Fox SB, Harris AL, Liu ET (2003) Breast cancer classification and prognosis based on gene expression profiles from a population-based study. Proc Natl Acad Sci 100(18):10393–10398. https://doi.org/10.1073/pnas.1732912100
3. Verhaak RG, Hoadley KA, Purdom E, Wang V, Qi Y, Wilkerson MD, Miller CR, Ding L, Golub T, Mesirov JP, Alexe G, Lawrence M, O'Kelly M, Tamayo P, Weir BA, Gabriel S, Winckler W, Gupta S, Jakkula L, Feiler HS, Hodgson JG, James CD, Sarkaria JN, Brennan C, Kahn A, Spellman PT, Wilson RK, Speed TP, Gray JW, Meyerson M, Getz G, Perou CM, Hayes DN, Cancer Genome Atlas Research N (2010) Integrated genomic analysis identifies clinically relevant subtypes of glioblastoma characterized by abnormalities in PDGFRA, IDH1, EGFR, and NF1. Cancer Cell 17(1):98–110. https://doi.org/10.1016/j.ccr.2009.12.020
4. The Cancer Genome Atlas Network (2012) Comprehensive molecular portraits of human breast tumours. Nature 490(7418):61–70. http://www.nature.com/nature/journal/v490/n7418/abs/nature11412.html#supplementary-information
5. The Cancer Genome Atlas Network (2015) Comprehensive, integrative genomic analysis of diffuse lower-grade gliomas. N Engl J Med 372(26):2481–2498. https://doi.org/10.1056/NEJMoa1402121
6. Cancer Genome Atlas Research N (2017) Integrated genomic and molecular characterization of cervical cancer. Nature 543(7645): 378–384. https://doi.org/10.1038/nature21386
7. Chalise P, Koestler DC, Bimali M, Yu Q, Fridley BL (2014) Integrative clustering methods for high-dimensional molecular data. Transl Cancer Res 3(3):202–216. https://doi.org/10.3978/j.issn.2218-676X.2014.06.03
8. Cancer Genome Atlas Research N (2012) Comprehensive genomic characterization of squamous cell lung cancers. Nature 489(7417):519–525. https://doi.org/10.1038/nature11404
9. Cancer Genome Atlas Research N (2013) Integrated genomic characterization of endometrial carcinoma. Nature 497(7447):67–73. https://doi.org/10.1038/nature12113
10. Cancer Genome Atlas Research N (2014) Comprehensive molecular profiling of lung adenocarcinoma. Nature 511(7511): 543–550. https://doi.org/10.1038/nature13385
11. Cancer Genome Atlas Research N (2014) Comprehensive molecular characterization of gastric adenocarcinoma. Nature 513(7517):

202–209. https://doi.org/10.1038/nature13480
12. Cancer Genome Atlas Research N (2015) The molecular taxonomy of primary prostate Cancer. Cell 163(4):1011–1025. https://doi.org/10.1016/j.cell.2015.10.025
13. Cancer Genome Atlas Research N (2017) Integrated genomic characterization of oesophageal carcinoma. Nature 541(7636): 169–175. https://doi.org/10.1038/nature20805
14. Chalise P, Fridley BL (2017) Integrative clustering of multi-level 'omic data based on non-negative matrix factorization algorithm. PLoS One 12(5):e0176278. https://doi.org/10.1371/journal.pone.0176278
15. Shen R, Olshen AB, Ladanyi M (2009) Integrative clustering of multiple genomic data types using a joint latent variable model with application to breast and lung cancer subtype analysis. Bioinformatics (Oxford, England) 25(22):2906–2912. https://doi.org/10.1093/bioinformatics/btp543
16. Wang B, Mezlini AM, Demir F, Fiume M, Tu Z, Brudno M, Haibe-Kains B, Goldenberg A (2014) Similarity network fusion for aggregating data types on a genomic scale. Nat Methods 11:333. https://doi.org/10.1038/nmeth.2810. https://www.nature.com/articles/nmeth.2810#supplementary-information
17. Cancer Genome Atlas N (2012) Comprehensive molecular portraits of human breast tumours. Nature 490(7418):61–70. https://doi.org/10.1038/nature11412
18. Lock EF, Dunson DB (2013) Bayesian consensus clustering. Bioinformatics 29(20): 2610–2616. https://doi.org/10.1093/bioinformatics/btt425
19. Kirk P, Griffin JE, Savage RS, Ghahramani Z, Wild DL (2012) Bayesian correlated clustering to integrate multiple datasets. Bioinformatics 28(24):3290–3297. https://doi.org/10.1093/bioinformatics/bts595
20. Mo Q, Shen R, Guo C, Vannucci M, Chan KS, Hilsenbeck SG (2018) A fully Bayesian latent variable model for integrative clustering analysis of multi-type omics data. Biostatistics 19(1): 71–86. https://doi.org/10.1093/biostatistics/kxx017
21. Mo Q, Wang S, Seshan VE, Olshen AB, Schultz N, Sander C, Powers RS, Ladanyi M, Shen R (2013) Pattern discovery and cancer gene identification in integrated cancer genomic data. Proc Natl Acad Sci U S A 110(11): 4245–4250. https://doi.org/10.1073/pnas.1208949110
22. Shen R, Wang S, Mo Q (2013) Sparse integrative clustering of multiple omics data sets. Ann Appl Stat 7(1):269–294. https://doi.org/10.1214/12-AOAS578
23. Rappoport N, Shamir R (2019) NEMO: cancer subtyping by integration of partial multiomic data. Bioinformatics 35(18):3348–3356. https://doi.org/10.1093/bioinformatics/btz058
24. Nguyen T, Tagett R, Diaz D, Draghici S (2017) A novel approach for data integration and disease subtyping. Genome Res 27:2025–2039. https://doi.org/10.1101/gr.215129.116
25. Curtis C, Shah SP, Chin SF, Turashvili G, Rueda OM, Dunning MJ, Speed D, Lynch AG, Samarajiwa S, Yuan Y, Graf S, Ha G, Haffari G, Bashashati A, Russell R, McKinney S, Group M, Langerod A, Green A, Provenzano E, Wishart G, Pinder S, Watson P, Markowetz F, Murphy L, Ellis I, Purushotham A, Borresen-Dale AL, Brenton JD, Tavare S, Caldas C, Aparicio S (2012) The genomic and transcriptomic architecture of 2,000 breast tumours reveals novel subgroups. Nature 486(7403):346–352. https://doi.org/10.1038/nature10983
26. The Cancer Genome Atlas Network (2015) Comprehensive genomic characterization of head and neck squamous cell carcinomas. Nature 517(7536):576–582. https://doi.org/10.1038/nature14129. http://www.nature.com/nature/journal/v517/n7536/abs/nature14129.html#supplementary-information
27. Cancer Genome Atlas Research Network. Electronic address edsc, Cancer Genome Atlas Research N (2017) Comprehensive and integrated genomic characterization of adult soft tissue sarcomas. Cell 171(4):950–965. e928. https://doi.org/10.1016/j.cell.2017.10.014
28. Network CGAR (2017) Comprehensive and integrative genomic characterization of hepatocellular carcinoma. Cell 169(7):1327–1341. e1323. https://doi.org/10.1016/j.cell.2017.05.046
29. Hmeljak J, Sanchez-Vega F, Hoadley KA, Shih J, Stewart C, Heiman D, Tarpey P, Danilova L, Drill E, Gibb EA, Bowlby R, Kanchi R, Osmanbeyoglu HU, Sekido Y, Takeshita J, Newton Y, Graim K, Gupta M, Gay CM, Diao L, Gibbs DL, Thorsson V, Iype L, Kantheti H, Severson DT, Ravegnini G, Desmeules P, Jungbluth AA, Travis WD, Dacic S, Chirieac LR, Galateau-Salle F, Fujimoto J, Husain AN, Silveira HC, Rusch

VW, Rintoul RC, Pass H, Kindler H, Zauderer MG, Kwiatkowski DJ, Bueno R, Tsao AS, Creaney J, Lichtenberg T, Leraas K, Bowen J, Network TR, Felau I, Zenklusen JC, Akbani R, Cherniack AD, Byers LA, Noble MS, Fletcher JA, Robertson AG, Shen R, Aburatani H, Robinson BW, Campbell P, Ladanyi M (2018) Integrative molecular characterization of malignant pleural mesothelioma. Cancer Discov 8(12):1548–1565. https://doi.org/10.1158/2159-8290.CD-18-0804

30. Mo Q, Wan L, Schell MJ, Jim H, Tworoger SS, Peng G (2021) Integrative analysis identifies multi-omics signatures that drive molecular classification of uveal melanoma. Cancers (Basel) 13(24). https://doi.org/10.3390/cancers13246168
31. Mo Q, Li R, Adeegbe DO, Peng G, Chan KS (2020) Integrative multi-omics analysis of muscle-invasive bladder cancer identifies prognostic biomarkers for frontline chemotherapy and immunotherapy. Commun Biol 3(1):784. https://doi.org/10.1038/s42003-020-01491-2
32. Hoadley KA, Yau C, Hinoue T, Wolf DM, Lazar AJ, Drill E, Shen R, Taylor AM, Cherniack AD, Thorsson V, Akbani R, Bowlby R, Wong CK, Wiznerowicz M, Sanchez-Vega F, Robertson AG, Schneider BG, Lawrence MS, Noushmehr H, Malta TM, Cancer Genome Atlas N, Stuart JM, Benz CC, Laird PW (2018) Cell-of-origin patterns dominate the molecular classification of 10,000 tumors from 33 types of cancer. Cell 173(2):291–304 e296. https://doi.org/10.1016/j.cell.2018.03.022
33. Lee DD, Seung HS (1999) Learning the parts of objects by non-negative matrix factorization. Nature 401(6755):788–791
34. Monti S, Tamayo P, Mesirov J, Golub T (2003) Consensus clustering: a resampling-based method for class discovery and visualization of gene expression microarray data. Mach Learn 52(1–2):91–118. https://doi.org/10.1023/A:1023949509487
35. Cantini L, Zakeri P, Hernandez C, Naldi A, Thieffry D, Remy E, Baudot A (2021) Benchmarking joint multi-omics dimensionality reduction approaches for the study of cancer. Nat Commun 12(1):124. https://doi.org/10.1038/s41467-020-20430-7
36. Paatero P, Tapper U (1994) Positive matrix factorization - a nonnegative factor model with optimal utilization of error-estimates of data values. Environmetrics 5(2):111–126. https://doi.org/10.1002/env.3170050203
37. Brunet JP, Tamayo P, Golub TR, Mesirov JP (2004) Metagenes and molecular pattern discovery using matrix factorization. Proc Natl Acad Sci U S A 101(12):4164–4169. https://doi.org/10.1073/pnas.0308531101
38. Berry MW, Browne M, Langville AN, Pauca VP, Plemmons RJ (2007) Algorithms and applications for approximate nonnegative matrix factorization. Comput Stat Data Anal 52(1):155–173. https://doi.org/10.1016/j.csda.2006.11.006
39. Lawson C, Hanson R (1995) Solving least squares problems. SIAM
40. Zhang S, Liu CC, Li W, Shen H, Laird PW, Zhou XJ (2012) Discovery of multi-dimensional modules by integrative analysis of cancer genomic data. Nucleic Acids Res 40(19):9379–9391. https://doi.org/10.1093/nar/gks725
41. Kaliki S, Shields CL (2017) Uveal melanoma: relatively rare but deadly cancer. Eye (Lond) 31(2):241–257. https://doi.org/10.1038/eye.2016.275
42. Amaro A, Gangemi R, Piaggio F, Angelini G, Barisione G, Ferrini S, Pfeffer U (2017) The biology of uveal melanoma. Cancer Metastasis Rev 36(1):109–140. https://doi.org/10.1007/s10555-017-9663-3
43. Collaborative Ocular Melanoma Study G (2001) Assessment of metastatic disease status at death in 435 patients with large choroidal melanoma in the Collaborative Ocular Melanoma Study (COMS): COMS report no. 15. Arch Ophthalmol 119(5):670–676. https://doi.org/10.1001/archopht.119.5.670
44. Lane AM, Kim IK, Gragoudas ES (2018) Survival rates in patients after treatment for metastasis from uveal melanoma. JAMA Ophthalmol 136(9):981–986. https://doi.org/10.1001/jamaophthalmol.2018.2466
45. Gupta MP, Lane AM, DeAngelis MM, Mayne K, Crabtree M, Gragoudas ES, Kim IK (2015) Clinical characteristics of uveal melanoma in patients with germline BAP1 mutations. JAMA Ophthalmol 133(8):881–887. https://doi.org/10.1001/jamaophthalmol.2015.1119
46. Harbour JW, Onken MD, Roberson ED, Duan S, Cao L, Worley LA, Council ML, Matatall KA, Helms C, Bowcock AM (2010) Frequent mutation of BAP1 in metastasizing uveal melanomas. Science 330(6009):1410–1413. https://doi.org/10.1126/science.1194472
47. White VA, Chambers JD, Courtright PD, Chang WY, Horsman DE (1998) Correlation

of cytogenetic abnormalities with the outcome of patients with uveal melanoma. Cancer 83(2):354–359

48. Damato B, Dopierala J, Klaasen A, van Dijk M, Sibbring J, Coupland SE (2009) Multiplex ligation-dependent probe amplification of uveal melanoma: correlation with metastatic death. Invest Ophthalmol Vis Sci 50(7): 3048–3055. https://doi.org/10.1167/iovs.08-3165

49. Robertson AG, Shih J, Yau C, Gibb EA, Oba J, Mungall KL, Hess JM, Uzunangelov V, Walter V, Danilova L, Lichtenberg TM, Kucherlapati M, Kimes PK, Tang M, Penson A, Babur O, Akbani R, Bristow CA, Hoadley KA, Iype L, Chang MT, Network TR, Cherniack AD, Benz C, Mills GB, Verhaak RGW, Griewank KG, Felau I, Zenklusen JC, Gershenwald JE, Schoenfield L, Lazar AJ, Abdel-Rahman MH, Roman-Roman S, Stern MH, Cebulla CM, Williams MD, Jager MJ, Coupland SE, Esmaeli B, Kandoth C, Woodman SE (2018) Integrative analysis identifies four molecular and clinical subsets in uveal melanoma. Cancer Cell 33(1):151. https://doi.org/10.1016/j.ccell.2017.12.013

50. Ceccarelli M, Barthel Floris P, Malta Tathiane M, Sabedot Thais S, Salama Sofie R, Murray Bradley A, Morozova O, Newton Y, Radenbaugh A, Pagnotta Stefano M, Anjum S, Wang J, Manyam G, Zoppoli P, Ling S, Rao Arjun A, Grifford M, Cherniack Andrew D, Zhang H, Poisson L, Carlotti Carlos G, da Tirapelli Daniela Pretti C, Rao A, Mikkelsen T, Lau Ching C, Yung WKA, Rabadan R, Huse J, Brat Daniel J, Lehman Norman L, Barnholtz-Sloan Jill S, Zheng S, Hess K, Rao G, Meyerson M, Beroukhim R, Cooper L, Akbani R, Wrensch M, Haussler D, Aldape Kenneth D, Laird Peter W, Gutmann David H, Anjum S, Arachchi H, Auman JT, Balasundaram M, Balu S, Barnett G, Baylin S, Bell S, Benz C, Bir N, Black Keith L, Bodenheimer T, Boice L, Bootwalla Moiz S, Bowen J, Bristow Christopher A, Butterfield Yaron SN, Chen Q-R, Chin L, Cho J, Chuah E, Chudamani S, Coetzee Simon G, Cohen Mark L, Colman H, Couce M, D'Angelo F, Davidsen T, Davis A, Demchok John A, Devine K, Ding L, Duell R, Elder JB, Eschbacher Jennifer M, Fehrenbach A, Ferguson M, Frazer S, Fuller G, Fulop J, Gabriel Stacey B, Garofano L, Gastier-Foster Julie M, Gehlenborg N, Gerken M, Getz G, Giannini C, Gibson William J, Hadjipanayis A, Hayes DN, Heiman David I, Hermes B, Hilty J, Hoadley Katherine A, Hoyle Alan P, Huang M, Jefferys Stuart R, Jones Corbin D, Jones Steven JM, Ju Z, Kastl A, Kendler A, Kim J, Kucherlapati R, Lai Phillip H, Lawrence Michael S, Lee S, Leraas Kristen M, Lichtenberg Tara M, Lin P, Liu Y, Liu J, Ljubimova Julia Y, Lu Y, Ma Y, Maglinte Dennis T, Mahadeshwar Harshad S, Marra Marco A, McGraw M, McPherson C, Meng S, Mieczkowski Piotr A, Miller CR, Mills Gordon B, Moore Richard A, Mose Lisle E, Mungall Andrew J, Naresh R, Naska T, Neder L, Noble Michael S, Noss A, O'Neill Brian P, Ostrom Quinn T, Palmer C, Pantazi A, Parfenov M, Park Peter J, Parker Joel S, Perou Charles M, Pierson Christopher R, Pihl T, Protopopov A, Radenbaugh A, Ramirez Nilsa C, Rathmell WK, Ren X, Roach J, Robertson AG, Saksena G, Schein Jacqueline E, Schumacher Steven E, Seidman J, Senecal K, Seth S, Shen H, Shi Y, Shih J, Shimmel K, Sicotte H, Sifri S, Silva T, Simons Janae V, Singh R, Skelly T, Sloan Andrew E, Sofia Heidi J, Soloway Matthew G, Song X, Sougnez C, Souza C, Staugaitis Susan M, Sun H, Sun C, Tan D, Tang J, Tang Y, Thorne L, Trevisan Felipe A, Triche T, Van Den Berg David J, Veluvolu U, Voet D, Wan Y, Wang Z, Warnick R, Weinstein John N, Weisenberger Daniel J, Wilkerson Matthew D, Williams F, Wise L, Wolinsky Y, Wu J, Xu Andrew W, Yang L, Yang L, Zack Travis I, Zenklusen Jean C, Zhang J, Zhang W, Zhang J, Zmuda E, Noushmehr H, Iavarone A, Verhaak RGW (2016) Molecular profiling reveals biologically discrete subsets and pathways of progression in diffuse glioma. Cell 164(3):550–563. https://doi.org/10.1016/j.cell.2015.12.028

Chapter 6

Analysis of Single-Cell RNA-seq Data

Xiaoru Dong and Rhonda Bacher

Abstract

As single-cell RNA sequencing experiments continue to advance scientific discoveries across biological disciplines, an increasing number of analysis tools and workflows for analyzing the data have been developed. In this chapter, we describe a standard workflow and elaborate on relevant data analysis tools for analyzing single-cell RNA sequencing data. We provide recommendations for the appropriate use of commonly used methods, with code examples and analysis interpretations.

Key words Single-cell RNA-seq, Gene expression, High-throughput sequencing, Quality control, Normalization

1 Introduction

Single-cell RNA-seq (scRNA-seq) is profoundly changing the study of heterogeneous cell populations and dynamic biological processes. However, there are challenges in analyzing the gene expression data due to sparsity, heterogeneous distributions, and technical noise [1]. To accommodate these features, novel statistical and computational methods have been developed or modified from those used in traditional RNA sequencing experiments (bulk RNA-seq). This chapter describes the typical workflow and statistical methods used in scRNA-seq gene expression analysis. We include brief examples using publicly available software and datasets accessible via GitHub or R/Bioconductor. A detailed script of all the codes used here is available on our GitHub: https://github.com/rhondabacher/scRNAseq_Analysis.

Brooke L. Fridley and Xuefeng Wang (eds.), *Statistical Genomics*,
Methods in Molecular Biology, vol. 2629, https://doi.org/10.1007/978-1-0716-2986-4_6,

2 Preliminaries

2.1 Code Setup

All examples will be carried out using the R statistical programming language. The R packages used throughout this chapter are available to install via Bioconductor or CRAN.

```
# BiocManager for installing Bioconductor packages:
if (!require("BiocManager", quietly = TRUE))
 install.packages("BiocManager")
## Bioconductor packages:
BiocManager::install("TENxPBMCData")
BiocManager::install("scater")
BiocManager::install("scran")
BiocManager::install("tradeSeq")
## CRAN packages:
install.pacakges("gridExtra")
install.pacakges("Seurat")
install.pacakges("dplyr")
install.pacakges("SeuratDisk")
install.pacakges("slingshot")
install.packages("RColorBrewer")
```

2.2 Data Description

We demonstrate a scRNA-seq data analysis on a moderate-sized dataset of peripheral blood mononuclear cells (PBMCs) from a healthy donor generated by 10X Genomics. The raw data was processed using Cell Ranger and resulted in a filtered dataset consisting of 4,340 cells and 33,694 gene features, which is available in the *TENxPBMCData* R/Bioconductor package [2].

3 Quality Control and Normalization

3.1 Read-Level Quality Control and Bioinformatic Processing

The raw single-cell sequencing data requires read-level quality control (QC), demultiplexing, alignment of reads, and transcript quantification. There are a number of bioinformatic pipelines for carrying out these steps, starting with the fastq files such as Cell-Ranger, scPipe [3], or scruff [4]. Many of these pipelines are specific to different scRNA-seq technologies and use a variety of read alignment procedures. An additional differentiating factor among preprocessing pipelines is how to identify and collapse the unique molecular identifier barcodes (UMIs). UMIs are strings of random nucleotides that allow for the identification of unique molecules after sequencing and bioinformatically removing PCR duplicates. Differences among preprocessing workflows do not appear to have major effects on clustering [5], but likely affect downstream analyses that heavily depend on accurate quantification [6].

Table 1
Additional R packages for scRNA-seq quality control processing

Name	Utility
DropletUtils [55]	Identify empty droplets
DropletQC [56]	Identify empty droplets or damaged cells
DoubletFinder [57]	Identify doublet cells
miQC [58]	Identify low-quality cells

For droplet-based scRNA-seq experiments, each droplet contains a unique barcode which is later used as a cell identifier. However, not all droplets will contain a single cell; many barcodes represent reads from ambient or background RNA, while others represent multiplets in which multiple cells were contained in a droplet. While some processing pipelines will automatically attempt to identify these barcodes, there are additional computational methods available for identifying problematic droplets (Table 1).

3.2 Cell-Level Quality Control

During sample processing, cells may become damaged or otherwise afflicted by technical artifacts. These "low-quality" cells are problematic in downstream analyses and must be identified. Three common metrics of cell quality are:

1. *The total number of counts present for each cell*: low total counts indicate insufficient or leaky capture of mRNA via a broken cell or an empty droplet with background mRNA [7]. Extraordinarily high counts may be caused by doublets or multiplets.
2. *The total number of genes detected in each cell*: cells with a small number of expressed genes may indicate the cell is damaged or dead [8]. If the number of expressed genes is large, it may be an indicator of doublets or multiplets [9].
3. *The fraction of reads mapped to mitochondrial genes*: the number of mitochondrial genes is generally higher in cells with necrosis or membrane rupture [10].

These features vary across biological tissues and cell types; thus, removing poor-quality cells while retaining biologically interesting cells requires a data-driven identification procedure. This is done by identifying cells that are outliers relative to the data at hand rather than setting absolute thresholds.

Below we walk through preprocessing the PBMC example using the R/Bioconductor package *scater* [11]. The *scater* package provides several tools for QC, including calculating and visualizing QC metrics to screen for low-quality cells. The R packages *scater* and *Seurat* [12] are among the most popular for performing standard cell-level quality control.

First, we load in the dataset:

```
library(TENxPBMCData)
sce <- TENxPBMCData(dataset = "pbmc4k")
```

Next, the function `perCellQCMetrics()` computes a number of QC metrics, including the total number of counts, total detected genes, and the mitochondrial count ratio. Cells with a high proportion of mitochondrial counts and large total counts are candidates for removal (Fig. 1a). We also examine the relationship between total counts and the number of genes detected to identify whether any cells fall outside of the overall trend (Fig. 1b).

```
library(gridExtra)
library(scater)
sce <- addPerCellQC(sce, subsets=list(Mito=grep
("^MT-",rowData(sce)$Symbol_TENx)))
p1 <- plotColData(sce, x = "sum", y = "subsets_Mi-
to_percent")
p2 <- plotColData(sce, x = "sum", y = "detected")
grid.arrange(p1, p2, ncol=2)
```

The distribution of QC features can also be examined univariately using violin plots (Fig. 2). A common threshold for mitochondrial read percentage is 10% or 20%; however, this is tissue- and experiment-specific [13]. Rather than removing these cells, it may be more beneficial to simply flag them and track them in downstream analyses.

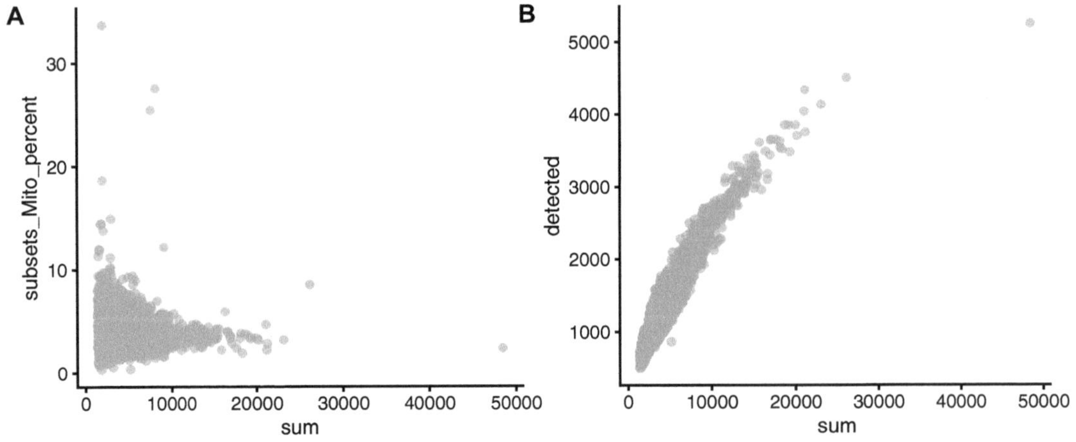

Fig. 1 (**a**) The total counts (sum) versus the percent of mitochondrial reads for every cell. (**b**) The total counts versus the number of detected genes for every cell

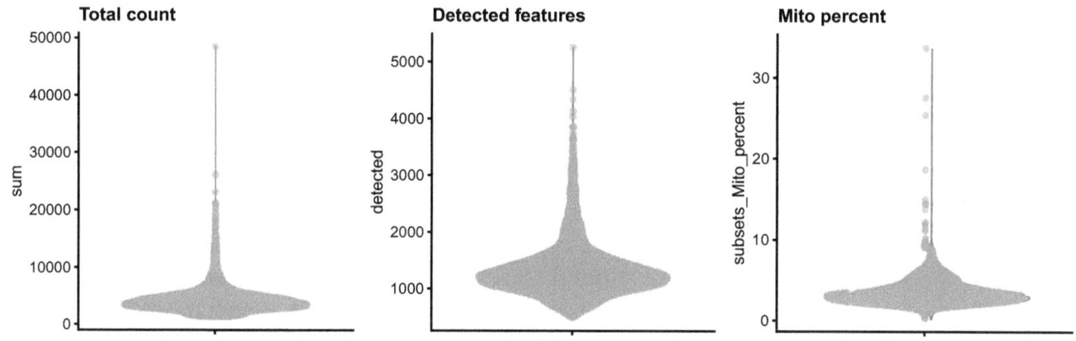

Fig. 2 Violin plots for the total number of counts (left), the total number of genes detected (middle), and the percent of mitochondrial reads (right)

```
p1 <- plotColData(sce, y="sum") + ggtitle("Total
count")
p2 <- plotColData(sce, y="detected") + ggtitle("De-
tected features")
p3 <- plotColData(sce, y="subsets_Mito_percent") +
ggtitle("Mito percent")
grid.arrange(p1, p2, p3, ncol = 3)
```

As mentioned, setting absolute thresholds is generally not recommended since these quality metrics will vary across cell types, states, and experimental parameters. Rather, we identify outlier cells using a dataset-specific threshold by using the `isOutlier()` function. The default here is to consider values larger or smaller than three median absolute deviations away from the sample median. The `type` parameter specifies in which direction outliers should be identified.

```
qc.lib2 <- isOutlier(colData(sce)$sum, type="-
both")
qc.nexprs2 <- isOutlier(colData(sce)$detected,
type="both")
qc.mito2 <- isOutlier(colData(sce)$subsets_Mito_-
percent, type="higher")
outliers <- qc.lib2 | qc.nexprs2 | qc.mito2
sce <- sce[, !outliers]
```

Rather than strictly removing low-quality cells, an alternative option is to flag these cells by creating a metadata column that can be used to track the cells at various stages of the analysis. This approach might help identify whether they are truly low-quality or potentially rare cell types.

```
colData(sce)$flag_lowquality <- 0
colData(sce)$flag_lowquality[outliers] <- 1
```

3.3 Normalization

Variations in library preparation, sequencing, and biological variability all contribute to differences in the total number of reads observed per cell in scRNA-seq data (e.g., sequencing depth or library size) [14]. The goal of normalization in scRNA-seq analysis is to remove these unwanted systematic variations to make the expression counts comparable across cells. Normalization methods originally developed for bulk RNA-seq aim to remove the variability in sequencing depth by estimating cell-specific scale factors; however, scaling approaches are unable to sufficiently remove technical artifacts in scRNA-seq data [15–17]. Below we describe normalization methods designed for scRNA data.

- *SCnorm* applies quantile regression to estimate gene-group scaling factors; gene groups are defined based on the estimated relationship between expression and sequencing depth [15]. SCnorm was developed for non-UMI-based methods, though it may be applied to UMI data with lower sparsity levels (e.g., less than 80% zeros).
- *Scran* estimates pool-based scale factors by summing expression counts in overlapping subsets of cells. Linear equations are used to deconvolve cell-specific scale factors. [16].
- *Sctransform* was designed for normalizing UMI counts [17]. Normalized counts are calculated as Pearson residuals obtained from a regularized negative binomial regression model. In addition to removing sequencing depth variations, the regularization approach shrinks the degree of expression variation due to technical noise.
- *Dino* normalizes UMI counts using posterior resampling of gene-specific means modeled by a hierarchical Gamma-Poisson distribution [18]. The Dino normalization approach is especially able to accommodate large variations in zeros across cells while preserving biological expression heterogeneity across cell types.

The PMBC dataset has a high degree of sparsity, with an average of 96% zeros per cell. We will demonstrate normalization by continuing with the *scran* R package on the cells remaining after quality control.

```
allz <- colMeans(counts(sce) == 0)
summary(allz)
   Min. 1st Qu.  Median    Mean 3rd Qu.    Max.
 0.9368  0.9582  0.9638  0.9634  0.9690  0.9852
```

The function `quickCluster()` is used to first roughly group cells based on similar expression levels, and then `computeSumFactors()` estimates the cell-specific factors for normalization and rescaling between the clusters. Since different cell types may have a large number of genes that are differentially expressed, the clustering ensures biological differences are not removed during normalization.

```
library(scran)
clusters <- quickCluster(sce)
sce <- computeSumFactors(sce, cluster=clusters)
```

Across multiple benchmarks and comparisons of scRNA-seq normalization methods, no method is universally superior. Data-driven normalization decisions are enabled by the R package, *scone,* which provides a framework to evaluate and compare normalization strategies [19]. In general, we advise users to select the normalization method most appropriate based on the single-cell sequencing platform and use exploratory data analysis to identify any potential post-normalization biases.

As an alternative to scaling normalization, downsampling has also been suggested for single-cell RNA-seq in order to normalize the entire distribution, including the zeros. The normalization method Dino is a compromise between scaling approaches which fail to handle discrepancies in the proportion of zeros across cells, but avoids the loss of information that occurs with downsampling.

3.4 Data Integration and Batch Effect Removal

Single-cell RNA-seq data are often generated across multiple experiments or run at different times. Datasets generated across experimental factors are referred to as "batches" and may contain unwanted heterogeneity. Simple batch effects may be removed using approaches developed for bulk RNA-seq such as ComBat [20] or included as a covariate in downstream analyses. However, complexities such as imbalances in cell-type proportions may be problematic. Methods for removing complex batch effects employ nonlinear or deep learning techniques and are referred to as "data integration." Caution should be used with data integration methods; removing batch effects can be helpful for identifying shared cell types, but it will make groups appear closer and possibly remove biology [21]. For this reason, data integration is typically only used at the preprocessing level to identify and annotate cell types. For statistical inference analyses, such as differential expression, the raw or normalized counts should be used as input (*see* Subheading 6).

4 Dimension Reduction

Single-cell RNA-seq data is high-dimensional, with the expression of tens of thousands of genes in thousands of cells [22]. Low counts, zero counts, and high correlation between genes are all characteristics of scRNA-seq data. These features combined with the high dimensionality introduce noise and redundant information into the analysis. Dimension reduction (DR) in both the gene space and the cell space improves the speed of analysis and our ability to disentangle cell heterogeneity signals [9].

4.1 Feature Selection

Reducing the gene space or the dimensionality in the numbers of features is referred to as feature selection. This is often done by selecting the highly variable genes (HVG), or those that provide the most information regarding cell-to-cell variations. Genes which display variability beyond that expected due to the mean-variance relationship or technical noise are considered highly variable [23, 24]. Additional approaches for identifying HVG include correlation-based and entropy-based approaches [25, 26].

4.2 Dimensionality Reduction

Dimensionality reduction algorithms in scRNA-seq project the high-dimensional cell space into a low-dimensional space. A reduced embedding in only two or three dimensions facilitates data visualization, clustering analysis, and trajectory analysis. In addition to principle component analysis (PCA) and multidimensional scaling (MDS), nonlinear dimension reduction methods such as t-distributed stochastic neighbor embedding (t-SNE) [27] and uniform approximation and projection method (UMAP) [28] are popular DR methods used in scRNA-seq methods. In comparison to PCA, nonlinear algorithms are stochastic and highly depend on the choice of parameters. However, they are widely used for generating engaging visualizations of cell clusters by optimizing the local structure of the data. Nonlinear methods also tend to distort global distances and direct interpretation should be avoided or performed with extreme caution [29]. Rather than avoiding these methods altogether, frameworks for evaluating and comparing the quality of dimension reduction should be considered [30].

Next, we demonstrate dimension reduction procedures on the PBMC dataset using the R package *Seurat*. We focus on visualization here; however, the reduced dimensions will be important for later analyses.

First, we convert our SingleCellExperiment object to a Seurat object.

```
library(Seurat)
colnames(sce) <- colData(sce)$Barcode
rownames(sce) <- rowData(sce)$Symbol_TENx
# need to convert the DelayedMatrix to a sparse
format to apply sctransform.
counts_sparse <- as(counts(sce), "dgCMatrix")
# create Seurat object
pbmc <- CreateSeuratObject(counts_sparse, project
= "pbmc4k")
```

Since we are again starting with the quality control processed data, we will use the `SCTransform()` function to normalize the data. This function also identifies the highly variable features and scales the data. By default, 3,000 HVGs are identified and retained for further analysis.

```
pbmc <- SCTransform(pbmc, verbose = FALSE)
```

Next, we will compare dimensionality reductions using PCA, t-SNE, and UMAP algorithms. PCA is a linear projection that maximizes the variance of the data at each principle component (PC). The function `RunPCA()` performs PCA and retains the top 50 PCs by default. The `DimPlot()` function is used to visualize the reduced cell space (Fig. 3a).

```
pbmc <- RunPCA(pbmc, verbose = FALSE)
DimPlot(pbmc, reduction = "pca")
```

PCA is used in data preprocessing and as input into nonlinear DR methods rather than as the final embedding. We next determine the dimensionality of the PBMC data and how many PCs to pass onto the t-SNE and UMAP methods. The elbow method is a common approach used to determine the number of PCs [9]. Each PC captures some biological signals, with the preceding PCs containing more information than later ones. We identify the PC at which the amount of variation appears to drop off. The `ElbowPlot()` function generates the elbow plot and we examine the first 50 PC dimensions (Fig. 3b). The standard deviation tapers off after PC30, which we will use to apply the UMAP and t-SNE algorithms (Fig. 4).

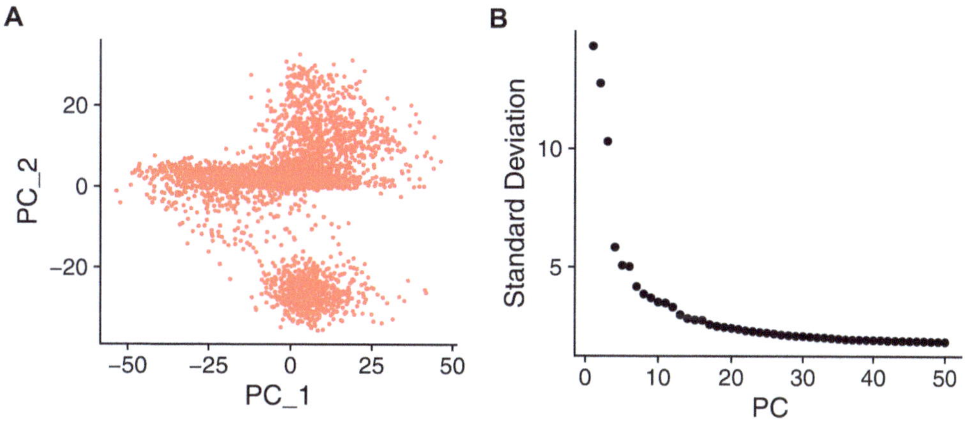

Fig. 3 (**a**) PCA projections of PBMC data. (**b**) An elbow plot visualizing the standard deviation of each PC

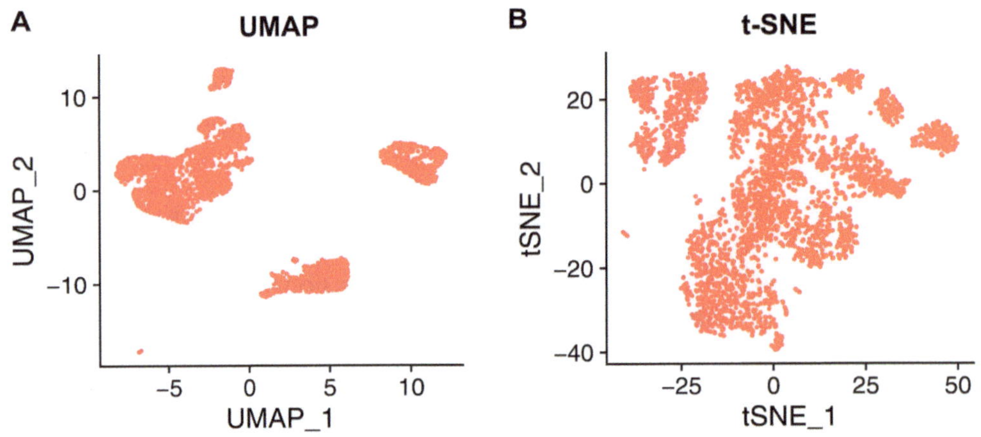

Fig. 4 (**a**) UMAP projection of PBMC data. (**b**) t-SNE projection of PBMC data

```
ElbowPlot(pbmc, reduction = "pca", ndims = 50)
# Run UMAP on first 30 PCs
pbmc <- RunUMAP(pbmc, reduction = "pca", dims = 1:
30, verbose = FALSE)
DimPlot(pbmc, reduction = "umap")
# Run t-SNE on first 30 PCs
pbmc <- RunTSNE(pbmc, reduction = "pca", dims = 1:
30, verbose = FALSE)
DimPlot(pbmc, reduction = "tsne")
```

As previously described, nonlinear methods are stochastic and rely on a number of hyperparameters; their performance also tends to be dataset-specific [31]. The topic of dimension reduction is still under debate; moving forward benchmarking and evaluation frameworks are critical to understanding method-specific characteristics [32].

5 Cell-Type Annotation

A major advantage of single-cell experiments is the ability to profile whole tissues or heterogenous cell mixtures without the need for cell sorting. Identifying the cell types is done computationally using one of two main approaches. The unsupervised approach first clusters cells into groups using the dimensionality-reduced representation based on the similarity of gene expression profiles—the groups typically correspond to specific cell types or states. From there, genes that are specific to a group or known to be marker genes are used to identify the cell cluster in a manual manner. The supervised approach annotates cells to reference populations, in which a model is typically trained on a reference dataset and then used for cell-type prediction.

The unsupervised approach is more discerning of new or rare cell types, but it is time-consuming and may require multiple rounds of clustering and annotating. Additionally, the results are subjective and require expert prior knowledge of marker genes. Supervised clustering is easier to implement and is generally accurate provided a sufficiently well-curated reference dataset. However, results may also be biased by the reference data, and rare or new cell types will not be detected if they are not in the reference set [33]. Head-to-head comparisons of these two approaches support the better overall performance of supervised methods, except in the case of rare cell identification and computational efficiency [34].

Next we apply a supervised clustering approach to the PBMC dataset using the reference-mapping approach Azimuth by *Seurat* [35]. Azimuth begins by identifying "anchors" between a reference dataset and the query scRNA-seq dataset. Anchors reflect similarities between cells of similar biological states such that the cell labels are transferred from the reference dataset to the query dataset.

We obtain a high-quality reference data of PBMCs from Azimuth [35]. It is saved in a compressed format.

```
remotes::install_github("mojaveazure/seurat-
disk")
library(SeuratDisk)
reference <- LoadH5Seurat("pbmc_multimodal.h5seur-
at")
```

The `FindTransferAnchors()` function is used to identify anchors between the reference dataset and the query dataset. The `MapQuery()` function transfers annotations from the reference dataset to the query dataset and projects the query dataset into UMAP coordinates.

```
anchors <- FindTransferAnchors(reference = refer-
ence, query = pbmc, normalization.method = "SCT",
reference.reduction = "spca", dims = 1:50)
pbmc <- MapQuery(anchorset = anchors, query = pbmc,
reference = reference,
refdata = list(celltype.level1 ="celltype.l1",
           celltype.level2 = "celltype.l2",
           celltype.level3 = "celltype.l3",
           predicted_ADT = "ADT"),
reference.reduction = "spca", reduction.model =
"wnn.umap")
```

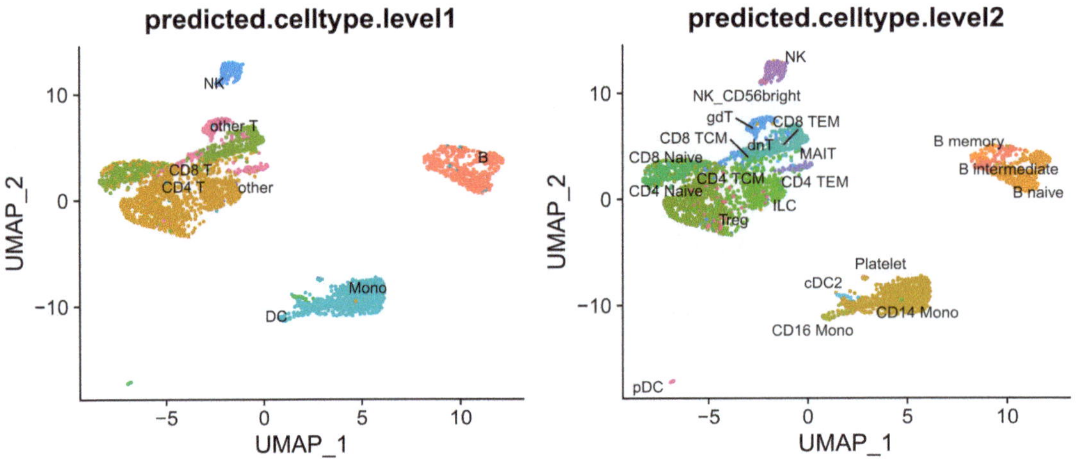

Fig. 5 UMAP projections of PBMC data colored by two different annotation granularity levels

The reference dataset here contains multiple granularity levels of annotations that we can examine on the PBMC dataset (Fig. 5).

```
p1 = DimPlot(pbmc, reduction = "umap", group.by =
"predicted.celltype.level1",
      label = TRUE, label.size = 3, repel = TRUE) +
NoLegend()
p2 = DimPlot(pbmc, reduction = "umap", group.by =
"predicted.celltype.level2",
      label = TRUE, label.size = 3 ,repel = TRUE) +
NoLegend()
p1 + p2
```

The level of annotation desired will be study-dependent, and some finer categories may need manual verification. For this example, we will save the broadest level for future analysis.

```
Idents(object = pbmc) <- "predicted.celltype.le-
vel1"
```

Supervised methods for cell annotation are still evolving, as well as the production of high-quality reference datasets through efforts such as the Human Cell Atlas [36] and HubMap [37]. The culmination of these efforts is expected to lead to improved accuracy of annotation for the scRNA-seq dataset and extension to annotating other single-cell omics datasets.

6 Downstream Statistical Analysis

Once the preprocessing and cell identification steps of the analysis workflow have been completed, downstream statistical analyses can be pursued. We focus on the two most common analyses: differential expression and trajectory inference.

6.1 Differential Expression Analysis

Differential expression (DE) analysis, a hallmark of traditional RNA-seq studies, compares gene expression differences under multiple conditions (e.g., species, phenotypes) and identifies condition-related genes [38]. In scRNA-seq, we can identify differentially expressed genes across cell types or across groups of the same cell type. Traditional RNA-seq DE methods include edgeR [39] and DESeq2 [40], which perform a statistical test of difference between groups for each gene using a negative binomial model. For scRNA-seq data, due to increased sparsity and heterogeneity, DE methods have been adapted using alternative model distributions such as the zero-inflated negative binomial model [41]. However, the choice of model is data-dependent and should be guided by examining the data distributions; often the standard negative binomial model without zero inflation is sufficient [42, 43].

Another consideration for differential expression analysis with scRNA-seq data is the appropriate sample size. In experiments with biological replicates, cells from the same individual or sample are correlated and should be modeled appropriately using either mixed models or pseudo-bulk approaches [44]. It should be noted that these approaches are not typically the default setting of standard single-cell analysis workflows; however, ignoring the correlation will lead to increased false-positive rates [44].

For the PBMC dataset, we only have a single sample, and we will identify cell-type-specific expression differences between cell clusters for demonstration. The `FindMarkers()` function in *Seurat* is used to calculate the differential expression. The default method uses the Wilcoxon Rank Sum test. Here we will identify DE genes between CD4+ T cells (CD4 T) and CD8+ T cells (CD8 T).

```
de.markers <- FindMarkers(pbmc, ident.1 = "CD4 T",
ident.2 = "CD8 T")
head(de.markers)
             p_val avg_log2FC pct.1 pct.2    p_val_adj
CD8B  2.904611e-220   -1.499543  0.124  0.821
4.310734e-216
CD8A  1.925126e-196   -1.180466  0.092  0.739
2.857079e-192
CTSW  3.187979e-125   -1.140770  0.239  0.740
4.731280e-121
```

```
CCL5   8.738088e-97   -2.682581 0.205 0.590
1.296820e-92
NKG7   1.088145e-90   -2.346978 0.197 0.585
1.614916e-86
CST7   6.501549e-86   -1.102272 0.089 0.449
9.648948e-82
```

To apply a pseudo-bulk approach in the case of multiple samples, we would simply sum the cell expression counts within each sample for all genes and then apply a traditional DE approach such as DESeq2. This can be specified in the `FindMarkers()` function by specifying `test.use = "DESeq2."` Alternatively, there are other options for differential expression analysis of scRNA-seq data for multiple individuals [45].

Using our list of differentially expressed genes, we can make a heatmap showing the expression differences between these two groups (Fig. 6).

```
de <- rownames(de.markers)[de.markers$p_va-
l_adj<0.05]
pbmc_sub <- subset(x = pbmc, subset = (predicted.
celltype.level1 == "CD4 T" | predicted.celltype.
level1 == "CD8 T"))
seurat_mat <- GetAssayData(pbmc_sub, slot = "scale.
data")
heatmap_mat <- seurat_mat[rownames(seurat_mat)[-
rownames(seurat_mat) %in% de],]
dis <- dist(heatmap_mat)
clusters <- hclust(dis)
DoHeatmap(pbmc_sub, features = clusters$labels
[clusters$order], size = 4, angle = 0, slot="s-
cale.data")
```

6.2 Trajectory Inference

Single-cell RNA-seq experiments also reflect underlying cellular dynamics such as those occurring during developmental stages and cell cycles. Rather than representing cells as clusters, the cells are computationally ordered along a trajectory to represent a continuous process. Trajectory inference (TI) analysis estimates each cell's relative position along the trajectory called the "pseudotime." The pseudo-temporal ordering places cells more similar to each other closer together, and those more divergent are placed further away, thus reflecting the underlying progression through biological processes.

TI methods typically begin by constructing a graph among cells in a low dimensional embedding on a reduced feature space, then aligning cells along the edges and branches of the graph [46]. The minimum spanning tree (MST) and nearest neighbor graph algorithms are most commonly used to construct trajectories [52]. The

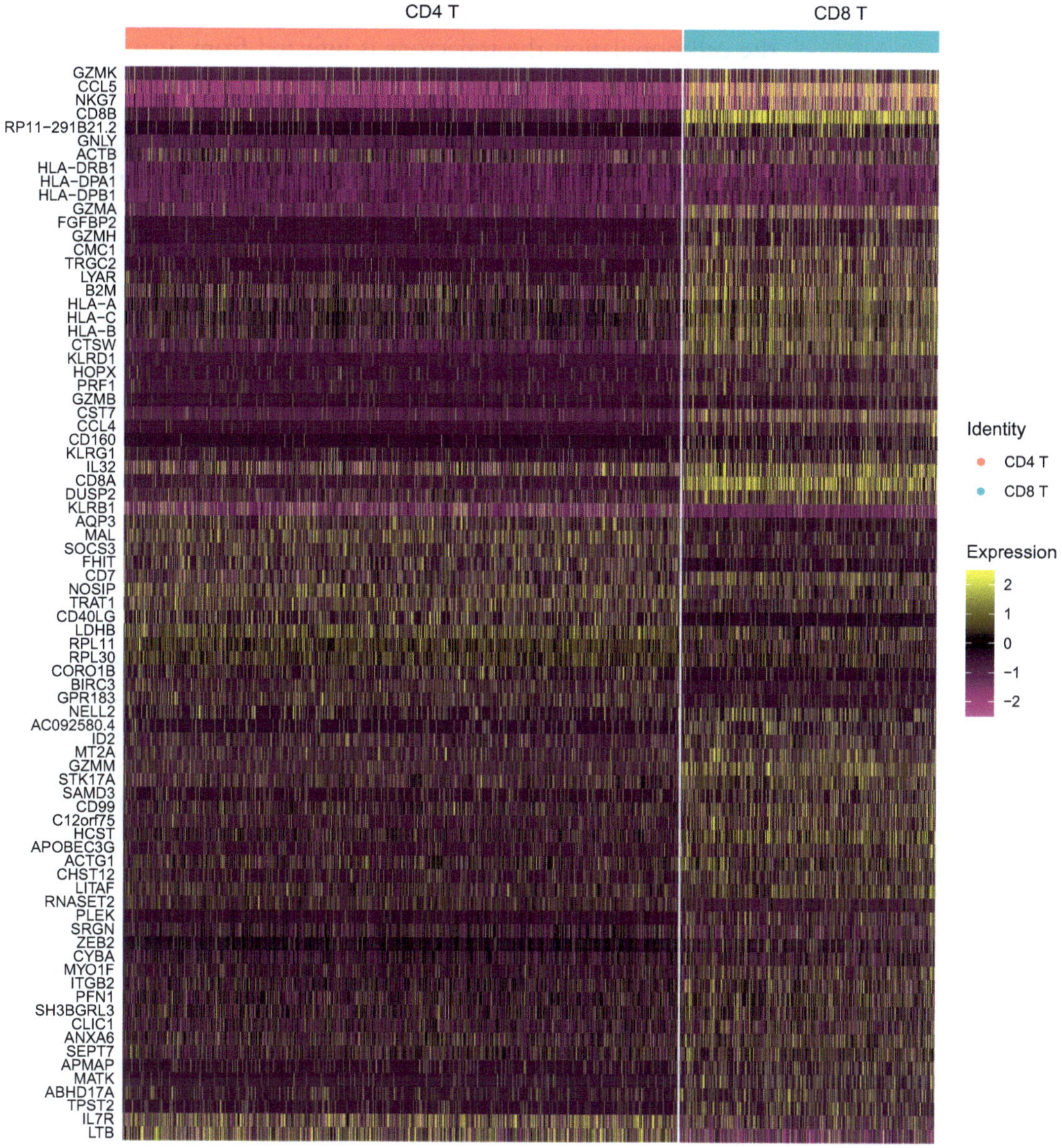

Fig. 6 Heatmap of genes differentially expressed between CD4+ T cells and CD8+ T cells (adjusted p-value < 0.05). The columns denote cells clustered by cell type

pseudotime for each cell along the trajectory is then used to identify genes that have dynamic expression patterns. A number of tools have been developed to construct cell transition trajectories, such as Wishbone [47], TSCAN [48], Slingshot [49], and Monocle3 [50]. No single method performs optimally for all datasets, and the choice of a proper trajectory analysis method is dependent on embeddings in low-dimensional space. Understanding the biological process of interest is important with regard to topology type (linear, bifurcation, tree, cycle, connected, and disconnected) and trajectory direction [51]. Consideration should also be given to the cell population the trajectory is inferred from; it is not appropriate to fit a single trajectory across discrete or biologically distinct cell clusters.

Using the PBMC dataset, we demonstrate using Slingshot for trajectory analysis. Slingshot typically performs well on linear, bifurcation, and tree topologies. We will extract the CD8+ T cell cluster for trajectory analysis. First we will rerun the dimension reduction on the cell subset and extract the embedding.

```
library(slingshot)
library(RColorBrewer)
pbmc_sub <- subset(x = pbmc, subset = (predicted.
celltype.l1 == "CD8 T"))
pbmc_sub <- RunPCA(pbmc_sub)
pbmc_sub <- RunUMAP(pbmc_sub, reduction = "pca",
dims = 1:30)
umap_coord <- Embeddings(pbmc_sub, reduction =
"umap")
```

Based on the reduced coordinate embedding, the `slingshot()` function builds the MST to determine the global topology and fits curves to illustrate the cell transition with pseudotime. The trajectory curve is extracted for each lineage using the `slingCurves()` function, and the `slingPseudotime()` function pulls the estimated pseudotime for cells in each lineage.

```
Sds <- slingshot(umap_coord)
line.ord <- slingCurves(sds)[["Lineage1"]][["s"]]
pse <- slingPseudotime(sds)
```

The trajectory can be visualized in low-dimensional space with a fitted curve with cells colored according to pseudotime, fading from red to green and finally to purple (Fig. 7a).

```
colors <- colorRampPalette(brewer.pal(11,'Spec-
tral')[-6])(100)
plotcol <- colors[cut(pse, breaks=100)]
plot(umap_coord[,1:2], pch = 16, cex = 0.5, col =
alpha(plotcol,0.7), main="CD8+ T cells")
lines(line.ord, lwd=2, col='black')
```

Using the trajectory, dynamic genes can be identified by modeling gene expression as a smooth function of pseudotime, typically using generalized additive models [52, 53]. We use tradeSeq to identify significantly dynamic genes within the CD8+ T cells and plot a single gene for demonstration (Fig. 7b).

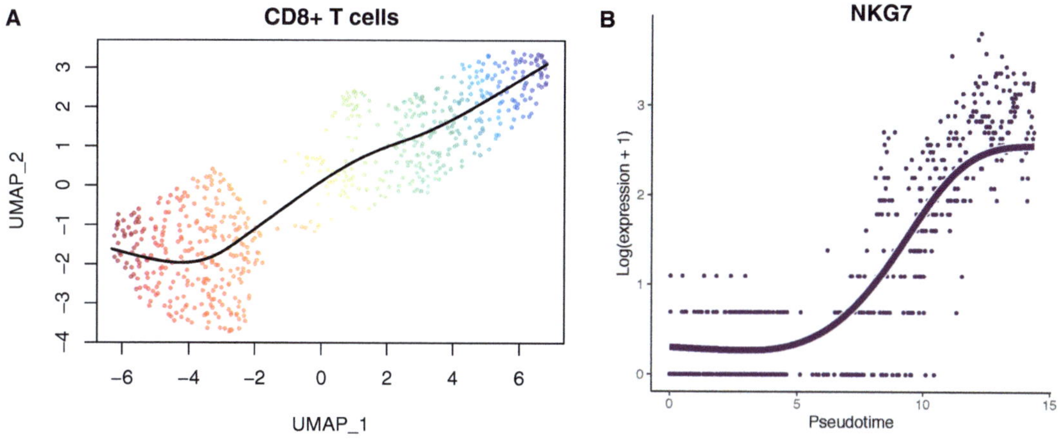

Fig. 7 (**a**) UMAP projection of CD8+ T cluster in the PBMC data colored by pseudotime. (**b**) Expression of the NKG7 gene over pseudotime

```
library(tradeSeq)
cellWeights <- slingCurveWeights(sds)
# Extract counts for 100 most variable features to
facilitate run-time.
counts <- pbmc_sub@assays$RNA@data[pbmc_sub@assays
$SCT@var.features[1:100],]
dyngenes <- fitGAM(counts = counts, pseudotime =
pse, cellWeights = cellWeights,
              nknots = 6, verbose = FALSE)
assoRes <- associationTest(dyngenes)
assoRes <- assoRes[order(assoRes$waldStat, de-
creasing = TRUE),]
head(assoRes)
        waldStat df pvalue meanLogFC
NKG7   1122.0939  5      0  1.645474
CCL5    851.3081  5      0  2.240000
GZMK    516.2338  5      0  1.880472
GZMH    440.2575  5      0  1.583884
FGFBP2  357.3571  5      0  1.130881
CMC1    279.2485  5      0  1.274895
plotSmoothers(dyngenes, counts, gene = "NKG7")
```

In addition to the traditional TI analysis, RNA velocity is a recent computational development used to order cells according to transient transcriptional kinetics. Using the ratios of reads mapping to exonic versus intronic regions, the transcriptional state of each gene can be inferred and used to fit a trajectory across cells [54]. This type of analysis does require modifying the preprocessing workflow in order to quantify the reads separately.

7 Summary

Single-cell RNA-seq technologies have only just begun to push the boundary in studying genome-wide expression dynamics and heterogeneity. With single-cell multi-omics and spatial transcriptomics on the horizon and datasets increasing exponentially in size, statistical and computational methodologies will continue to evolve in order to facilitate high-quality data analysis.

References

1. Bacher R, Kendziorski C (2016) Design and computational analysis of single-cell RNA-sequencing experiments. Genome Biol 17:63
2. Hansen K, Risso D, Hicks S (2021) TENxPBMCData: PBMC data from 10X Genomics
3. Tian L, Su S, Dong X et al (2018) scPipe: a flexible R/Bioconductor preprocessing pipeline for single-cell RNA-sequencing data. PLOS Comput Biol 14:e1006361
4. Wang Z, Hu J, Johnson WE et al (2019) Scruff: an R/Bioconductor package for preprocessing single-cell RNA-sequencing data. BMC Bioinf 20:222
5. You Y, Tian L, Su S et al (2021) Benchmarking UMI-based single-cell RNA-seq preprocessing workflows. Genome Biol 22:339
6. Soneson C, Srivastava A, Patro R et al (2021) Preprocessing choices affect RNA velocity results for droplet scRNA-seq data. PLoS Comput Biol 17:e1008585
7. Amezquita RA, Lun ATL, Becht E et al (2020) Orchestrating single-cell analysis with Bioconductor. Nat Methods 17:137–145
8. Hong R, Koga Y, Bandyadka S et al (2022) Comprehensive generation, visualization, and reporting of quality control metrics for single-cell RNA sequencing data. Nat Commun 13: 1688
9. Luecken MD, Theis FJ (2019) Current best practices in single-cell RNA-seq analysis: a tutorial. Mol Syst Biol 15:e8746
10. Chen G, Ning B, Shi T (2019) Single-cell RNA-seq technologies and related computational data analysis. Front Genet 10:317
11. McCarthy DJ, Campbell KR, Lun ATL et al (2017) Scater: pre-processing, quality control, normalization and visualization of single-cell RNA-seq data in R. Bioinformatics 33:1179–1186
12. Stuart T, Butler A, Hoffman P et al (2019) Comprehensive integration of single-cell data. Cell 177:1888–1902.e21
13. Galow A-M, Kussauer S, Wolfien M et al (2021) Quality control in scRNA-Seq can discriminate pacemaker cells: the mtRNA bias. Cell Mol Life Sci 78:6585–6592
14. Bacher R, Chu L-F, Argus C et al (2022) Enhancing biological signals and detection rates in single-cell RNA-seq experiments with cDNA library equalization. Nucleic Acids Res 50:e12–e12
15. Bacher R, Chu L-F, Leng N et al (2017) SCnorm: robust normalization of single-cell RNA-seq data. Nat Methods 14:584–586
16. L Lun AT, Bach K, Marioni JC (2016) Pooling across cells to normalize single-cell RNA sequencing data with many zero counts. Genome Biol 17:75
17. Hafemeister C, Satija R (2019) Normalization and variance stabilization of single-cell RNA--seq data using regularized negative binomial regression. Genome Biol 20:296
18. Brown J, Ni Z, Mohanty C et al (2021) Normalization by distributional resampling of high throughput single-cell RNA-sequencing data. Bioinformatics 37:4123–4128
19. Cole MB, Risso D, Wagner A et al (2019) Performance assessment and selection of normalization procedures for single-cell RNA-seq. Cell Syst 8:315–328.e8
20. Johnson WE, Li C, Rabinovic A (2007) Adjusting batch effects in microarray expression data using empirical Bayes methods. Biostatistics 8: 118–127
21. Luecken MD, Büttner M, Chaichoompu K et al (2022) Benchmarking atlas-level data integration in single-cell genomics. Nat Methods 19:41–50

22. Ding J, Condon A, Shah SP (2018) Interpretable dimensionality reduction of single cell transcriptome data with deep generative models. Nat Commun 9:2002
23. Vallejos CA, Risso D, Scialdone A et al (2017) Normalizing single-cell RNA sequencing data: challenges and opportunities. Nat Methods 14: 565–571
24. Brennecke P, Anders S, Kim JK et al (2013) Accounting for technical noise in single-cell RNA-seq experiments. Nat Methods 10: 1093–1095
25. Lall S, Ghosh A, Ray S et al (2022) Sc-REnF: an entropy guided robust feature selection for single-cell RNA-seq data. Brief Bioinform 23: bbab517
26. Ranjan B, Sun W, Park J et al (2021) DUBStepR is a scalable correlation-based feature selection method for accurately clustering single-cell data. Nat Commun 12:5849
27. van der Maaten L, Hinton G (2008) Visualizing data using t-SNE. J Mach Learn Res 9: 2579–2605
28. McInnes L, Healy J, Melville J (2020) UMAP: uniform manifold approximation and projection for dimension reduction. ArXiv180203426 Cs Stat
29. Chari T, Banerjee J, Pachter L (2021) The specious art of single-cell genomics. bioRxiv 2021.08.25.457696; https://doi.org/10.1101/2021.08.25.457696
30. Johnson EM, Kath W, Mani M (2022) EMBEDR: Distinguishing signal from noise in single-cell omics data. Patterns 3(3):100443
31. Kobak D, Berens P (2019) The art of using t-SNE for single-cell transcriptomics. Nat Commun 10:5416
32. Dong X, Bacher R (2022) Data-driven assessment of dimension reduction quality for single-cell omics data. Patterns 3:100465
33. Ranjan B, Schmidt F, Sun W et al (2021) scConsensus: combining supervised and unsupervised clustering for cell type identification in single-cell RNA sequencing data. BMC Bioinf 22:186
34. Sun X, Lin X, Li Z et al (2022) A comprehensive comparison of supervised and unsupervised methods for cell type identification in single-cell RNA-seq. Brief Bioinform 23: bbab567
35. Hao Y, Hao S, Andersen-Nissen E et al (2021) Integrated analysis of multimodal single-cell data. Cell 184:3573–3587.e29
36. Regev A, Teichmann SA, Lander ES et al (2017) The human cell atlas. elife 6:e27041
37. HuBMAP Consortium, Writing Group, Snyder MP et al (2019) The human body at cellular resolution: the NIH human biomolecular atlas program. Nature 574:187–192
38. Costa-Silva J, Domingues D, Lopes FM (2017) RNA-Seq differential expression analysis: an extended review and a software tool. PLoS One 12:e0190152
39. Robinson MD, McCarthy DJ, Smyth GK (2010) edgeR: a Bioconductor package for differential expression analysis of digital gene expression data. Bioinformatics 26:139–140
40. Love MI, Huber W, Anders S (2014) Moderated estimation of fold change and dispersion for RNA-seq data with DESeq2. Genome Biol 15:550
41. Risso D, Perraudeau F, Gribkova S et al (2018) A general and flexible method for signal extraction from single-cell RNA-seq data. Nat Commun 9:284
42. Svensson V (2020) Droplet scRNA-seq is not zero-inflated. Nat Biotechnol 38:147–150
43. Soneson C, Robinson MD (2018) Bias, robustness and scalability in single-cell differential expression analysis. Nat Methods 15: 255–261
44. Squair JW, Gautier M, Kathe C et al (2021) Confronting false discoveries in single-cell differential expression. Nat Commun 12:5692
45. Zhang M, Liu S, Miao Z et al (2022) IDEAS: individual level differential expression analysis for single-cell RNA-seq data. Genome Biol 23(1):33
46. Chervov A, Bac J, Zinovyev A (2020) Minimum spanning vs. principal trees for structured approximations of multi-dimensional datasets. Entropy 22:1274
47. Setty M, Tadmor MD, Reich-Zeliger S et al (2016) Wishbone identifies bifurcating developmental trajectories from single-cell data. Nat Biotechnol 34:637–645
48. Ji Z, Ji H (2016) TSCAN: pseudo-time reconstruction and evaluation in single-cell RNA-seq analysis. Nucleic Acids Res 44:e117–e117
49. Street K, Risso D, Fletcher RB et al (2018) Slingshot: cell lineage and pseudotime inference for single-cell transcriptomics. BMC Genomics 19:477
50. Cao J, Spielmann M, Qiu X et al (2019) The single-cell transcriptional landscape of mammalian organogenesis. Nature 566:496–502
51. Saelens W, Cannoodt R, Todorov H et al (2019) A comparison of single-cell trajectory inference methods. Nat Biotechnol 37:547–554
52. Van den Berge K, Roux de Bézieux H, Street K et al (2020) Trajectory-based differential expression analysis for single-cell sequencing data. Nat Commun 11:1201

53. Trapnell C, Cacchiarelli D, Grimsby J et al (2014) The dynamics and regulators of cell fate decisions are revealed by pseudotemporal ordering of single cells. Nat Biotechnol 32: 381–386
54. La Manno G, Soldatov R, Zeisel A et al (2018) RNA velocity of single cells. Nature 560:494–498
55. Lun ATL, Riesenfeld S, Andrews T et al (2019) EmptyDrops: distinguishing cells from empty droplets in droplet-based single-cell RNA sequencing data. Genome Biol 20:63
56. Muskovic W, Powell JE (2021) DropletQC: improved identification of empty droplets and damaged cells in single-cell RNA-seq data. Genome Biol 22:329
57. McGinnis CS, Murrow LM, Gartner ZJ (2019) DoubletFinder: doublet detection in single-cell RNA sequencing data using artificial nearest neighbors. Cell Syst 8:329–337.e4
58. Hippen AA, Falco MM, Weber LM et al (2021) miQC: an adaptive probabilistic framework for quality control of single-cell RNA-sequencing data. PLOS Comput Biol 17: e1009290

Chapter 7

A Primer on Preprocessing, Visualization, Clustering, and Phenotyping of Barcode-Based Spatial Transcriptomics Data

Oscar Ospina, Alex Soupir, and Brooke L. Fridley

Abstract

Recent developments in spatially resolved transcriptomics (ST) have resulted in a large number of studies characterizing the architecture of tissues, the spatial distribution of cell types, and their interactions. Furthermore, ST promises to enable the discovery of more accurate drug targets while also providing a better understanding of the etiology and evolution of complex diseases. The analysis of ST brings similar challenges as seen in other gene expression assays such as scRNA-seq; however, there is the additional spatial information that warrants the development of suitable algorithms for the quality control, preprocessing, visualization, and other discovery-enabling approaches (e.g., clustering, cell phenotyping). In this chapter, we review some of the existing algorithms to perform these analytical tasks and highlight some of the unmet analytical challenges in the analysis of ST data. Given the diversity of available ST technologies, we focus this chapter on the analysis of barcode-based RNA quantitation techniques.

Key words Spatially resolved transcriptomics, Gene expression, Spatial dependency, Tissue microenvironment, Cell phenotyping, Cell-type deconvolution, Clustering

1 Introduction

Recent years have seen an increased interest in spatially resolved transcriptomics (ST) [1, 2], which is evidenced by the explosion in the number of publications on this topic (Fig. 1). The meaning of the term ST is broad and may refer to a number of experimental approaches that aim to estimate the quantity and physical location of RNA transcripts within a tissue sample. Hence, the ultimate goal of ST is to provide the spatial resolution that other gene expression assays lack (i.e., bulk and single-cell RNA-seq), (Fig. 2). The various ST methods range from computational reconstruction of the spatial arrangement of cells from nonspatial data to in situ capturing or labeling of RNA transcripts [3]. For the purposes of this book chapter, we will focus on bioinformatic analysis methods for

Brooke L. Fridley and Xuefeng Wang (eds.), *Statistical Genomics*,
Methods in Molecular Biology, vol. 2629, https://doi.org/10.1007/978-1-0716-2986-4_7,

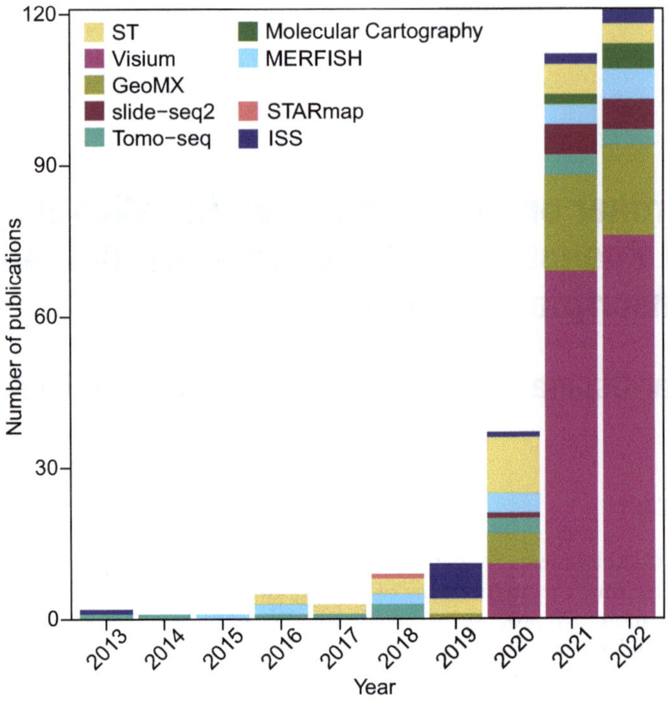

Fig. 1 The number of publications in spatial transcriptomics per year, for some of the most common ST methods in the literature. Includes peer-reviewed publications and preprints (as of July 2022). (Data taken from The Museum of Spatial Transcriptomics [1])

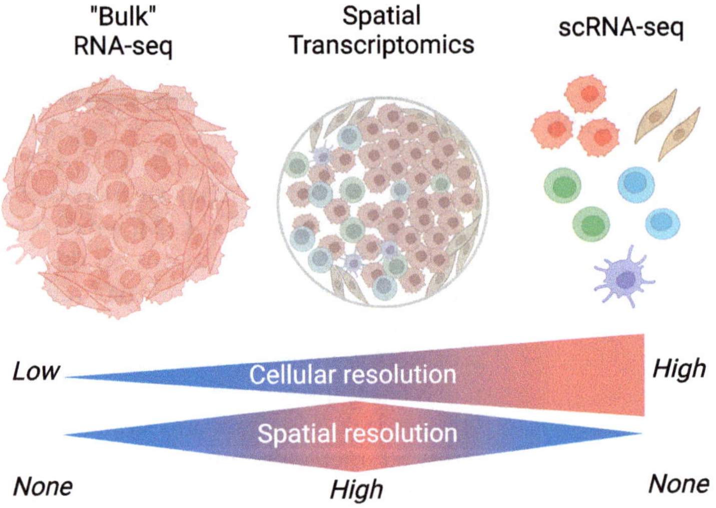

Fig. 2 Spatial transcriptomics (ST) provides the spatial resolution that bulk RNA-seq and single-cell RNA-seq (scRNA-seq) lack. Although some ST technologies do not have a cellular resolution as high as scRNA-seq, the newest ST methods provide data at the single-cell level

technologies that capture transcripts in situ (i.e., spatial barcoding), with special focus on "spot"- or "region of interest" (ROI)-based methods. This chapter's aim is to provide a starting point for potential ST users, by introducing some of the possibilities for statistical and bioinformatic analyses including visualization, clustering, and deconvolution (i.e., spot or ROI phenotyping).

Although the last few years have been characterized by major advances in ST data generation and cellular resolution [4], methods that register the spatial location of RNA transcripts are not new. The development of fluorescent in situ hybridization (FISH) was and continues to be a cornerstone in the study of transcriptomics in space [5]. In modern ST, advances in FISH technology now allows for the profiling of thousands of RNA species in a tissue at single-cell resolution (e.g., MERFISH [6], CosMx–SMI [7], Molecular Cartography [8]). In assays where RNA is captured and subsequently sequenced (e.g., ST/Visium [9], GeoMx-DSP [10], slide-seq2 [11], Seq-Scope [12]), FISH can be used to guide tissue niche detection (see [13] for an example). In all cases, the advantage of modern ST over traditional FISH is the high multiplexing capability, which can reach data outputs at the whole-transcriptome level.

The use of ST already has advanced the study of spatial biology across many organs and species (although its use on non-model species continues to be very limited [1]). Most studies using ST have addressed the differences in spatial tissue architecture between normal and diseased tissues, in human and murine model, with many publicly available data sets generated from brain [14–17], gut [18], liver [19], and a number of cancerous tumors and other tissues [9, 20–29]. Another area of study involves the assessment of expression patterns within infected tissues (e.g., [30]). These studies have not only elucidated the diversity of gene expression regions within tissues but also explored cell-to-cell communication in situ [31–33]. For example, the use of ST allows for studies to determine the functionality of immune cells and how they interact with cancerous cells [34–36]. Such knowledge will in turn become valuable for the identification of druggable targets for many diseases [26, 37, 38].

However, there are some challenges when it comes to using ST technologies. From the data analysis standpoint, ST impose similar challenges to studies involving the use of single-cell transcriptomics, such as zero inflation due to gene dropout [39, 40], large dimensionality, and, depending on the research question, detection of cell-type-specific markers and/or lack of appropriate cell-type reference atlases. In addition to these challenges, ST data present the additional challenge of spatial dependency among sampling units (i.e., capture spots, ROIs, fluorescent cells), a factor often neglected in the statistical analysis approaches used by researchers. The spatial dependency can have important effects on the observed gene expression, as neighboring sampling units (e.g., spots) are

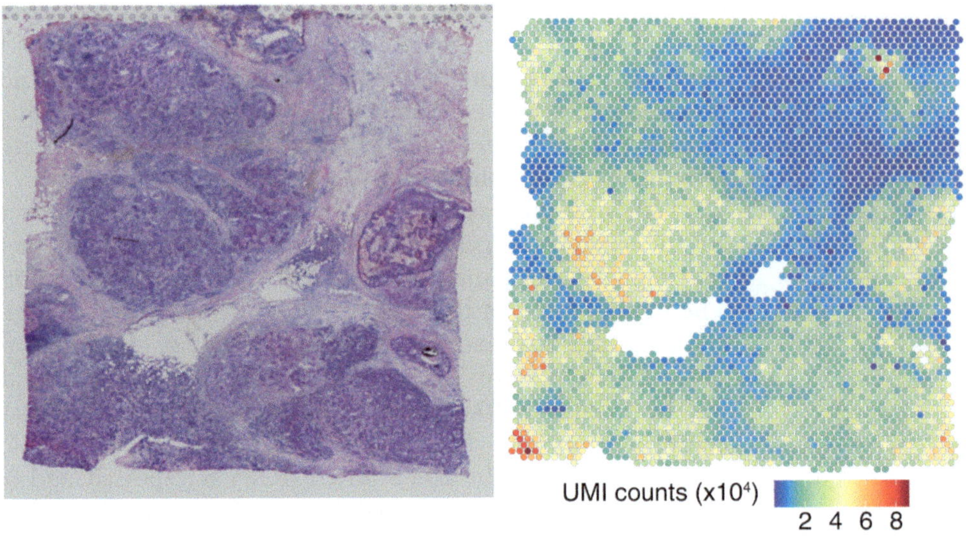

Fig. 3 As opposed to scRNA-seq, differences in data output do not necessarily reflect technical artifacts but rather can reflect true tissue biology. In this breast cancer sample, tissue architecture corresponds to the number of UMI counts observed. Areas of higher cellularity (dark purple on the left) show higher UMI counts (yellow/red on the right). (Data from 10X Genomics)

predicted to be transcriptomically more similar among them, than to distant sampling units. This spatial dependency in ST data is a driving factor for gene expression patterns observed in tissues [41, 42]. Nonetheless, our understanding of how the spatial autocorrelation decays with distance within biological tissues is still very poor and rarely assessed in ST studies. Understanding the effect of spatial dependency on the gene expression patterns will become even more necessary as the range of sizes of assayed tissue increaseson (i.e., from a few millimeters to a whole organ or individual [43]). Similarly, bioinformatic methods that explicitly account for spatial dependency in ST are relatively scarce [44]. Although spatial dependency is an important factor to consider in ST, the variation in tissue architecture or the spatial arrangement of 'tissue niches' also have major impacts on gene expression. Neighboring sampling units exhibit spatial autocorrelation, but as the assessed region becomes larger, the tissue compartmentalization (or niches) can break the spatial dependency between spots. The latter is especially true for highly structured tissues, such as cancerous tissues with tumor and stroma compartments. Similarly, tissue regions with lower cellularity (i.e., fewer cells per area or spot) would yield lower gene counts than regions with dense cellularity [45] (Fig. 3). In single-cell applications, cells with low counts are often discarded, but care needs to be taken in ST to avoid losing valuable information related to the tissue architecture.

In this chapter, we aim to describe general approaches for the bioinformatic and statistical analysis of ST data with a focus on spot-/ROI-based technologies such as Visium and GeoMx. The goal of this chapter is to provide the reader with an overview of the opportunities for ST data analysis and associated challenges. To promote further interest in the analysis of ST, we have made available the code that generated the figures presented in this chapter (https://github.com/FridleyLab/ST_book_chapter).

2 Pre-processing and Quality Control

Data formats from ST transcriptomic experiments are as diverse as the platforms that generate the data. Fortunately, many software solutions have been developed to ease data input. The packages Seurat [46], Giotto [47], STUtility [48], spatialGE [49], SpatialExperiment, and Squidpy [50], among others, directly accept files generated by the various platforms. Files in DCC format from Nanostring's GeoMx platform can be acquired easily by GeomxTools and spatialGE. Interoperability between data structures is generally straightforward, allowing analysis of the same data set by different software packages. Most software can take tables with gene count data and spatial coordinates, and 10X Visium HDF5 files can also be loaded into the R package Seurat via the Load10X_Spatial function. In the case of the spatialGE R package, the sparse matrices generated by Space Ranger can be provided for analysis.

Just like with any high-dimensional technology, performing data quality control prior to any data exploration is paramount. One of the major challenges with ST is the ability to dissect technical effects from the tissue architecture. Researchers might be tempted to remove sampling units that yield low gene counts; however, this could result in the elimination of biologically important information. For example, in some cancer tissues, stromal areas may show lower cellular density than tumor areas, resulting in lower gene counts from stromal regions. In those cases, removing spots with low gene counts will inevitably result in the loss of useful spatial and biological information. As a result, most ST studies do not eliminate sampling units from the data [22, 27, 51–53].

The practice of removing spots with high relative abundance of mitochondrial transcripts still can be considered an appropriate way to deal with technical effects, although this filtering is rarely seen in the published literature (e.g., [34, 54]). The approach is valid under the assumption that cells (or groups of cells as in spot-/ROI-based methods) exhibit high expression of mitochondrial genes when they are under stress or during cell death [55]. Another approach to adjusting for technical artifacts is the use of regression models, whereby the mitochondrial percent is regressed out (e.g., SCTransform [56], Fig. 4). The use of residuals from regression models to reduce technical artifacts is common in the ST literature [29, 34, 57–60], especially for those involving spatial barcoding technologies (e.g., Visium), and is implemented as part of the

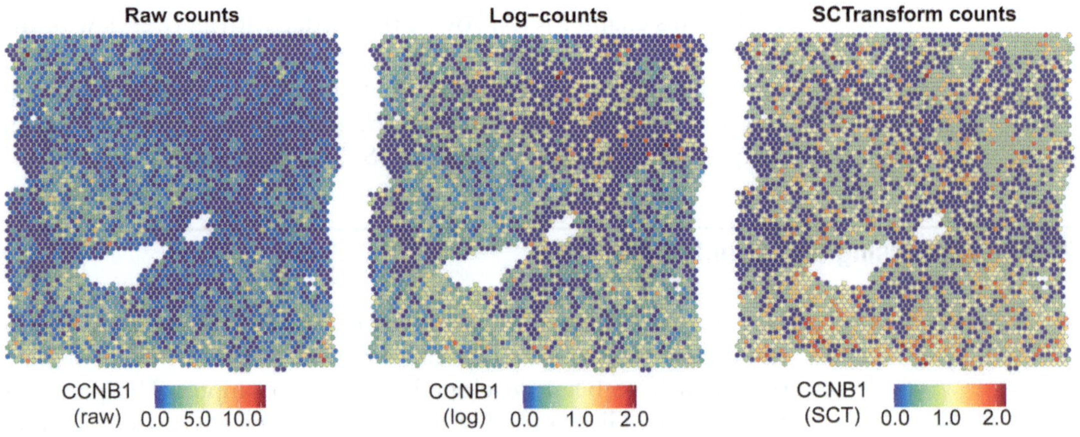

Fig. 4 Spatial differences in gene counts may obscure tissue architecture (left), hence normalization is necessary. Logarithmic transformation of library-sized scaled counts mitigates the tissue architecture differences (center), but more elaborated approaches such as regression against a percentage of mitochondrial counts (right) can further help reduce technical effects. (Data from 10X Genomics)

normalization of gene counts in the R package Seurat. It is also common to use the logarithmic transformation of gene counts scaled by the corresponding library sizes (total number of counts in the sampling unit) to normalize ST data [49, 56, 61, 62]. Other approaches include L2 normalization (e.g., MERFISH and Slide-seq [63]) or third quartile normalization (e.g., GeoMx [64–66]). An important area requiring method development is the harmonization of data sets from different experiments [67, 68]. Integration of data sets via the SCTransform function in Seurat is often carried out to reduce potential batch effects (e.g., [69]).

Some ST data sets exhibit marked gene dropout in which many genes have zero counts for the level of gene expression. The most common approach to deal with this challenge is to remove genes with an excess of zeroes. Nonetheless, filtering based on zero content often leads to significant reduction in the number of genes available for downstream analysis (i.e., hundreds to a few thousand genes removed from the analysis). An alternative option to removing zero-count genes is the application of zero-inflated negative binomial models and nonparametric models, which are beginning to become available for the analysis of ST data [70–72]. On the other side of the spectrum are genes showing high expression. Mitochondrial and ribosomal genes are usually part of this set of highly expressed genes, especially when tissue is undergoing necrosis or cellular stress [73]. Although regression analysis can help adjust for these differences in gene counts, another solution is to discard all mitochondrial and ribosomal genes (e.g., [16]), which may remove some of the noise contributed by necrosis and cellular stress.

3 Visualization

In ST experiments, gene expression and associated spatial information are deconstructed into numerous file formats. Hence, in order to visualize ST data, most analytical software includes methods for the reconstruction of the spatial context of gene expression, in many cases offering the possibility to overlay gene expression on the corresponding image of the stained tissue (e.g., H&E images). Many ST platforms have developed their own web-based analytical suites for the visualization of the generated data (e.g., GeoMx Analysis Suite, AtoMX, MERSCOPE™ Vizualizer, 10X Loupe), most of them using a point-and-click, user-friendly interface.

For a more flexible option, users can create custom plots and complete more complex downstream analysis by using a number of command-line tools for ST visualization. One of the most complete analytical software for ST analysis is the Giotto Suite [47], which allows for the visualization of gene expression along with corresponding tissue images; it is also compatible with multiple ST platforms. For comparative analyses, most software packages allow for the plotting of multiple genes simultaneously. For example, Seurat allows for not only the visualization of gene expression at each sampling unit but also for the creation of visual summaries (e.g., scatterplots, violin plots). Manual annotations of spatial features, as opposed to data-driven clustering, are also possible in STUtility. In this case, annotations are made via a browser window where the user manually selects sampling units. This information is saved as metadata for a specific sample, and gene expression visual summaries that show differences among the manually annotated features can also be generated (Fig. 5).

Another option for the visualization of ST data is the estimation of a "gene expression surface" (a.k.a. activity map). The gene expression surfaces allow for an easier detection of expression patterns matching tissue niches, as opposed to directly displaying the gene expression values of each sampling unit (e.g., Seurat's spatial feature plots or spatialGE's quilt plots). Several methods have been used to perform spatial data interpolation, including splines, inverse distance weighting, and kriging [74]. Spatial interpolation or kriging infers the gene expression values of unsampled areas based on the gene expression values of neighboring sampling units. An example of these plots are the kriging plots generated by spatialGE (Fig. 6). However, in ST technologies where sampling is not dense, such as GeoMx, spatial interpolation may yield inaccurate gene expression surfaces.

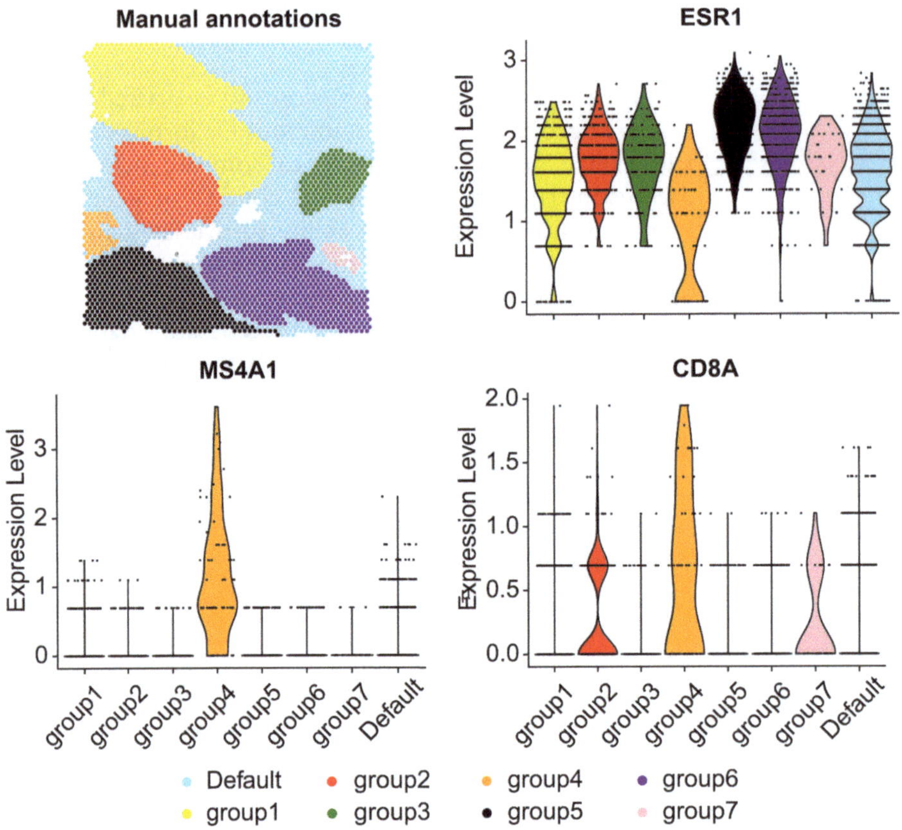

Fig. 5 Manual annotation of spots is possible with STUtility. The expression of specific genes can be compared across annotations. In this sample of ductal breast carcinoma, the black (group 5) and purple (group 6) colored areas express more estrogen receptors than the rest of the tissue. On the other hand, the orange (group 4) colored area is potentially a tertiary lymphoid structure with abundant B and T cells, given the high expression of *MS4A1* and *CD8A*, respectively

Because GeoMx experiments often consist of a relatively small number of ROIs (e.g., tens of ROIs), the visualization of gene expression is often better achieved by using traditional heatmaps supplemented with ROI annotations. The GeoMx platform allows for the quantification of gene expression from tissue areas with previously known attributes. For example, in cancer studies, researchers often pick ROIs based on their locations within larger tissue compartments such as tumor and stroma regions. Selection of GeoMx ROIs from specific areas is possible by the application of fluorescent probes to tissues, which assists pathologists during the ROI selection process. Furthermore, finer resolution of spatial gene expression can be obtained by segmenting the ROIs, which allows, for example, the study of tumor/stroma interfaces. Gene counts tables and ROI annotations can be obtained from Nanostring's GeoMx Analysis Suite. Then, the use of R packages that allow for

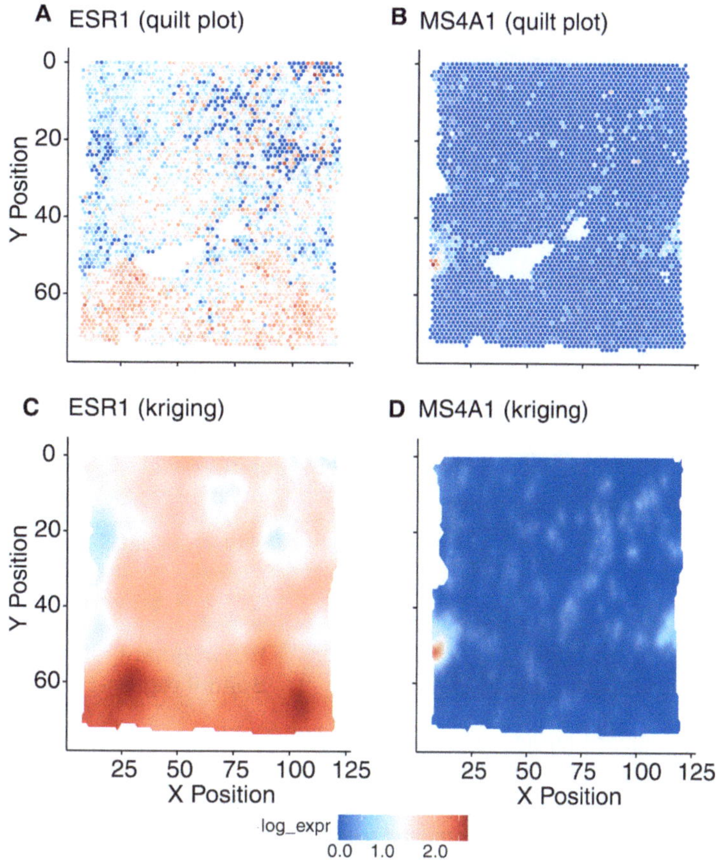

Fig. 6 Spatial expression of a given gene can be visualized via quilt plots in spatialGE (**a**, **b**). For a more "smoothed" visualization, kriging can be applied to produce a gene expression surface. In this case, agreement with the manual annotations done previously is observed (Fig. 5). The estrogen receptor (**c**) is being expressed more in the areas corresponding to groups 5, 6, and 7. Conversely, *MS4A1* is only expressed in a reduced area corresponding to group 4 (**d**)

flexible construction of heatmaps (such as ComplexHeatmap) can be leveraged to create custom visualizations (Fig. 7).

4 Clustering

Approaches for clustering of sampling units are abundant and continue to be an area of active research and development. As expected, many methods employ workflows similar to those used in the scRNA-seq analysis (e.g., principal component analysis (PCA) and Louvain clustering). Others have explored solutions rooted in Bayesian statistics or modern machine learning. Additionally, there is diversity in the approaches to account for spatial dependency. Some methods disregard the spatial information,

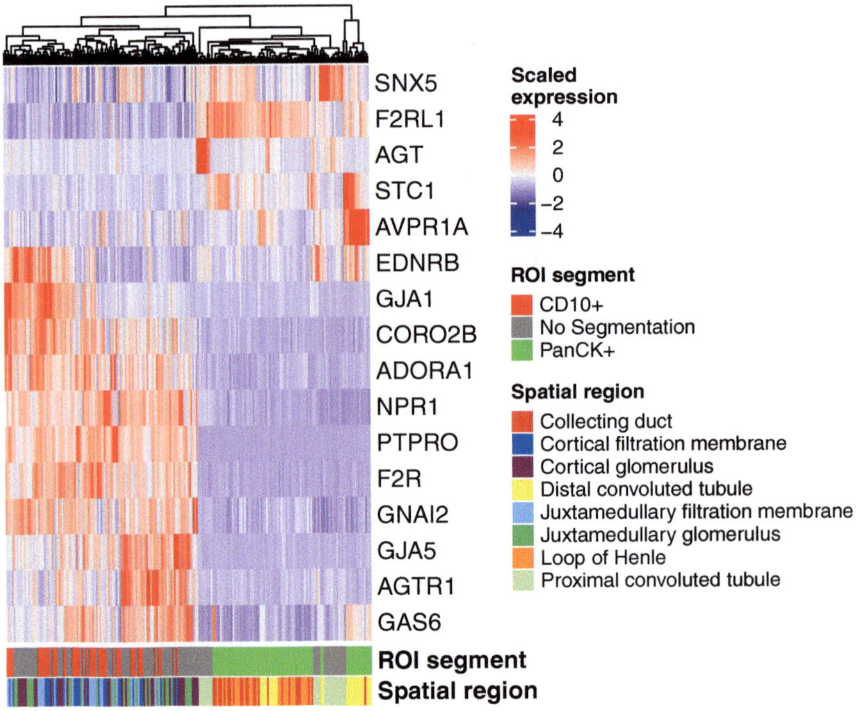

Fig. 7 Heatmaps provide an easy visualization of ROI gene expression for GeoMx experiments. Here, the heatmap shows the expression of genes involved in the regulation of renal function (obtained from KEGG), as quantified in the glomerular and tubular regions of the kidney samples. Each column in this heatmap represents an ROI, with ROIs grouped together based on their gene expression similarity. The ROI segmentation and selection are facilitated by the use of fluorescent markers (e.g., PanCK and CD10). As shown in the plot, proximal and distal tubules are differentiated by their PanCK staining (proximal tubules are PanCK-). Filtration membranes can be identified by CD10+ staining. Further pathology annotation for these samples shows that the gene *F2RL1*, involved in blood vessel dilation, is expressed in the Loops of Henle, collecting ducts, and distal tubules. This pattern of expression is expected due to the fact that the aforementioned structures are responsible for the regulation of urine/blood electrolytes and water homeostasis. The data set in this heatmap was downloaded from Nanostring's Spatial Organ Atlas website and consists of 203 ROIs taken from four kidney samples

while others account for not only the locations of sampling units, but also other data sources (e.g., stained tissue imaging data).

Clustering methods such as those implemented in Seurat are commonly used in ST studies [69, 75, 76]. Seurat uses Louvain clustering to detect "communities" of sampling units [32, 77]. However, it does not account for the spatial dependency of ST data. A Python alternative, Squidpy [50], uses spatial coordinates to create a graph and identify nearest neighbors for clustering. Even though spatial dependency is an important feature in ST data, recovering the overall similarities among sampling units does not always require explicit incorporation of the spatial coordinates, with most methods usually agreeing on the identification of most tissue niches (Fig. 8). An advantage of Louvain clustering, as implemented in Seurat, is that it is fast to compute simultaneously over different parameter values. One of the most important parameters

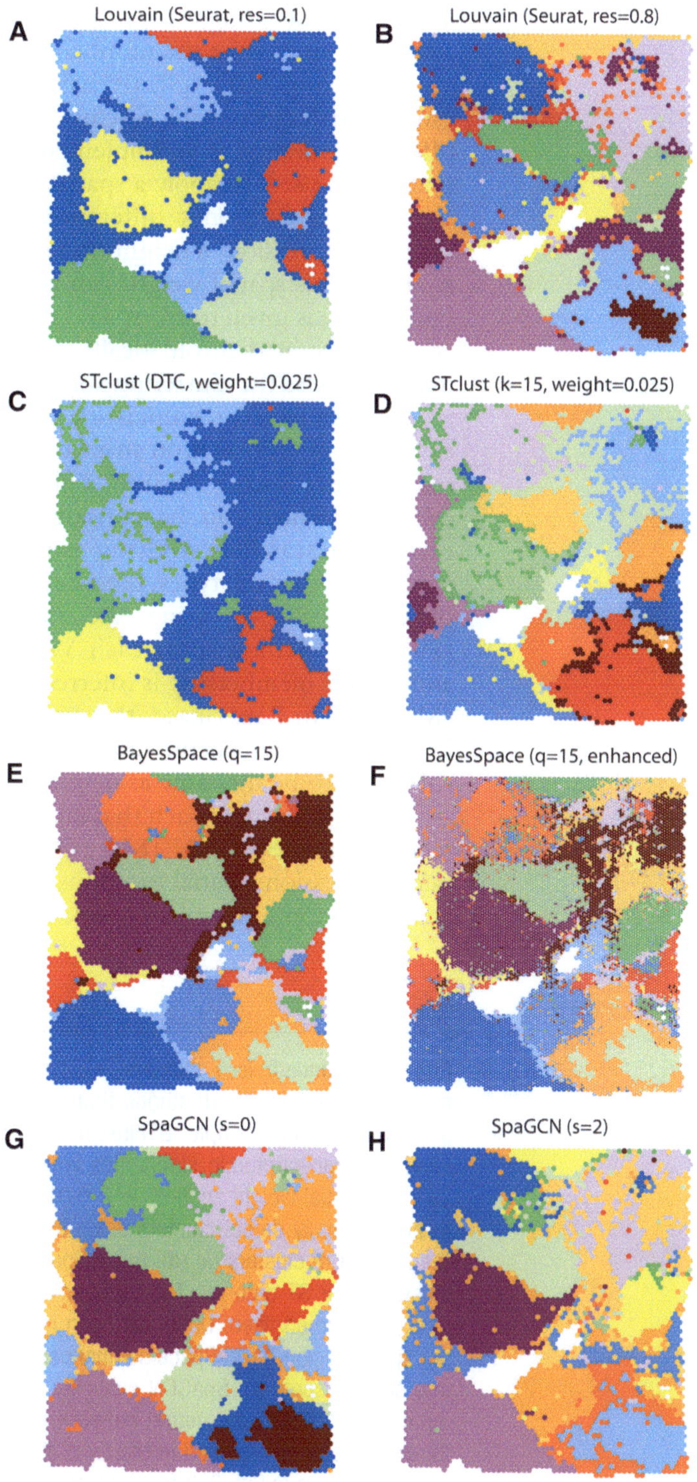

Fig. 8 Different methods to cluster sampling units applied to ST data generated with Visium technology. One of the most commonly used methods is Louvain clustering, as implemented in Seurat, which allows tuning of the number of clusters by varying the resolution parameter (res). Lower resolution results in

in Louvain clustering is the resolution of clustering, which provides a rough control over the number of communities.

Clustering that explicitly accounts for the spatial information can be performed with STclust from the R package spatialGE [49]. The method is an implementation of hierarchical clustering that allows users to apply a spatial weight to "shrink" or "regularize" the transcriptomic differences between sampling units, with the gene expression of neighboring spots being shrunk towards one another. This spatial weight can also be interpreted as how much importance is given to the spatial positions. The number of clusters (k) is determined automatically via dynamic tree cuts [78], or the user can assess the clustering solutions over a range of k values.

Using a Bayesian framework, the R package BayesSpace aims to extract more information from each spot. As in the case of many technologies, such as Visium, a given spot contains more than one cell type, which can be considered a "mini-bulk" sample. For example, researchers studying astrocytes, which often are accompanied by other cell types, used BayesSpace to ascertain the location of those cells by using the algorithm's enhanced clustering [79]. The method splits each Visium spot into six "sub-spots," and cluster membership is inferred via Markov chain Monte Carlo (MCMC). In BayesSpace, the spatial information is accounted for in the model by "encouraging" cluster membership for spots and sub-spots to be the same as neighboring spots [80].

A clustering approach that takes advantage of machine learning techniques is SpaGCN [81]. This Python-based method passes gene expression, spatial locations, and histology image features to a convolutional neural network to predict cluster memberships.

Fig. 8 (continued) less clusters (**a**), while larger resolution results in more clusters (**b**). Alternative methods that explicitly account for spatial information are available. For example, STclust from spatialGE uses hierarchical clustering based on transcriptomic similarities weighted by spatial distances. Users can vary the spatial weight applied ($w = 0.025$ in **c** and **d**). STclust also allows users to automatically detect clusters using dynamic tree cuts (DTC, **c**) or specifying the number of clusters (k) (**d**). BayesSpace offers a novel Bayesian approach to clustering for ST data (**e**). The number of clusters (q) can be inferred from the data or assumed/fixed. Users also can infer more than one cluster assignment per spot by using BayesSpace's enhanced clustering, which partitions each spot into six sub-spots (**f**). SpaGCN explicitly accounts for spatial information as an input of a convolutional neural network, in addition to gene expression data. A third input can be provided in the form of a histology image. Here, we used the histology image in Fig. 3. When no weight is assumed for the image information ($s = 0$, **g**), the clustering is driven entirely by gene expression and spatial locations. If $s = 2$ (**h**), then the clustering reflects a higher degree based on the histology image

Features from the histology image are extracted by tuning two parameters that control the size of the image area to extract data from and the weight given to those features. SpaGCN is currently one of the most sophisticated clustering methods developed for ST data, but as with the majority of methods, empirical studies, as well as additional benchmarking, are needed to validate their utility.

5 Phenotyping of Sampling Units

Researchers using ST are likely to be interested in uncovering the biological identity of their sampling units (i.e., spot or ROI "phenotyping"). This aspect of the ST data analysis workflow has sparked the development of many deconvolution methods. The currently developed approaches vary in implementation, but just like for scRNA-seq analysis, a major difference arises from the use or not of an annotated single-cell transcriptome (i.e., 'reference' data set). In some cases, a reference data set is not required by the method. Instead, a database containing gene expression signatures is necessary to infer the cell-type identities. An important consideration, however, is that the ST methods on which we have focused in this chapter do not provide single-cell-level data. As a consequence, inferred cell types from deconvolution algorithms applied to spot or ROIs (i.e., mini-bulk data) should be generally interpreted as the predominant cell type in the sampling unit.

The speed at which ST deconvolution methods are being developed outpaces the number of methods used in empirical research. As opposed to the use of deconvolution methods, many studies opt for the use of canonical markers to identify cell types (e.g., [58, 69, 82]). Others classify sampling units at the tissue niche level (e.g., [69, 83, 84]). One potential reason for this disconnect is that the various ST phenotyping methods have only recently appeared in the literature. Another reason is that most deconvolution methods require a single-cell reference transcriptome, which may not be available for specific tissue types. When a single-cell reference is available, the algorithm SPOTlight [85] has been used to phenotype Visium spots from the kidney (Fig. 9, [59]) and cancerous tissues [86]. SPOTlight uses nonnegative matrix factorization (NMF) to infer "topics," which are representations of the cell types. The NMF analysis is initialized using marker genes for each cell type in the single-cell reference. It is a computationally intensive method, and a subset of cells in the reference is often used [85].

Another reference-based method is Tangram [87], a Python module that uses deep learning to deconvolve cell mixtures from tissue regions that have available corresponding single-cell data. Tangram has been used to predict the locations of cells in specific regions of the mouse embryo for the validation of a new ST

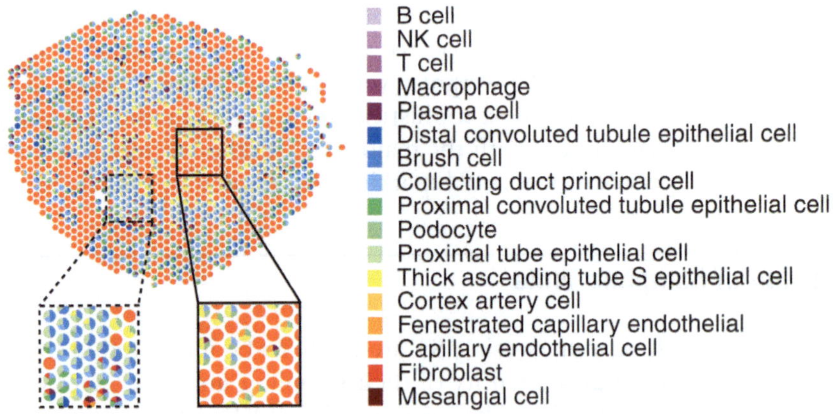

Fig. 9 SPOTlight spot-level deconvolution of a 10X Visium mouse kidney sample. The single-cell RNA-seq reference was subset to at most 100 cells per type to reduce computation time. SPOTlight and other deconvolution methods provide proxies for cell proportions at each sampling unit (spots in this case). In this example, the kidney tissue area within the dashed square was predicted to contain mostly distal and proximal tubule epithelial cells. Conversely, the area within the solid square is almost entirely composed of capillary endothelial cells

approach, Stereo-seq [88]. Among other methods requiring a single-cell reference transcriptome, users can find RCTD [89], Stereoscope [90], and Ecotyper [91]; these methods have been used in a variety of tissue types (Table 1). In the case of EcoTyper, bulk RNA-seq can be provided in lieu of scRNA-seq data, with the deconvolved cell types from bulk RNA-seq being used as the reference data set. In spite of the abundance of deconvolution methods, studies providing systematic validation and benchmarking are limited (but see [92]). Of equal importance and necessity are studies that assess how the selection of a single-cell reference affects the methods' performance. This is especially true in the case of highly heterogeneous tissues such as tumors, where publicly available scRNA-seq reference data may not be comparable with newly generated spatial data [93].

Deconvolution methods that do not require a scRNA-seq reference are also available (Table 1); however, their use in research has been much more limited. These methods often rely on the use of signature matrices. The objective is to find similarities between the gene expression patterns of each cell type in the signature matrix and the gene expression measured in the sampling units. In the case of SpatialDecon [93], built-in "cell profile" (signature) matrices have been developed for a variety of tissue types. Users also have the possibility of creating custom signature matrices from single-cell data or prior biological knowledge. Of note is that SpatialDecon was optimized for the GeoMx platform; however, SpatialDecon authors also tested the algorithm on Visium data [93]. A more technology-agnostic method is STdeconvolve [94], which predicts "topic" (i.e., cell type) proportions for each sampling unit or spot. These topics can be annotated by looking at

Table 1
A summary of some deconvolution methods for spatial transcriptomics

Method	Description	Requires scRNA reference	Publications which used the method
SPOTlight [85]	Nonnegative matrix factorization initialized with cell-type gene markers to detect cell-type proportions within sampling units	Yes	Tumor-associated immune cells [107]; kidney [59, 60]; melanoma [86], psoriatic skin [108]; glioblastoma [109]; zebrafish embryo [110]; colon adenocarcinoma [111]; multiple sclerosis brain [112]
Tangram [87]	Uses deep learning to map/infer cell types to their spatial locations in mini-bulk[a] and FISH-based ST[b]	Yes, specific to the tissue niche to deconvolve	Mouse embryo [88]
RCTD [89] (spacexr)	Supervised learning method using maximum likelihood framework applicable to mini-bulk[a] and FISH-based ST[b]	Yes	Tumor-associated immune cells [107]; renal cell cancer [113]; mouse cerebellum [71]; mouse liver metastasis [114]; multiple sclerosis brain [112]; monkey brain [115]
Stereoscope [90]	Avoids the use of marker genes for deconvolution. Assumes negative binomial distribution	Yes	Breast cancer [25, 116]; adipose tissue [24]
DSTG [61]	Pseudo-ST data generated from scRNA-seq and neural network implementation to classify real ST data	Yes	–
spatialDWLS [117] (Giotto)	Dampened weighted least squares (DWLS) to infer cell fractions	Yes, to construct signature matrix	Multiple sclerosis brain [112]
cell2location [118]	Cell signatures built via negative binomial regression	Yes, to construct signature matrix	Liver, thymus, spleen, gut [119]
EcoTyper [91]	NMF to define cell types from scRNA-seq or CIBERSORTx deconvolved bulk RNA-seq. Then, finds similarities between those cell types and ST data	Yes, scRNA-seq or bulk RNA-seq to define cell types	Human lymph node [120]

(continued)

Table 1 (continued)

Method	Description	Requires scRNA reference	Publications which used the method
STdeconvolve [94]	Latent Dirichlet allocation (LDA) to infer "cell topic" proportions within sampling units	No, cell-type identification requires knowledge of marker genes	–
SpatialDecon [93]	Uses log-normal regression to reduce the skewness of data. Provides cell-type proportions per sampling unit	No, uses pre-built signature matrices. It also allows for the creation of custom signature matrices from scRNA-seq. Enables GeoMx ROI deconvolution	Soft-tissue sarcomas [121]; Colonic mucosa [122]

[a]In this chapter, "mini-bulk" refers to those ST platforms that yield transcriptomes comprised of a small number of cells (i.e., Visium, GeoMx). The cellular resolution in mini-bulk platforms lies in between scRNA-seq and traditional bulk RNA-seq

[b]FISH-based ST refers to methods that yield high-throughput, often subcellular resolution fluorescent imaging data. Methods such as MERFISH and CODEX fall into this category

genes with high expression in a topic compared to the other topics (i.e., manual, marker-based annotation), or gene-set enrichment analysis (GSEA).

6 Final Remarks

This chapter has focused on ST technologies that do not deliver data at the single-cell level (i.e., mini-bulk). Even as these technologies become more affordable and common in the literature, there are already ST technologies that produce single-cell level spatial data (i.e., MERFISH, CosMx-SMI, Molecular Cartography, Seq-Scope). Development of methods for clustering and phenotyping that efficiently handle these large data outputs is warranted. However, progress in that area will be limited to the extent that output formats from the different technologies are standardized. These single-cell-level ST technologies will be key to comprehensively understand cell-cell interactions, tumor/stroma interactions, neuronal function, drug targets, and many other elusive topics important in the study of spatial biology.

Regarding methods for spatial cell-cell communication, many approaches are being developed. Solutions that have been tested on FISH data include CytoTalk [95], DIALOGUE [96], SVCA [97], GCNG [98], and Giotto [47]. Methods that test for cell interactions on mini-bulk spatial data are still being developed (but see for example [99]). In contrast, more development has been directed to detect spatial heterogeneity patterns. Detecting a lack of uniformity in expression within a tissue sample is of major interest as it provides the first exploration of possible tissue domains. Many packages employ spatial autocorrelation metrics such as Moran's I [100] (e.g., Seurat, MERINGUE [101], spatialGE), Gaussian process regression [41], and variogram analysis [102]. Additionally, approaches that perform non-parametric testing [71] and neural networks [81, 103] are also available.

Another area of development for ST analytical tools is data integration. To date, work has been done to supplement the lack of cellular resolution in ST, by integrating scRNA-seq data. A good example of such advances is the many phenotyping tools available to the scientific community. Nonetheless, more efforts are needed to integrate other popular spatial omics data, such as proteomics and metabolomics. For example, the Giotto suite can analyze CODEX [104] and MIBI [105] spatial proteomics data in addition to ST. One limiting factor is the possibility of extracting different data types from a single specimen. The GeoMx-DSP platform can provide both proteome and transcriptome profiling, which can be imported for analysis by R packages (GeomxTools and NanoStringNCTools). As more integrative technologies such as the recent Spatial-CITE-seq [106] become available, perhaps more tools for their analysis will be developed.

References

1. Moses L, Pachter L (2022) Museum of spatial transcriptomics. Nat Methods 19(5): 534–546. https://doi.org/10.1038/s41592-022-01409-2
2. Maniatis S, Petrescu J, Phatnani H (2021) Spatially resolved transcriptomics and its applications in cancer. Curr Opin Genet Dev 66:70–77. https://doi.org/10.1016/j.gde.2020.12.002
3. Asp M, Bergenstrahle J, Lundeberg J (2020) Spatially resolved transcriptomes-next generation tools for tissue exploration. BioEssays 42(10):e1900221. https://doi.org/10.1002/bies.201900221
4. Burgess DJ (2019) Spatial transcriptomics coming of age. Nat Rev Genet 20(6):317. https://doi.org/10.1038/s41576-019-0129-z
5. Lein E, Borm LE, Linnarsson S (2017) The promise of spatial transcriptomics for neuroscience in the era of molecular cell typing. Science 358(6359):64–69. https://doi.org/10.1126/science.aan6827
6. Chen KH, Boettiger AN, Moffitt JR, Wang S, Zhuang X (2015) RNA imaging. Spatially resolved, highly multiplexed RNA profiling in single cells. Science 348(6233):aaa6090. https://doi.org/10.1126/science.aaa6090
7. He S, Bhatt R, Brown C, Brown EA, Buhr DL, Chantranuvatana K, Danaher P, Dunaway D, Garrison RG, Geiss G, Gregory MT, Hoang ML, Khafizov R, Killingbeck EE, Kim D, Kim TK, Kim Y, Klock A, Korukonda M, Kutchma A, Lee E, Lewis ZR, Liang Y, Nelson JS, Ong GT, Perillo EP, Phan JC, Phan-Everson T, Piazza E, Rane T, Reitz Z, Rhodes M, Rosenbloom A, Ross D, Sato H, Wardhani AW, Williams-Wietzikoski CA, Wu L, Beechem JM (2022) High-plex imaging of RNA and proteins at subcellular resolution in fixed tissue by spatial molecular imaging. Nature Biotechnology 40:1794–1806. https://doi.org/10.1038/s41587-022-01483-z
8. Groiss S, Pabst D, Faber C, Meier A, Bogdoll A, Unger C, Nilges B, Strauss S, Föderl-Höbenreich E, Hardt M, Geipel A, Reinecke F, Korfhage C, Zatloukal K (2021) Highly resolved spatial transcriptomics for detection of rare events in cells. bioRxiv. https://doi.org/10.1101/2021.10.11.463936

9. Stahl PL, Salmen F, Vickovic S, Lundmark A, Navarro JF, Magnusson J, Giacomello S, Asp M, Westholm JO, Huss M, Mollbrink A, Linnarsson S, Codeluppi S, Borg A, Ponten F, Costea PI, Sahlen P, Mulder J, Bergmann O, Lundeberg J, Frisen J (2016) Visualization and analysis of gene expression in tissue sections by spatial transcriptomics. Science 353(6294):78–82. https://doi.org/10.1126/science.aaf2403
10. Zollinger DR, Lingle SE, Sorg K, Beechem JM, Merritt CR (2020) GeoMx RNA assay: high multiplex, digital, spatial analysis of RNA in FFPE tissue. Methods Mol Biol 2148:331–345. https://doi.org/10.1007/978-1-0716-0623-0_21
11. Stickels RR, Murray E, Kumar P, Li J, Marshall JL, Di Bella DJ, Arlotta P, Macosko EZ, Chen F (2021) Highly sensitive spatial transcriptomics at near-cellular resolution with slide-seqV2. Nat Biotechnol 39(3):313–319. https://doi.org/10.1038/s41587-020-0739-1
12. Cho CS, Xi J, Si Y, Park SR, Hsu JE, Kim M, Jun G, Kang HM, Lee JH (2021) Microscopic examination of spatial transcriptome using Seq-scope. Cell 184(13):3559–3572 e3522. https://doi.org/10.1016/j.cell.2021.05.010
13. Guilliams M, Bonnardel J, Haest B, Vanderborght B, Wagner C, Remmerie A, Bujko A, Martens L, Thone T, Browaeys R, De Ponti FF, Vanneste B, Zwicker C, Svedberg FR, Vanhalewyn T, Goncalves A, Lippens S, Devriendt B, Cox E, Ferrero G, Wittamer V, Willaert A, Kaptein SJF, Neyts J, Dallmeier K, Geldhof P, Casaert S, Deplancke B, Ten Dijke P, Hoorens A, Vanlander A, Berrevoet F, Van Nieuwenhove Y, Saeys Y, Saelens W, Van Vlierberghe H, Devisscher L, Scott CL (2022) Spatial proteogenomics reveals distinct and evolutionarily conserved hepatic macrophage niches. Cell 185(2):379–396 e338. https://doi.org/10.1016/j.cell.2021.12.018
14. Ortiz C, Navarro JF, Jurek A, Martin A, Lundeberg J, Meletis K (2020) Molecular atlas of the adult mouse brain. Sci Adv 6(26):eabb3446. https://doi.org/10.1126/sciadv.abb3446
15. Chen WT, Lu A, Craessaerts K, Pavie B, Sala Frigerio C, Corthout N, Qian X, Lalakova J, Kuhnemund M, Voytyuk I, Wolfs L, Mancuso R, Salta E, Balusu S, Snellinx A, Munck S, Jurek A, Fernandez Navarro J, Saido TC, Huitinga I, Lundeberg J, Fiers M, De Strooper B (2020) Spatial transcriptomics and in situ sequencing to study Alzheimer's disease. Cell 182(4):976–991 e919. https://doi.org/10.1016/j.cell.2020.06.038
16. Maynard KR, Collado-Torres L, Weber LM, Uytingco C, Barry BK, Williams SR, Catallini JL 2nd, Tran MN, Besich Z, Tippani M, Chew J, Yin Y, Kleinman JE, Hyde TM, Rao N, Hicks SC, Martinowich K, Jaffe AE (2021) Transcriptome-scale spatial gene expression in the human dorsolateral prefrontal cortex. Nat Neurosci 24(3):425–436. https://doi.org/10.1038/s41593-020-00787-0
17. Hasel P, Rose IVL, Sadick JS, Kim RD, Liddelow SA (2021) Neuroinflammatory astrocyte subtypes in the mouse brain. Nat Neurosci 24(10):1475–1487. https://doi.org/10.1038/s41593-021-00905-6
18. Fawkner-Corbett D, Antanaviciute A, Parikh K, Jagielowicz M, Geros AS, Gupta T, Ashley N, Khamis D, Fowler D, Morrissey E, Cunningham C, Johnson PRV, Koohy H, Simmons A (2021) Spatiotemporal analysis of human intestinal development at single-cell resolution. Cell 184(3):810–826 e823. https://doi.org/10.1016/j.cell.2020.12.016
19. Wu R, Guo W, Qiu X, Wang S, Sui C, Lian Q, Wu J, Shan Y, Yang Z, Yang S, Wu T, Wang K, Zhu Y, Wang S, Liu C, Zhang Y, Zheng B, Li Z, Zhang Y, Shen S, Zhao Y, Wang W, Bao J, Hu J, Wu X, Jiang X, Wang H, Gu J, Chen L (2021) Comprehensive analysis of spatial architecture in primary liver cancer. Sci Adv 7(51):eabg3750. https://doi.org/10.1126/sciadv.abg3750
20. Theocharidis G, Thomas BE, Sarkar D, Mumme HL, Pilcher WJR, Dwivedi B, Sandoval-Schaefer T, Sirbulescu RF, Kafanas A, Mezghani I, Wang P, Lobao A, Vlachos IS, Dash B, Hsia HC, Horsley V, Bhasin SS, Veves A, Bhasin M (2022) Single cell transcriptomic landscape of diabetic foot ulcers. Nat Commun 13(1):181. https://doi.org/10.1038/s41467-021-27801-8
21. He B, Bergenstrahle L, Stenbeck L, Abid A, Andersson A, Borg A, Maaskola J, Lundeberg J, Zou J (2020) Integrating spatial gene expression and breast tumour morphology via deep learning. Nat Biomed Eng 4(8):827–834. https://doi.org/10.1038/s41551-020-0578-x
22. Stur E, Corvigno S, Xu M, Chen K, Tan Y, Lee S, Liu J, Ricco E, Kraushaar D, Castro P, Zhang J, Sood AK (2022) Spatially resolved transcriptomics of high-grade serous ovarian carcinoma. iScience 25(3):103923. https://doi.org/10.1016/j.isci.2022.103923

23. Berglund E, Maaskola J, Schultz N, Friedrich S, Marklund M, Bergenstrahle J, Tarish F, Tanoglidi A, Vickovic S, Larsson L, Salmen F, Ogris C, Wallenborg K, Lagergren J, Stahl P, Sonnhammer E, Helleday T, Lundeberg J (2018) Spatial maps of prostate cancer transcriptomes reveal an unexplored landscape of heterogeneity. Nat Commun 9(1):2419. https://doi.org/10.1038/s41467-018-04724-5
24. Backdahl J, Franzen L, Massier L, Li Q, Jalkanen J, Gao H, Andersson A, Bhalla N, Thorell A, Ryden M, Stahl PL, Mejhert N (2021) Spatial mapping reveals human adipocyte subpopulations with distinct sensitivities to insulin. Cell Metab 33(9):1869–1882 e1866. https://doi.org/10.1016/j.cmet.2021.07.018
25. Andersson A, Larsson L, Stenbeck L, Salmen F, Ehinger A, Wu SZ, Al-Eryani G, Roden D, Swarbrick A, Borg A, Frisen J, Engblom C, Lundeberg J (2021) Spatial deconvolution of HER2-positive breast cancer delineates tumor-associated cell type interactions. Nat Commun 12(1):6012. https://doi.org/10.1038/s41467-021-26271-2
26. Tavares-Ferreira D, Shiers S, Ray PR, Wangzhou A, Jeevakumar V, Sankaranarayanan I, Cervantes AM, Reese JC, Chamessian A, Copits BA, Dougherty PM, Gereau RW, Burton MD, Dussor G, Price TJ (2022) Spatial transcriptomics of dorsal root ganglia identifies molecular signatures of human nociceptors. Sci Transl Med 14(632):eabj8186. https://doi.org/10.1126/scitranslmed.abj8186
27. Dhainaut M, Rose SA, Akturk G, Wroblewska A, Nielsen SR, Park ES, Buckup M, Roudko V, Pia L, Sweeney R, Le Berichel J, Wilk CM, Bektesevic A, Lee BH, Bhardwaj N, Rahman AH, Baccarini A, Gnjatic S, Pe'er D, Merad M, Brown BD (2022) Spatial CRISPR genomics identifies regulators of the tumor microenvironment. Cell 185(7):1223–1239 e1220. https://doi.org/10.1016/j.cell.2022.02.015
28. Garcia-Alonso L, Handfield LF, Roberts K, Nikolakopoulou K, Fernando RC, Gardner L, Woodhams B, Arutyunyan A, Polanski K, Hoo R, Sancho-Serra C, Li T, Kwakwa K, Tuck E, Lorenzi V, Massalha H, Prete M, Kleshchevnikov V, Tarkowska A, Porter T, Mazzeo CI, van Dongen S, Dabrowska M, Vaskivskyi V, Mahbubani KT, Park JE, Jimenez-Linan M, Campos L, Kiselev VY, Lindskog C, Ayuk P, Prigmore E, Stratton MR, Saeb-Parsy K, Moffett A, Moore L, Bayraktar OA, Teichmann SA, Turco MY, Vento-Tormo R (2021) Mapping the temporal and spatial dynamics of the human endometrium in vivo and in vitro. Nat Genet 53(12):1698–1711. https://doi.org/10.1038/s41588-021-00972-2
29. Chen H, Murray E, Sinha A, Laumas A, Li J, Lesman D, Nie X, Hotaling J, Guo J, Cairns BR, Macosko EZ, Cheng CY, Chen F (2021) Dissecting mammalian spermatogenesis using spatial transcriptomics. Cell Rep 37(5): 109915. https://doi.org/10.1016/j.celrep.2021.109915
30. Delorey TM, Ziegler CGK, Heimberg G, Normand R, Yang Y, Segerstolpe A, Abbondanza D, Fleming SJ, Subramanian A, Montoro DT, Jagadeesh KA, Dey KK, Sen P, Slyper M, Pita-Juarez YH, Phillips D, Biermann J, Bloom-Ackermann Z, Barkas N, Ganna A, Gomez J, Melms JC, Katsyv I, Normandin E, Naderi P, Popov YV, Raju SS, Niezen S, Tsai LT, Siddle KJ, Sud M, Tran VM, Vellarikkal SK, Wang Y, Amir-Zilberstein L, Atri DS, Beechem J, Brook OR, Chen J, Divakar P, Dorceus P, Engreitz JM, Essene A, Fitzgerald DM, Fropf R, Gazal S, Gould J, Grzyb J, Harvey T, Hecht J, Hether T, Jane-Valbuena J, Leney-Greene M, Ma H, McCabe C, McLoughlin DE, Miller EM, Muus C, Niemi M, Padera R, Pan L, Pant D, Pe'er C, Pfiffner-Borges J, Pinto CJ, Plaisted J, Reeves J, Ross M, Rudy M, Rueckert EH, Siciliano M, Sturm A, Todres E, Waghray A, Warren S, Zhang S, Zollinger DR, Cosimi L, Gupta RM, Hacohen N, Hibshoosh H, Hide W, Price AL, Rajagopal J, Tata PR, Riedel S, Szabo G, Tickle TL, Ellinor PT, Hung D, Sabeti PC, Novak R, Rogers R, Ingber DE, Jiang ZG, Juric D, Babadi M, Farhi SL, Izar B, Stone JR, Vlachos IS, Solomon IH, Ashenberg O, Porter CBM, Li B, Shalek AK, Villani AC, Rozenblatt-Rosen O, Regev A (2021) COVID-19 tissue atlases reveal SARS-CoV-2 pathology and cellular targets. Nature 595(7865):107–113. https://doi.org/10.1038/s41586-021-03570-8
31. Rao A, Barkley D, Franca GS, Yanai I (2021) Exploring tissue architecture using spatial transcriptomics. Nature 596(7871): 211–220. https://doi.org/10.1038/s41586-021-03634-9
32. Longo SK, Guo MG, Ji AL, Khavari PA (2021) Integrating single-cell and spatial transcriptomics to elucidate intercellular tissue dynamics. Nat Rev Genet 22(10): 627–644. https://doi.org/10.1038/s41576-021-00370-8

33. Marshall JL, Noel T, Wang QS, Chen H, Murray E, Subramanian A, Vernon KA, Bazua-Valenti S, Liguori K, Keller K, Stickels RR, McBean B, Heneghan RM, Weins A, Macosko EZ, Chen F, Greka A (2022) High resolution slide-seqV2 spatial transcriptomics enables discovery of disease-specific cell neighborhoods and pathways. iScience:104097. https://doi.org/10.1016/j.isci.2022.104097
34. Lv J, Shi Q, Han Y, Li W, Liu H, Zhang J, Niu C, Gao G, Fu Y, Zhi R, Wu K, Li S, Gu F, Fu L (2021) Spatial transcriptomics reveals gene expression characteristics in invasive micropapillary carcinoma of the breast. Cell Death Dis 12(12):1095. https://doi.org/10.1038/s41419-021-04380-6
35. Ji AL, Rubin AJ, Thrane K, Jiang S, Reynolds DL, Meyers RM, Guo MG, George BM, Mollbrink A, Bergenstrahle J, Larsson L, Bai Y, Zhu B, Bhaduri A, Meyers JM, Rovira-Clave X, Hollmig ST, Aasi SZ, Nolan GP, Lundeberg J, Khavari PA (2020) Multimodal analysis of composition and spatial architecture in human squamous cell carcinoma. Cell 182(2):497–514 e422. https://doi.org/10.1016/j.cell.2020.05.039
36. Jin S, Ramos R (2022) Computational exploration of cellular communication in skin from emerging single-cell and spatial transcriptomic data. Biochem Soc Trans 50(1): 297–308. https://doi.org/10.1042/BST20210863
37. Joshi N, Watanabe S, Verma R, Jablonski RP, Chen CI, Cheresh P, Markov NS, Reyfman PA, McQuattie-Pimentel AC, Sichizya L, Lu Z, Piseaux-Aillon R, Kirchenbuechler D, Flozak AS, Gottardi CJ, Cuda CM, Perlman H, Jain M, Kamp DW, Budinger GRS, Misharin AV (2020) A spatially restricted fibrotic niche in pulmonary fibrosis is sustained by M-CSF/M-CSFR signalling in monocyte-derived alveolar macrophages. Eur Respir J 55(1). https://doi.org/10.1183/13993003.00646-2019
38. Su S, Li X (2021) Dive into single, seek out multiple: probing cancer metastases via single-cell sequencing and imaging techniques. Cancers (Basel) 13(5). https://doi.org/10.3390/cancers13051067
39. Li Q, Zhang M, Xie Y, Xiao G (2021) Bayesian modeling of spatial molecular profiling data via Gaussian process. Bioinformatics. https://doi.org/10.1093/bioinformatics/btab455
40. Zhao P, Zhu J, Ma Y, Zhou X (2022) Modeling zero inflation is not necessary for spatial transcriptomics. Genome Biol 23(1):118. https://doi.org/10.1186/s13059-022-02684-0
41. Svensson V, Teichmann SA, Stegle O (2018) SpatialDE: identification of spatially variable genes. Nat Methods 15(5):343–346. https://doi.org/10.1038/nmeth.4636
42. Fornito A, Arnatkeviciute A, Fulcher BD (2019) Bridging the gap between connectome and transcriptome. Trends Cogn Sci 23(1):34–50. https://doi.org/10.1016/j.tics.2018.10.005
43. Srivatsan SR, Regier MC, Barkan E, Franks JM, Packer JS, Grosjean P, Duran M, Saxton S, Ladd JJ, Spielmann M, Lois C, Lampe PD, Shendure J, Stevens KR, Trapnell C (2021) Embryo-scale, single-cell spatial transcriptomics. Science 373(6550): 111–117. https://doi.org/10.1126/science.abb9536
44. Hu J, Schroeder A, Coleman K, Chen C, Auerbach BJ, Li M (2021) Statistical and machine learning methods for spatially resolved transcriptomics with histology. Comput Struct Biotechnol J 19:3829–3841. https://doi.org/10.1016/j.csbj.2021.06.052
45. Saiselet M, Rodrigues-Vitoria J, Tourneur A, Craciun L, Spinette A, Larsimont D, Andry G, Lundeberg J, Maenhaut C, Detours V (2020) Transcriptional output, cell-type densities, and normalization in spatial transcriptomics. J Mol Cell Biol 12(11): 906–908. https://doi.org/10.1093/jmcb/mjaa028
46. Hao Y, Hao S, Andersen-Nissen E, Mauck WM 3rd, Zheng S, Butler A, Lee MJ, Wilk AJ, Darby C, Zager M, Hoffman P, Stoeckius M, Papalexi E, Mimitou EP, Jain J, Srivastava A, Stuart T, Fleming LM, Yeung B, Rogers AJ, McElrath JM, Blish CA, Gottardo R, Smibert P, Satija R (2021) Integrated analysis of multimodal single-cell data. Cell 184(13):3573–3587 e3529. https://doi.org/10.1016/j.cell.2021.04.048
47. Dries R, Zhu Q, Dong R, Eng CL, Li H, Liu K, Fu Y, Zhao T, Sarkar A, Bao F, George RE, Pierson N, Cai L, Yuan GC (2021) Giotto: a toolbox for integrative analysis and visualization of spatial expression data. Genome Biol 22(1):78. https://doi.org/10.1186/s13059-021-02286-2
48. Bergenstrahle J, Larsson L, Lundeberg J (2020) Seamless integration of image and molecular analysis for spatial transcriptomics

workflows. BMC Genomics 21(1):482. https://doi.org/10.1186/s12864-020-06832-3
49. Ospina OE, Wilson CM, Soupir AC, Berglund A, Smalley I, Tsai KY, Fridley B (2022) spatialGE: quantification and visualization of the tumor microenvironment heterogeneity using spatial transcriptomics. Bioinformatics 38(9):2645–2647. https://doi.org/10.1093/bioinformatics/btac145
50. Palla G, Spitzer H, Klein M, Fischer D, Schaar AC, Kuemmerle LB, Rybakov S, Ibarra IL, Holmberg O, Virshup I, Lotfollahi M, Richter S, Theis FJ (2022) Squidpy: a scalable framework for spatial omics analysis. Nat Methods 19(2):171–178. https://doi.org/10.1038/s41592-021-01358-2
51. Calvanese V, Capellera-Garcia S, Ma F, Fares I, Liebscher S, Ng ES, Ekstrand S, Aguadé-Gorgorió J, Vavilina A, Lefaudeux D, Nadel B, Li JY, Wang Y, Lee LK, Ardehali R, Iruela-Arispe ML, Pellegrini M, Stanley EG, Elefanty AG, Schenke-Layland K, Mikkola HKA (2022) Mapping human haematopoietic stem cells from haemogenic endothelium to birth. Nature 604(7906):534–540. https://doi.org/10.1038/s41586-022-04571-x
52. Kadur Lakshminarasimha Murthy P, Sontake V, Tata A, Kobayashi Y, Macadlo L, Okuda K, Conchola AS, Nakano S, Gregory S, Miller LA, Spence JR, Engelhardt JF, Boucher RC, Rock JR, Randell SH, Tata PR (2022) Human distal lung maps and lineage hierarchies reveal a bipotent progenitor. Nature 604(7904):111–119. https://doi.org/10.1038/s41586-022-04541-3
53. Liu D, Li H, Dong H, Qu M, Yang L, Chen L, Li Y, Wang H, He Y (2022) Spatial multiomics analysis reveals only minor genetic and epigenetic changes in human liver cancer stem-like cells compared with other tumor parenchymal cells. Front Cell Dev Biol 10:810687. https://doi.org/10.3389/fcell.2022.810687
54. Li YH, Cao Y, Liu F, Zhao Q, Adi D, Huo Q, Liu Z, Luo JY, Fang BB, Tian T, Li XM, Liu D, Yang YN (2021) Visualization and analysis of gene expression in Stanford type a aortic dissection tissue section by spatial Transcriptomics. Front Genet 12:698124. https://doi.org/10.3389/fgene.2021.698124
55. Lun AT, McCarthy DJ, Marioni JC (2016) A step-by-step workflow for low-level analysis of single-cell RNA-seq data with bioconductor. F1000Res 5:2122. https://doi.org/10.12688/f1000research.9501.2
56. Hafemeister C, Satija R (2019) Normalization and variance stabilization of single-cell RNA-seq data using regularized negative binomial regression. Genome Biol 20(1):296. https://doi.org/10.1186/s13059-019-1874-1
57. Parigi SM, Larsson L, Das S, Ramirez Flores RO, Frede A, Tripathi KP, Diaz OE, Selin K, Morales RA, Luo X, Monasterio G, Engblom C, Gagliani N, Saez-Rodriguez J, Lundeberg J, Villablanca EJ (2022) The spatial transcriptomic landscape of the healing mouse intestine following damage. Nat Commun 13(1):828. https://doi.org/10.1038/s41467-022-28497-0
58. Kiss T, Nyul-Toth A, DelFavero J, Balasubramanian P, Tarantini S, Faakye J, Gulej R, Ahire C, Ungvari A, Yabluchanskiy A, Wiley G, Garman L, Ungvari Z, Csiszar A (2022) Spatial transcriptomic analysis reveals inflammatory foci defined by senescent cells in the white matter, hippocampi and cortical grey matter in the aged mouse brain. Geroscience 44(2):661–681. https://doi.org/10.1007/s11357-022-00521-7
59. Melo Ferreira R, Sabo AR, Winfree S, Collins KS, Janosevic D, Gulbronson CJ, Cheng YH, Casbon L, Barwinska D, Ferkowicz MJ, Xuei X, Zhang C, Dunn KW, Kelly KJ, Sutton TA, Hato T, Dagher PC, El-Achkar TM, Eadon MT (2021) Integration of spatial and single-cell transcriptomics localizes epithelial cell-immune cross-talk in kidney injury. JCI Insight 6(12). https://doi.org/10.1172/jci.insight.147703
60. Dixon EE, Wu H, Muto Y, Wilson PC, Humphreys BD (2022) Spatially resolved transcriptomic analysis of acute kidney injury in a female murine model. J Am Soc Nephrol 33(2):279–289. https://doi.org/10.1681/ASN.2021081150
61. Song Q, Su J (2021) DSTG: deconvoluting spatial transcriptomics data through graph-based artificial intelligence. Brief Bioinform 22(5). https://doi.org/10.1093/bib/bbaa414
62. Shengquan C, Boheng Z, Xiaoyang C, Xuegong Z, Rui J (2021) stPlus: a reference-based method for the accurate enhancement of spatial transcriptomics. Bioinformatics 37 (Suppl_1):i299–i307. https://doi.org/10.1093/bioinformatics/btab298
63. Xu Y, McCord RP (2021) CoSTA: unsupervised convolutional neural network learning for spatial transcriptomics analysis. BMC Bioinform 22(1):397. https://doi.org/10.1186/s12859-021-04314-1

64. Park J, Foox J, Hether T, Danko DC, Warren S, Kim Y, Reeves J, Butler DJ, Mozsary C, Rosiene J, Shaiber A, Afshin EE, MacKay M, Rendeiro AF, Bram Y, Chandar V, Geiger H, Craney A, Velu P, Melnick AM, Hajirasouliha I, Beheshti A, Taylor D, Saravia-Butler A, Singh U, Wurtele ES, Schisler J, Fennessey S, Corvelo A, Zody MC, Germer S, Salvatore S, Levy S, Wu S, Tatonetti NP, Shapira S, Salvatore M, Westblade LF, Cushing M, Rennert H, Kriegel AJ, Elemento O, Imielinski M, Rice CM, Borczuk AC, Meydan C, Schwartz RE, Mason CE (2022) System-wide transcriptome damage and tissue identity loss in COVID-19 patients. Cell Rep Med 3(2):100522. https://doi.org/10.1016/j.xcrm.2022.100522
65. Rendeiro AF, Ravichandran H, Bram Y, Chandar V, Kim J, Meydan C, Park J, Foox J, Hether T, Warren S, Kim Y, Reeves J, Salvatore S, Mason CE, Swanson EC, Borczuk AC, Elemento O, Schwartz RE (2021) The spatial landscape of lung pathology during COVID-19 progression. Nature 593(7860):564–569. https://doi.org/10.1038/s41586-021-03475-6
66. Brady L, Kriner M, Coleman I, Morrissey C, Roudier M, True LD, Gulati R, Plymate SR, Zhou Z, Birditt B, Meredith R, Geiss G, Hoang M, Beechem J, Nelson PS (2021) Inter- and intra-tumor heterogeneity of metastatic prostate cancer determined by digital spatial gene expression profiling. Nat Commun 12(1):1426. https://doi.org/10.1038/s41467-021-21615-4
67. Dries R, Chen J, Del Rossi N, Khan MM, Sistig A, Yuan GC (2021) Advances in spatial transcriptomic data analysis. Genome Res 31(10):1706–1718. https://doi.org/10.1101/gr.275224.121
68. Liu B, Li Y, Zhang L (2021) Analysis and visualization of spatial transcriptomic data. Front Genet 12:785290. https://doi.org/10.3389/fgene.2021.785290
69. Hildebrandt F, Andersson A, Saarenpaa S, Larsson L, Van Hul N, Kanatani S, Masek J, Ellis E, Barragan A, Mollbrink A, Andersson ER, Lundeberg J, Ankarklev J (2021) Spatial Transcriptomics to define transcriptional patterns of zonation and structural components in the mouse liver. Nat Commun 12(1):7046. https://doi.org/10.1038/s41467-021-27354-w
70. BinTayyash N, Georgaka S, John ST, Ahmed S, Boukouvalas A, Hensman J, Rattray M (2021) Non-parametric modelling of temporal and spatial counts data from RNA-seq experiments. Bioinformatics 37(21):3788–3795. https://doi.org/10.1093/bioinformatics/btab486
71. Zhu J, Sun S, Zhou X (2021) SPARK-X: non-parametric modeling enables scalable and robust detection of spatial expression patterns for large spatial transcriptomic studies. Genome Biol 22(1):184. https://doi.org/10.1186/s13059-021-02404-0
72. Li K, Yan C, Li C, Chen L, Zhao J, Zhang Z, Bao S, Sun J, Zhou M (2022) Computational elucidation of spatial gene expression variation from spatially resolved transcriptomics data. Mol Ther Nucleic Acids 27:404–411. https://doi.org/10.1016/j.omtn.2021.12.009
73. Yu X, Abbas-Aghababazadeh F, Chen YA, Fridley BL (2021) Statistical and bioinformatics analysis of data from bulk and single-cell RNA sequencing experiments. Methods Mol Biol 2194:143–175. https://doi.org/10.1007/978-1-0716-0849-4_9
74. Diggle PJ, Ribeiro PJ (2007) An overview of model-based geostatistics. In: Diggle PJ, Ribeiro PJ (eds) Model-based geostatistics, Springer series in statistics. Springer, New York, pp 134–156. https://doi.org/10.1007/978-0-387-48536-2_6
75. Gerber T, Loureiro C, Schramma N, Chen S, Jain A, Weber A, Weigert A, Santel M, Alim K, Treutlein B, Camp JG (2022) Spatial transcriptomic and single-nucleus analysis reveals heterogeneity in a gigantic single-celled syncytium. elife:11. https://doi.org/10.7554/eLife.69745
76. Luo W, Lin GN, Song W, Zhang Y, Lai H, Zhang M, Miao J, Cheng X, Wang Y, Li W, Wei W, Gao WQ, Yang R, Wang J (2021) Single-cell spatial transcriptomic analysis reveals common and divergent features of developing postnatal granule cerebellar cells and medulloblastoma. BMC Biol 19(1):135. https://doi.org/10.1186/s12915-021-01071-8
77. Butler A, Hoffman P, Smibert P, Papalexi E, Satija R (2018) Integrating single-cell transcriptomic data across different conditions, technologies, and species. Nat Biotechnol 36(5):411–420. https://doi.org/10.1038/nbt.4096
78. Langfelder P, Zhang B, Horvath S (2008) Defining clusters from a hierarchical cluster tree: the dynamic tree cut package for R. Bioinformatics 24(5):719–720. https://doi.org/10.1093/bioinformatics/btm563
79. Sadick JS, O'Dea MR, Hasel P, Dykstra T, Faustin A, Liddelow SA (2022) Astrocytes and oligodendrocytes undergo subtype-

specific transcriptional changes in Alzheimer's disease. Neuron 110(11):1788–1805 e1710. https://doi.org/10.1016/j.neuron.2022.03.008

80. Zhao E, Stone MR, Ren X, Guenthoer J, Smythe KS, Pulliam T, Williams SR, Uytingco CR, Taylor SEB, Nghiem P, Bielas JH, Gottardo R (2021) Spatial transcriptomics at subspot resolution with BayesSpace. Nat Biotechnol 39(11):1375–1384. https://doi.org/10.1038/s41587-021-00935-2
81. Hu J, Li X, Coleman K, Schroeder A, Ma N, Irwin DJ, Lee EB, Shinohara RT, Li M (2021) SpaGCN: integrating gene expression, spatial location and histology to identify spatial domains and spatially variable genes by graph convolutional network. Nat Methods 18(11):1342–1351. https://doi.org/10.1038/s41592-021-01255-8
82. Peng L, Jin X, Li BY, Zeng X, Liao BH, Jin T, Chen JW, Gao XS, Wang W, He Q, Chen G, Gong LN, Shen H, Wang KJ, Li H, Luo DY (2022) Integrating single-cell RNA sequencing with spatial transcriptomics reveals immune landscape for interstitial cystitis. Signal Transduct Target Ther 7(1):161. https://doi.org/10.1038/s41392-022-00962-8
83. Tower RJ, Busse E, Jaramillo J, Lacey M, Hoffseth K, Guntur AR, Simkin J, Sammarco MC (2022) Spatial transcriptomics reveals metabolic changes underly age-dependent declines in digit regeneration. elife 11. https://doi.org/10.7554/eLife.71542
84. Moncada R, Barkley D, Wagner F, Chiodin M, Devlin JC, Baron M, Hajdu CH, Simeone DM, Yanai I (2020) Integrating microarray-based spatial transcriptomics and single-cell RNA-seq reveals tissue architecture in pancreatic ductal adenocarcinomas. Nat Biotechnol 38(3):333–342. https://doi.org/10.1038/s41587-019-0392-8
85. Elosua-Bayes M, Nieto P, Mereu E, Gut I, Heyn H (2021) SPOTlight: seeded NMF regression to deconvolute spatial transcriptomics spots with single-cell transcriptomes. Nucleic Acids Res 49(9):e50. https://doi.org/10.1093/nar/gkab043
86. Pascual G, Dominguez D, Elosua-Bayes M, Beckedorff F, Laudanna C, Bigas C, Douillet D, Greco C, Symeonidi A, Hernandez I, Gil SR, Prats N, Bescos C, Shiekhattar R, Amit M, Heyn H, Shilatifard A, Benitah SA (2021) Dietary palmitic acid promotes a prometastatic memory via Schwann cells. Nature 599(7885): 485–490. https://doi.org/10.1038/s41586-021-04075-0
87. Biancalani T, Scalia G, Buffoni L, Avasthi R, Lu Z, Sanger A, Tokcan N, Vanderburg CR, Segerstolpe A, Zhang M, Avraham-Davidi I, Vickovic S, Nitzan M, Ma S, Subramanian A, Lipinski M, Buenrostro J, Brown NB, Fanelli D, Zhuang X, Macosko EZ, Regev A (2021) Deep learning and alignment of spatially resolved single-cell transcriptomes with tangram. Nat Methods 18(11):1352–1362. https://doi.org/10.1038/s41592-021-01264-7
88. Chen A, Liao S, Cheng M, Ma K, Wu L, Lai Y, Qiu X, Yang J, Xu J, Hao S, Wang X, Lu H, Chen X, Liu X, Huang X, Li Z, Hong Y, Jiang Y, Peng J, Liu S, Shen M, Liu C, Li Q, Yuan Y, Wei X, Zheng H, Feng W, Wang Z, Liu Y, Wang Z, Yang Y, Xiang H, Han L, Qin B, Guo P, Lai G, Munoz-Canoves P, Maxwell PH, Thiery JP, Wu QF, Zhao F, Chen B, Li M, Dai X, Wang S, Kuang H, Hui J, Wang L, Fei JF, Wang O, Wei X, Lu H, Wang B, Liu S, Gu Y, Ni M, Zhang W, Mu F, Yin Y, Yang H, Lisby M, Cornall RJ, Mulder J, Uhlen M, Esteban MA, Li Y, Liu L, Xu X, Wang J (2022) Spatiotemporal transcriptomic atlas of mouse organogenesis using DNA nanoball-patterned arrays. Cell 185(10):1777–1792 e1721. https://doi.org/10.1016/j.cell.2022.04.003
89. Cable DM, Murray E, Zou LS, Goeva A, Macosko EZ, Chen F, Irizarry RA (2022) Robust decomposition of cell type mixtures in spatial transcriptomics. Nat Biotechnol 40(4):517–526. https://doi.org/10.1038/s41587-021-00830-w
90. Andersson A, Bergenstrahle J, Asp M, Bergenstrahle L, Jurek A, Fernandez Navarro J, Lundeberg J (2020) Single-cell and spatial transcriptomics enables probabilistic inference of cell type topography. Commun Biol 3(1):565. https://doi.org/10.1038/s42003-020-01247-y
91. Luca BA, Steen CB, Matusiak M, Azizi A, Varma S, Zhu C, Przybyl J, Espin-Perez A, Diehn M, Alizadeh AA, van de Rijn M, Gentles AJ, Newman AM (2021) Atlas of clinically distinct cell states and ecosystems across human solid tumors. Cell 184(21): 5482–5496 e5428. https://doi.org/10.1016/j.cell.2021.09.014
92. Li B, Zhang W, Guo C, Xu H, Li L, Fang M, Hu Y, Zhang X, Yao X, Tang M, Liu K, Zhao X, Lin J, Cheng L, Chen F, Xue T, Qu K (2022) Benchmarking spatial and single-cell transcriptomics integration methods for transcript distribution prediction and cell type deconvolution. Nat Methods 19(6):

662–670. https://doi.org/10.1038/s41592-022-01480-9

93. Danaher P, Kim Y, Nelson B, Griswold M, Yang Z, Piazza E, Beechem JM (2022) Advances in mixed cell deconvolution enable quantification of cell types in spatial transcriptomic data. Nat Commun 13(1):385. https://doi.org/10.1038/s41467-022-28020-5
94. Miller BF, Huang F, Atta L, Sahoo A, Fan J (2022) Reference-free cell type deconvolution of multi-cellular pixel-resolution spatially resolved transcriptomics data. Nat Commun 13(1):2339. https://doi.org/10.1038/s41467-022-30033-z
95. Hu Y, Peng T, Gao L, Tan K (2021) CytoTalk: De novo construction of signal transduction networks using single-cell transcriptomic data. Sci Adv 7(16). https://doi.org/10.1126/sciadv.abf1356
96. Jerby-Arnon L, Regev A (2022) DIALOGUE maps multicellular programs in tissue from single-cell or spatial transcriptomics data. Nat Biotechnol. https://doi.org/10.1038/s41587-022-01288-0
97. Arnol D, Schapiro D, Bodenmiller B, Saez-Rodriguez J, Stegle O (2019) Modeling cell-cell interactions from spatial molecular data with spatial variance component analysis. Cell Reports 29(1):202–211.e206. https://doi.org/10.1016/j.celrep.2019.08.077
98. Yuan Y, Bar-Joseph Z (2020) GCNG: graph convolutional networks for inferring gene interaction from spatial transcriptomics data. Genome Biol 21(1):300. https://doi.org/10.1186/s13059-020-02214-w
99. Shao X., Li C., Yang H., Lu X., Liao J., Qian J., Wang K., Cheng J., Yang P, Chen, H., Xu X., & Fan X. (2022). Knowledge-graph-based cell-cell communication inference for spatially resolved transcriptomic data with SpaTalk. Nat Commun 13, 4429
100. Moran PA (1950) Notes on continuous stochastic phenomena. Biometrika 37(1/2): 17–23
101. Miller BF, Bambah-Mukku D, Dulac C, Zhuang X, Fan J (2021) Characterizing spatial gene expression heterogeneity in spatially resolved single-cell transcriptomic data with nonuniform cellular densities. Genome Res 31(10):1843–1855. https://doi.org/10.1101/gr.271288.120
102. Edsgard D, Johnsson P, Sandberg R (2018) Identification of spatial expression trends in single-cell gene expression data. Nat Methods 15(5):339–342. https://doi.org/10.1038/nmeth.4634
103. Bae S, Choi H, Lee DS (2021) Discovery of molecular features underlying the morphological landscape by integrating spatial transcriptomic data with deep features of tissue images. Nucleic Acids Res 49(10):e55. https://doi.org/10.1093/nar/gkab095
104. Goltsev Y, Samusik N, Kennedy-Darling J, Bhate S, Hale M, Vazquez G, Black S, Nolan GP (2018) Deep profiling of mouse splenic architecture with CODEX multiplexed imaging. Cell 174(4):968–981 e915. https://doi.org/10.1016/j.cell.2018.07.010
105. Angelo M, Bendall SC, Finck R, Hale MB, Hitzman C, Borowsky AD, Levenson RM, Lowe JB, Liu SD, Zhao S, Natkunam Y, Nolan GP (2014) Multiplexed ion beam imaging of human breast tumors. Nat Med 20(4):436–442. https://doi.org/10.1038/nm.3488
106. Liu Y, DiStasio M, Su G, Asashima H, Enninful A, Qin X, Deng Y, Bordignon P, Cassano M, Tomayko M, Xu M, Halene S, Craft JE, Hafler D, Fan R (2022) Spatial-CITE-seq: spatially resolved high-plex protein and whole transcriptome co-mapping. https://doi.org/10.1101/2022.04.01.486788
107. Nieto P, Elosua-Bayes M, Trincado JL, Marchese D, Massoni-Badosa R, Salvany M, Henriques A, Nieto J, Aguilar-Fernandez S, Mereu E, Moutinho C, Ruiz S, Lorden P, Chin VT, Kaczorowski D, Chan CL, Gallagher R, Chou A, Planas-Rigol E, Rubio-Perez C, Gut I, Piulats JM, Seoane J, Powell JE, Batlle E, Heyn H (2021) A single-cell tumor immune atlas for precision oncology. Genome Res 31(10):1913–1926. https://doi.org/10.1101/gr.273300.120
108. Zhu Y, Wu Z, Yan W, Shao F, Ke B, Jiang X, Gao J, Guo W, Lai Y, Ma H, Chen D, Xu Q, Sun Y (2022) Allosteric inhibition of SHP2 uncovers aberrant TLR7 trafficking in aggravating psoriasis. EMBO Mol Med 14(3):e14455. https://doi.org/10.15252/emmm.202114455
109. Ravi VM, Neidert N, Will P, Joseph K, Maier JP, Kuckelhaus J, Vollmer L, Goeldner JM, Behringer SP, Scherer F, Boerries M, Follo M, Weiss T, Delev D, Kernbach J, Franco P, Schallner N, Dierks C, Carro MS, Hofmann UG, Fung C, Sankowski R, Prinz M, Beck J, Salie H, Bengsch B, Schnell O, Heiland DH (2022) T-cell dysfunction in the glioblastoma microenvironment is mediated by myeloid cells releasing interleukin-10. Nat Commun 13(1):925. https://doi.org/10.1038/s41467-022-28523-1

110. Liu C, Li R, Li Y, Lin X, Zhao K, Liu Q, Wang S, Yang X, Shi X, Ma Y, Pei C, Wang H, Bao W, Hui J, Yang T, Xu Z, Lai T, Berberoglu MA, Sahu SK, Esteban MA, Ma K, Fan G, Li Y, Liu S, Chen A, Xu X, Dong Z, Liu L (2022) Spatiotemporal mapping of gene expression landscapes and developmental trajectories during zebrafish embryogenesis. Dev Cell 57(10):1284–1298 e1285. https://doi.org/10.1016/j.devcel.2022.04.009
111. Zhang R, Feng Y, Ma W, Guo Y, Luo M, Li Y, Zang Y, Dong X, Lu S, Guo Q, Xu Q, Chen H, Li Y, Liu L, Chen A, Chen G, Xu X (2022) Spatial transcriptome unveils a discontinuous inflammatory pattern in proficient mismatch repair colorectal adenocarcinoma. Fundam Res. https://doi.org/10.1016/j.fmre.2022.01.036
112. Kaufmann M, Schaupp AL, Sun R, Coscia F, Dendrou CA, Cortes A, Kaur G, Evans HG, Mollbrink A, Navarro JF, Sonner JK, Mayer C, DeLuca GC, Lundeberg J, Matthews PM, Attfield KE, Friese MA, Mann M, Fugger L (2022) Identification of early neurodegenerative pathways in progressive multiple sclerosis. Nat Neurosci 25(7):944–955. https://doi.org/10.1038/s41593-022-01097-3
113. Meylan M, Petitprez F, Becht E, Bougouin A, Pupier G, Calvez A, Giglioli I, Verkarre V, Lacroix G, Verneau J, Sun CM, Laurent-Puig P, Vano YA, Elaidi R, Mejean A, Sanchez-Salas R, Barret E, Cathelineau X, Oudard S, Reynaud CA, de Reynies A, Sautes-Fridman C, Fridman WH (2022) Tertiary lymphoid structures generate and propagate anti-tumor antibody-producing plasma cells in renal cell cancer. Immunity 55(3):527–541 e525. https://doi.org/10.1016/j.immuni.2022.02.001
114. Zhao T, Chiang ZD, Morriss JW, LaFave LM, Murray EM, Del Priore I, Meli K, Lareau CA, Nadaf NM, Li J, Earl AS, Macosko EZ, Jacks T, Buenrostro JD, Chen F (2022) Spatial genomics enables multi-modal study of clonal heterogeneity in tissues. Nature 601(7891):85–91. https://doi.org/10.1038/s41586-021-04217-4
115. Kamath T, Abdulraouf A, Burris SJ, Langlieb J, Gazestani V, Nadaf NM, Balderrama K, Vanderburg C, Macosko EZ (2022) Single-cell genomic profiling of human dopamine neurons identifies a population that selectively degenerates in Parkinson's disease. Nat Neurosci 25(5):588–595. https://doi.org/10.1038/s41593-022-01061-1
116. Wu SZ, Al-Eryani G, Roden DL, Junankar S, Harvey K, Andersson A, Thennavan A, Wang C, Torpy JR, Bartonicek N, Wang T, Larsson L, Kaczorowski D, Weisenfeld NI, Uytingco CR, Chew JG, Bent ZW, Chan CL, Gnanasambandapillai V, Dutertre CA, Gluch L, Hui MN, Beith J, Parker A, Robbins E, Segara D, Cooper C, Mak C, Chan B, Warrier S, Ginhoux F, Millar E, Powell JE, Williams SR, Liu XS, O'Toole S, Lim E, Lundeberg J, Perou CM, Swarbrick A (2021) A single-cell and spatially resolved atlas of human breast cancers. Nat Genet 53(9):1334–1347. https://doi.org/10.1038/s41588-021-00911-1
117. Dong R, Yuan GC (2021) SpatialDWLS: accurate deconvolution of spatial transcriptomic data. Genome Biol 22(1):145. https://doi.org/10.1186/s13059-021-02362-7
118. Kleshchevnikov V, Shmatko A, Dann E, Aivazidis A, King HW, Li T, Elmentaite R, Lomakin A, Kedlian V, Gayoso A, Jain MS, Park JS, Ramona L, Tuck E, Arutyunyan A, Vento-Tormo R, Gerstung M, James L, Stegle O, Bayraktar OA (2022) Cell2location maps fine-grained cell types in spatial transcriptomics. Nat Biotechnol 40(5):661–671. https://doi.org/10.1038/s41587-021-01139-4
119. Suo C, Dann E, Goh I, Jardine L, Kleshchevnikov V, Park JE, Botting RA, Stephenson E, Engelbert J, Tuong ZK, Polanski K, Yayon N, Xu C, Suchanek O, Elmentaite R, Dominguez Conde C, He P, Pritchard S, Miah M, Moldovan C, Steemers AS, Mazin P, Prete M, Horsfall D, Marioni JC, Clatworthy MR, Haniffa M, Teichmann SA (2022) Mapping the developing human immune system across organs. Science, 376 (6597):eabo0510. https://doi.org/10.1126/science.abo0510
120. Steen CB, Luca BA, Esfahani MS, Azizi A, Sworder BJ, Nabet BY, Kurtz DM, Liu CL, Khameneh F, Advani RH, Natkunam Y, Myklebust JH, Diehn M, Gentles AJ, Newman AM, Alizadeh AA (2021) The landscape of tumor cell states and ecosystems in diffuse large B cell lymphoma. Cancer Cell 39(10):1422–1437 e1410. https://doi.org/10.1016/j.ccell.2021.08.011
121. Italiano A, Bessede A, Pulido M, Bompas E, Piperno-Neumann S, Chevreau C, Penel N, Bertucci F, Toulmonde M, Bellera C, Guegan JP, Rey C, Sautes-Fridman C, Bougouin A, Cantarel C, Kind M, Spalato M, Dadone-Montaudie B, Le Loarer F, Blay JY, Fridman WH (2022) Pembrolizumab in soft-tissue

sarcomas with tertiary lymphoid structures: a phase 2 PEMBROSARC trial cohort. Nat Med 28(6):1199–1206. https://doi.org/10.1038/s41591-022-01821-3

122. Gonzalez CG, Mills RH, Kordahi MC, Carrillo-Terrazas M, Secaira-Morocho H, Widjaja CE, Tsai MS, Mittal Y, Yee BA, Vargas F, Weldon K, Gauglitz JM, Delaroque C, Sauceda C, Rossitto LA, Ackermann G, Humphrey G, Swafford AD, Siegel CA, Buckey JC Jr, Raffals LE, Sadler C, Lindholm P, Fisch KM, Valaseck M, Suriawinata A, Yeo GW, Ghosh P, Chang JT, Chu H, Dorrestein P, Zhu Q, Chassaing B, Knight R, Gonzalez DJ, Dulai PS (2022) The host-microbiome response to hyperbaric oxygen therapy in ulcerative colitis patients. Cell Mol Gastroenterol Hepatol 14(1):35–53. https://doi.org/10.1016/j.jcmgh.2022.03.008

Chapter 8

Statistical Analysis of Multiplex Immunofluorescence and Immunohistochemistry Imaging Data

Julia Wrobel, Coleman Harris, and Simon Vandekar

Abstract

Advances in multiplexed single-cell immunofluorescence (mIF) and multiplex immunohistochemistry (mIHC) imaging technologies have enabled the analysis of cell-to-cell spatial relationships that promise to revolutionize our understanding of tissue-based diseases and autoimmune disorders. Multiplex images are collected as multichannel TIFF files; then denoised, segmented to identify cells and nuclei, normalized across slides with protein markers to correct for batch effects, and phenotyped; and then tissue composition and spatial context at the cellular level are analyzed. This chapter discusses methods and software infrastructure for image processing and statistical analysis of mIF/mIHC data.

Key words Multiplex imaging, Bioimage analysis, Spatial statistics, Cell phenotyping, Image normalization

1 Introduction

The development of multiplexed immunofluorescence (mIF) and multiplex immunohistochemistry (mIHC) imaging has revolutionized researchers' ability to study tissue structure and function at a cellular level while preserving the original spatial context of the sample. Functionally, mIF/mIHC is a hybrid between multiplex single-cell techniques like flow and mass cytometry [1, 2], which allow for the staining of multiple antibody markers but require the dissolution of tissue in a suspended solution, and traditional immunohistochemistry, which preserves the original tissue structure but staining is limited to 2–3 antibodies [3]. Along with similar single-cell imaging technologies, including multiplexed ion beam imaging (MIBI) [4, 5] and imaging mass cytometry (IMC) [6, 7], mIF platforms have particular promise in the field of cancer immunology, where they have been used to characterize the tumor immune microenvironment (TIME). This chapter surveys statistical methods for preprocessing and analysis of mIF data using examples

Brooke L. Fridley and Xuefeng Wang (eds.), *Statistical Genomics*,
Methods in Molecular Biology, vol. 2629, https://doi.org/10.1007/978-1-0716-2986-4_8,

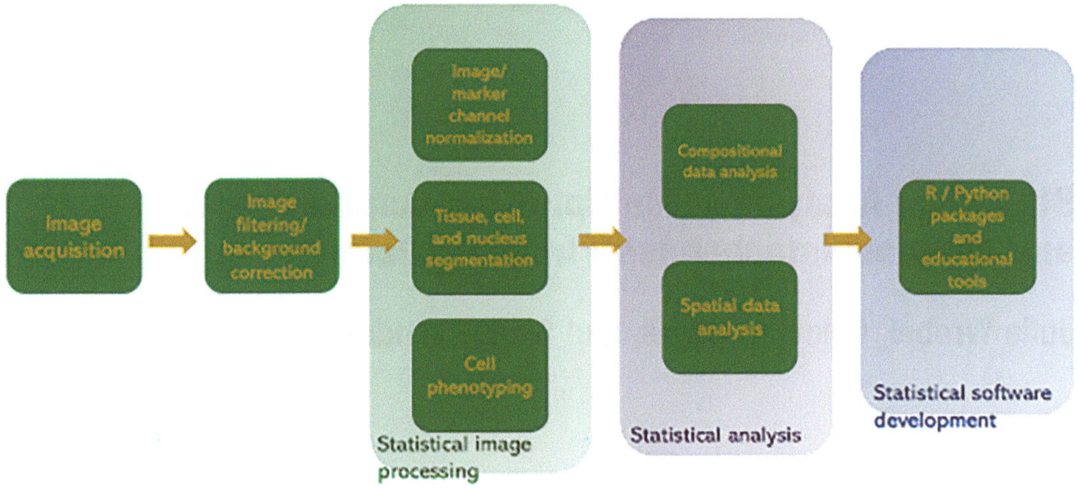

Fig. 1 Typical image acquisition, processing, and analysis workflow for mIF/mIHC data

motivated by scientific questions about the TIME—specifically, the study of abundance, spatial location, and protein expression of immune cells in and around tumors, and how immune profiling relates to improved prognosis for cancer patients. However, the methods highlighted can be directly applied or adapted for mIF data from many other research areas; recent mIF advances in non-cancer domains include the study of neurodegenerative diseases [8, 9], type 1 diabetes [10], inflammatory bowel disease [11], and multiple sclerosis [12].

The collection of mIF data follows a complex data acquisition, image processing, and analysis pipeline depicted at a high level in Fig. 1. First, tissue is placed on a slide. An advantage of mIF imaging is that it can be applied to regions of interest (ROIs) of a whole tissue slide or to tissue microarrays (TMAs), allowing for analysis across large tissue areas or a large number of samples [13]. The slide is then stained with multiple, or *multiplex*, primary antibodies, referred to as *markers*, that attach to biomarkers of interest. Markers can be broadly categorized as either *phenotypic markers*, which are used to categorize cell type, or *functional markers*, which inform cell function and can be present across multiple cell types. The marker antibodies are attached to fluorophores, which emit light at specific wavelengths when imaged, and the signal of light intensity at a particular wavelength is used to quantify biomarker abundance. Then the intensity spectra must be unmixed to give separate intensity values for each marker. Resulting images are stored as multichannel tiff files where each channel provides the tissue intensities for a particular marker.

These images are then segmented to identify tissue compartments (e.g., tumor vs. stroma), cells, and nuclei. Single-cell segmentation is an important and rapidly advancing area of research that benefits from recent success in neural networks and deep learning. Many algorithms have been developed as part of proprietary software packages such as Inform and HALO but are also being developed by academic research groups [14–16]. These algorithms are often tissue-specific and can be influenced by the image production process, so they sometimes must be retrained for new data. Research in this area works to make the algorithms more general and reduce bias. Because research in single-cell segmentation is broad in itself, the topic is not discussed further in this chapter.

After segmentation, cells are *phenotyped*, or given biological label(s), based on their average marker intensities. Segmentation and phenotyping steps convert images into a tabular dataset where each row is a cell and columns are features identifying cell spatial location, tissue type, phenotype, marker intensities, and other cell or image characteristics. Finally, the datasets are analyzed, potentially in combination with patient-level outcomes such as survival time, disease subtype, or cancer stage. While examples in this chapter focus on cell-level data, the images can also be analyzed directly using the same methodology, using pixels rather than cells as the unit of observation. In our more detailed discussion below, we focus on steps of the mIF/mIHC pipeline downstream of image acquisition that can be applied and improved upon by statisticians and bioinformaticians.

2 Ovarian and Lung Cancer Datasets

With 184,000 deaths per year, globally, ovarian cancer is the most lethal reproductive cancer [17]. In addition, 75% of patients with high-grade serous ovarian cancer have tumor recurrence and develop therapy-resistant disease [18]. Lung cancer is even more deadly, making up 25% of all cancer deaths [19]. There is mounting evidence of the prognostic significance of the tumor-immune composition in lung and ovarian cancers [20, 21]. However, the biological significance of spatial relationships between cell types in the TIME of these cancers is not yet well understood, and mIF imaging may provide key scientific insights. To address gaps in our understanding of the biology of these cancers, a high-grade serous ovarian cancer (HGSOC) and a non-small cell lung carcinoma (NSCLC) mIHC dataset were collected at the University of Colorado Anschutz Medical Campus and will be used to illustrate the methods in this chapter.

Images were collected on Vectra 3.0™ and Vectra Polaris™ instruments for the ovarian and lung datasets, respectively, and image processing was performed using InForm software [22–25]. The Vectra 3.0™ and Vectra Polaris™ by Akoya Biosciences are currently the most widely used mIF/mIHC platforms, and InForm is the accompanying proprietary software that provides a graphical user interface (GUI) for image processing steps including spectral unmixing, segmentation, and phenotyping. For our lung and ovarian datasets, InForm was used for spectral unmixing, nucleus, cell, and tissue segmentation, and cell phenotyping; however, it is also common to use a mixture of Inform and other GUI-based software such as HALO and QuPATH to convert images into tabular datasets [26]. Both datasets are freely available in *R* in a tabular post-segmentation format as part of the Bioconductor ExperimentHub package VectraPolarisData [27]. Specifically, these data are stored as SpatialExperiment class objects [28], and the code used to reproduce figures in this chapter using these data is available at https://github.com/julia-wrobel/mIF_chapter_code. Each dataset provides cell spatial locations and marker channel intensities for over 1.5 million cells, as well as patient-level clinical information.

The HGSOC ovarian dataset comes from a spatial analysis of a tumor microarray of 128 subjects with ovarian cancer collected at the University of Colorado [29] and contains one image per subject. Images were stained for DAPI, five phenotypic markers (CD3, CD8, CD68, CD19, cytokeratin), and two functional markers (phospho-stat3 and IER3), allowing for the identification of CD4+, CD8+, CD68+, and CD19+ immune cells. Patient-level variables include age, cancer stage at diagnosis, BRCA mutation status, debulking status, survival time, survival status, and time-to-recurrence.

The NSCLC lung data was collected to analyze the relationship between tumor-infiltrating immune cells and major histocompatibility II expressing cancer cells in the NSCLC tumor immune microenvironment and consists of 761 mIF-imaged regions of interest (ROIs) from 153 patients [30]. Images were stained for DAPI, five phenotypic markers (CD3, CD8, CD14, CD19, cytokeratin), and one functional marker (MHCII), allowing for identification of CD4+, CD8+, CD14+, and CD19+ immune cells. Patient-level variables include age, gender, cancer stage at diagnosis, pack-years of cigarette smoking, survival time, survival status, and time-to-recurrence. Because each sample consists of multiple ROIs that are not spatially adjacent, these images cannot be stitched together and treated as one image. Instead, repeated measures approaches must be considered when analyzing image data in models for patient-level outcomes. Figure 2 (far left) shows an example image from this dataset which combines intensities from

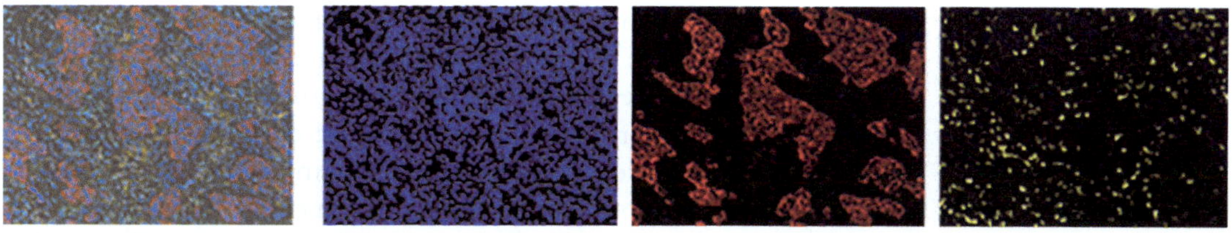

Fig. 2 One image from the NSCLC dataset. Left panel is the composite image combining all eight channels collected. The next three images are three individual channels from the same image, shown, from left to right, the nucleus (DAPI) channel, the cytokeratin channel, and the CD8 channel

all 8 channels. The three right columns are the DAPI, cytokeratin, and CD8 channels for this image, respectively; these are used to identify cell nuclei, tumor areas, and CD8 T cells.

3 Methods

Multiplex images are typically provided as multichannel tiff files and transformed into tabular sets after a cell segmentation step. In the following sections, we direct our focus towards analyzing the cell-level tabular data after single-cell segmentation. These methods applied at the cell level can also be applied directly to pixels of a non-tabular mIF image, but at a greater computational cost. We divide our review of methods into the following subsections that address different areas of the processing pipeline depicted in Fig. 1. In Subheading 3.1, we cover marker channel processing steps that must be addressed before data is appropriate for answering questions pertaining to scientific hypotheses. In Subheading 3.2, we cover methods for cell phenotyping, the process of classifying cell types. In Subheading 3.3, we discuss the analysis of image summary data using statistics that capture the cellular composition of tissue samples. An increasing focus of the literature on mIF data is dedicated to spatial analysis, and in Subheading 3.4, we detail existing methods; in particular, we emphasize approaches grounded in point process theory because this is an area of mIF where we believe expertise from statisticians and bioinformaticians can have the greatest impact. In Subheading 3.5, we conclude with a review of available open-source software tools for the visualization, preprocessing, and analysis of multiplex imaging data.

3.1 Image Transformation, Normalization, and Batch Correction

Image transformation, normalization, and batch correction are, in general, used to make the data more appropriate for downstream analysis by adjusting intensity distributions or removing systematic biases due to processing variables. Image transformations (e.g., log, arcsinh, or root) are used to make the data more amenable to modeling, but do not explicitly adjust for systematic effects due to image acquisition variables. Image normalization is the process of making the same tissue type appear similar in images that were acquired across different sites, times, imaging parameters, or platforms [31–35]. Normalization adjusts the distribution of marker intensities in each slide, batch, channel, or image separately to make the distributions appear more similar. Batch correction (often called harmonization in the imaging literature) is the process of removing systematic effects in groups of cells or images that are due to processing steps or other non-biological sources [36]. Batch correction methods use a single model across slides to account for covariates while estimating and removing batch effects across slides. Transformation, normalization, and batch correction methods are applied to each channel of the mIF data, so it is important to have procedures that are effective for a wide range of different marker types, including both phenotypic markers and functional markers. The top two rows of Fig. 3 show evidence of potential batch effects in the NSCLC data. In this figure, the distributions of unnormalized marker intensities are shown for three subjects and two immune cell markers (CD14 and CD19). Each subject has densities plotted for multiple ROIs (dotted lines) as well as the subject-specific mean density (solid lines). For this dataset, each slide contains data for a single subject, and the range of intensity values differs substantially for each slide.

Normalization and batch correction algorithms can be applied at the image or cell level [37]. The goals of the two procedures are methodologically similar, and batch correction methods from genomics have been very effectively applied to imaging datasets [34, 38, 39]. There is an emerging literature on batch correction in mIF data [31, 32, 34, 37, 40, 41], and this research area is likely to grow as the effects of imaging parameters and platforms are better understood. It is well-accepted that there are effects of slide, staining round, and batch on the distribution of mIF intensities at the image and cell level. mIF data are unique from magnetic resonance imaging (where batch correction is being widely applied) because the tissue populations differ significantly in their composition across samples, so close alignment of the image histograms is not always an indication of effective normalization (Fig. 2, top rows). Batch adjustment methods developed for other genomic assays, such as sc-RNAseq are promising, but cannot be directly applied due to greater overlap in the distributions of expressed and unexpressed cells. For these reasons, mIF normalization and batch correction is a rapidly growing area of research. Here, we provide a brief overview of recently proposed methods and software for normalization and batch correction in mIF data.

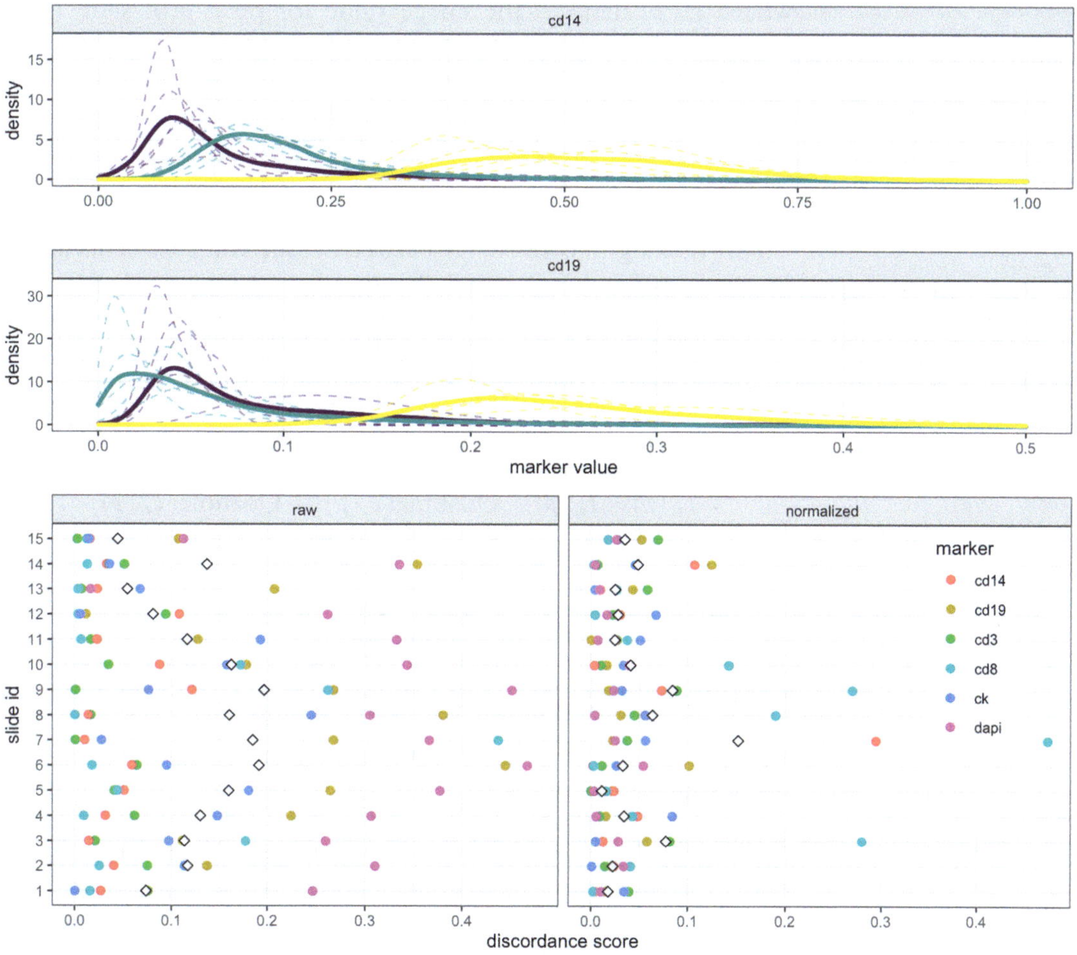

Fig. 3 Batch effects and results of normalization for the NSCLC data. Top and middle rows: unnormalized marker intensity distributions for the immune cell markers CD14 (top row) and CD19 (middle row). Densities are plotted for three randomly selected subjects, each shown in a different color. Each subject has densities plotted for multiple ROIs (dotted lines) as well as the subject-specific mean density (solid lines). Large differences in the intensity ranges across subjects may indicate slide-level batch effects. Bottom row: plots of discordance scores of phenotypic marker channels for 15 subjects in the NSCLC data before and after normalization. Simple log10 and mean division transform-based normalization was applied using the *mxnorm* package. Smaller discordance scores in the normalized data indicate a reduction in batch effects

3.1.1 Image Normalization

Raza et al. were one of the earliest groups to explore normalization methods for mIF data applied at the image level [31]. They compared a linear normalization,

$$\tilde{I}_{ic}(p) = \frac{I_{ic}(p) - \min_p I_{ic}(p)}{\max_p I_{ic}(p) - \min_p I_{ic}(p)},$$

where $I_{ic}(p)$ denotes the image value for pixel p of slide i and channel c, and $\tilde{I}_{ic}(p)$ denotes the normalized data, and a tan$h$ transformation after applying mean division normalization,

$$\tilde{I}_{ic}(p) = \tan h\left\{\frac{I_{ic}(p)}{2 \times \bar{I}_{ic}}\right\},$$

where $\bar{I}_{ic}$ is the slide mean. The methods were evaluated based on expert ranking in consistency between tissue compartments across images and KL-divergence within and between tissue classes. Linear normalization almost always scored better across their evaluation procedures.

Graf et al. proposed several quantile-based normalization methods and applied the methods at the pixel and segmented cell levels of analysis. They compared linear

$$\tilde{I}_{ic}(p) = I_{ic}(p) + \text{Quantile}(I_i(p)) - \text{Quantile}(I_{ic}(p))$$

and multiplicative normalizations

$$\tilde{I}_{ic}(p) = \frac{I_{ic}(p) \times \text{Quantile}(I_c(p))}{\text{Quantile}(I_{ic}(p))},$$

where Quantile($I_c(p)$) is the quantile across all images and Quantile ($I_{ic}(p)$) is the quantile in slide i only. They considered the median for the linear transformations and the median and 75th quantile for the multiplicative normalization as well as several other quantile-based normalizations (see Table 1 of [37]). They used a novel approach to evaluate normalization by leveraging the DAPI channel, which was repeatedly stained through multiple rounds and serves as a "virtual" ground truth because it marks the nucleus and should be identical across staining rounds. Acquisition parameters were varied through the rounds, but the same tissues were used, so they could assume (in the absence of batch effects) that the tissue should have consistent image intensity through the rounds of staining. Their normalization procedure and methods assumed control regions that are expected to be consistent across the staining rounds, and they used the coefficient of variation to quantify batch effects. The authors found that the multiplicative normalization applied on log-transformed data had the best performance (the lowest coefficient of variation).

Other normalization methods have relied on control regions for estimation [32] or evaluation [40] of batch effects. These assume that adjacent control sections are used across slides and that alignment of these control regions implies alignment of the tissue of interest. These are not possible if a good control region is used and may not be effective if sections of the control tissue are not adjacent sections. In general, prior to transforming the data, normalization by dividing by the mean, quantile/median, or range of the data seems to work, but more evaluation across a greater range of data sets is merited.

3.1.2 Batch Correction

Harris et al. compared several normalization batch correction methods to adjust for slide effects in mIF data at the cell level [34]. They compared an adaptation of the widely used ComBat method [36, 38], and a functional data registration method to perform histogram matching across slides [34, 42] as well as several transformations and a mean division normalization. To assess the methods in the absence of a ground truth, they established a framework for evaluating batch effects for mIF data that includes (a) quantification of misalignment of the expression histograms for each marker channel, (b) quantification of discordance between slide-level phenotyping (no batch effects present) and across slide phenotyping (batch effects may be present) before and after normalization, (c) quantification of proportion of variance due to slide, and (d) visualization of batch and biological effects in UMAP embedding, across these metrics, applying mean division, mean division with a log transformation, and using functional data registration all performed well. The authors recommend mean division and mean division with log transformation as a balance between efficacy and simplicity.

3.1.3 Future Research and Software Development

Based on the current research, normalization is demonstrated to be reliable and effective; however, the simplistic procedures currently used are unlikely to remain the top performers as we learn more about sources of systematic variation in mIF data and more public data sets are released that can be used for evaluation [43, 44]. Future directions may more rigorously consider batch adjustment methods. For example, ComBat has been demonstrated to fail with mIF data due to zero inflation [34], but a recently developed method using a negative binomial model may hold more promise [45]. In addition, the normalization methods used currently are robust across channels, but do not explicitly leverage information about the joint distribution across channels, the staining round, or the spatial information available in the image. These features may be used to better understand and quantify systematic effects in mIF.

Software designed for other assays (e.g., Seurat, Giotto) can be used for mIF normalization or batch correction. In addition, most of the recently proposed normalization procedures provide code to implement their methods [37, 40, 41]. In particular, *mxnorm* is a recently published *R* package devoted to normalization of cell-level mIF data [41]. This package provides the normalization and batch correction algorithms used in [34] and also allows users to specify new algorithms which can be compared using the authors' proposed evaluation framework. The third row of Fig. 3 shows plots of discordance scores of phenotypic marker channels for 15 subjects in the NSCLC data before and after normalization. Simple log10 and mean division transform-based normalization was applied using the *mxnorm* package. The discordance score is a novel evaluation metric proposed by [34] that quantifies the difference between threshold-based phenotyping at the slide and population level. Smaller discordance scores indicate a reduction in batch effects, and smaller discordance is seen for the normalized values as compared to the raw marker intensity values.

3.2 Cell Phenotyping

Cell phenotyping is the process of identifying discrete cell types from marker expression values. Phenotyping can use unsupervised algorithms but often requires expert intervention to label and validate cell annotations. This step in the analysis pipeline is not exclusive to mIF and has been well studied in other fields, including flow and mass cytometry [46, 47]. Typical preprocessing before phenotyping for other assays includes batch correction, transformation (e.g., log or arcsinh), Z-scaling, and dimension reduction. These steps have not been rigorously evaluated in mIF data. For example, dimension reduction may not be necessary due to the relatively small number of image marker channels (10 s as opposed to 100 s), but data whitening may be beneficial. Here, we provide a brief overview of the approaches used and proposed for cell phenotyping using mIF data. These include marker gating and unsupervised or semisupervised clustering methods.

3.2.1 Marker Gating

Marker gating is a widely used procedure for determining a threshold to define marker-positive cells that have expression values above the given threshold [32, 43, 48]. This procedure is performed manually by visualizing univariate or bivariate histograms of cell expression values and image overlays to determine a threshold that separates modes in the data. If three or more channels are necessary to determine cell phenotypes, marker gating can be performed iteratively for each channel or pair of channels [49–51]. Software with features for performing marker gating of mIF data includes Cellprofiler Analyst [50, 51], histoCAT [52], CyLinter [53], and Advanced Cell Classifier [54]. Marker gating is useful for phenotyping cells of a known type that can be identified based on a subset of known marker channels. Otherwise, unsupervised algorithms are more convenient for identifying known and unknown cell classes.

3.2.2 Unsupervised Clustering Algorithms

K-means, Louvain, and Leiden algorithms are unsupervised clustering/community detection algorithms that have been widely implemented for single-cell assays because they are computationally efficient in large data sets [55–58] and are included in most software programs for single-cell analysis, such as Seurat, Giotto, and PhenoGraph [15, 57, 59, 60]. After clustering, manual annotations are assigned for each cluster by visualizing the clusters in a dimension-reduced projection (e.g., UMAP, tSNE), visualizing the cluster memberships on the slides, and visualizing expression profiles (Fig. 4). K-means is a distance-based clustering; in contrast, Louvain and Leiden clustering are popular graph-based clustering

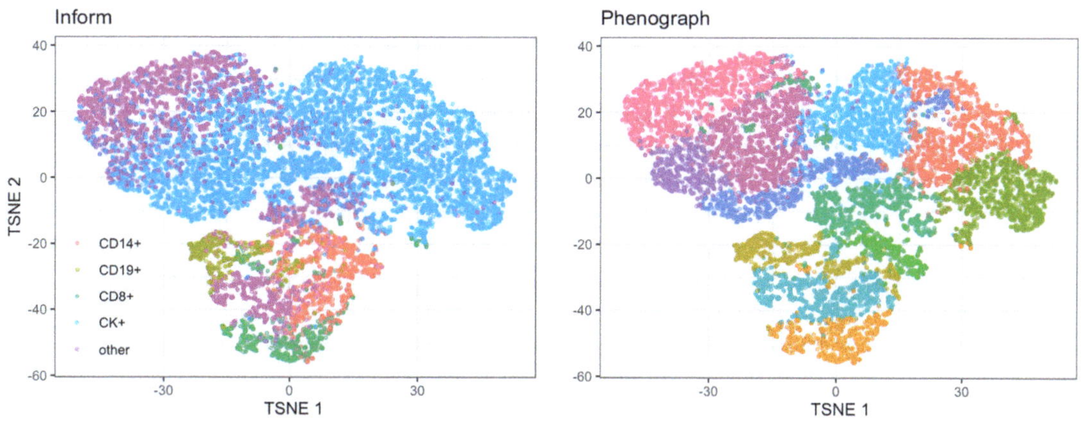

Fig. 4 Clustering results applied to five ROIs from one subject in the NSCLC dataset. TSNE was first applied to the raw marker values to reduce dimensions for visualization purposes. The cells for this subject were plotted along the first two TSNE directions and colored by phenotypes calculated using Inform software (left) and by clusters produced by applying PhenoGraph to log10 mean division normalized marker intensities (right). PhenoGraph produced 12 clusters for the normalized data, more than are identifiable by the markers collected in the dataset, which must be combined to identify clusters comparable with those obtained by Inform

methods that group cells based on the similarity of the cells in a local neighborhood, where the neighborhoods are determined by the marker expression values.

Figure 4 shows clustering results applied to five ROIs from one subject in the NSCLC dataset. TSNE was first applied to the raw marker values to reduce dimensions for visualization purposes. The cells for this subject were plotted along the first two TSNE directions and colored by phenotypes calculated using Inform software (left), and colored by clusters produced by applying PhenoGraph to log10 mean division normalized marker intensities (right). PhenoGraph produced 12 clusters for the normalized data, more than are identifiable by the markers collected in the dataset. However, many map on to clusters labeled in Inform.

These clustering algorithms were not specifically developed for single-cell analysis, and adaptations have been developed to address characteristic features of single-cell data such as dimensionality and rare cell types. For example, RaceID uses outlier detection to revise K-means cluster identities because K-means clustering tends to form clusters of similar sizes, so rare cell types are often mixed with discrete cell types [61]. In addition, methodological and computational modifications to the analysis process have been developed to speed computation in large data sets (>1 million cells) [57, 62–64]. Two examples are PhenoGraph and FastPG for clustering cells that are scalable to large datasets [57, 62]. Both make modifications to the k-nearest neighbors computation, weighting estimation, and computational processing of the Louvain algorithm to improve processing speed. Many of the clustering methods mentioned here have been evaluated in single-cell RNA sequencing [46, 47], but these algorithms have not been rigorously evaluated using mIF data. Software tools specifically designed for mIF or other spatial assays may include the above-mentioned phenotyping algorithms [59, 65]; however, it is likely that the phenotyping pipelines for other single-cell assays are not directly applicable here, given differences in acquisition and processing [66].

Recently, several phenotyping algorithms have been proposed specifically for microscopy data that can be applied to mIF data at the cell or image-pixel level. Astir is a cell-level, scalable, semisupervised phenotyping algorithm using a probabilistic model and variational inference with a neural network that was specifically designed for mIF data. As input, Astir takes the mIF expression matrix and a user-specified marker file, which defines combinations of marker values attributed to each cell type. Astir can be used to broadly classify cells before applying unsupervised clustering on the subpopulations or to identify known cell types automatically. Cell phenotyping has been performed using morphological summary features in addition to median image intensities within each cell to perform unsupervised clustering [14, 67]. These methods still rely on single-cell segmentation as a preprocessing step but use spatial information through cell shape, size, composition, and texture. When single-cell segmentation is difficult, then phenotyping can be applied directly to the images in order to identify tissue classes or neighborhoods in mIF analysis that are spatially contiguous [67–72]. Some of these methods, which have been developed in other similar fields such as imaging mass spectrometry, will likely be useful in tissues where cell segmentation is difficult. After defining neighborhoods, they can be used to study spatial interactions between these tissue compartments.

To summarize, phenotyping in mIF data is not well evaluated, but it is possible that algorithms effective in other single-cell assays will be useful in mIF data with methodological modifications to the phenotyping analysis pipeline. Quality assessment and visualization tools that allow users to evaluate cell phenotypes in situ are critical for quality assessment [15, 53]. Given the variability in possible markers, unsupervised or semisupervised methods are likely to be most applicable. Finally, mIF data accommodate image-level clustering, which allows analysts to circumvent single-cell segmentation, which could be biased or difficult depending on the tissue type. Cell and image-level phenotypes can then be passed on to downstream spatial analytic methods.

3.3 Analysis of Cellular Composition and Marker Expression

As a first step of probing the tissue microenvironment, it is common to analyze the cellular composition of the overall sample and tissue sub-compartments. One typical approach is estimating the number, percentage, or density of cell subtypes in different tissue regions; for example, summarizing immune cell proportions within tumor and stromal regions [73]. It is also common to characterize cell type composition across levels of a patient-level covariate, such as responders and nonresponders to chemotherapy. Summary statistics including the percentage or proportion of positive cells for each phenotypic form can be easily calculated in image analysis software such as Halo and Inform. However, these quantities are not normally distributed, even after normality transformations, so modeling choices need to be carefully considered [13]. In addition, zero inflation is common in mIF studies where rare cell populations contribute to zero cell counts for some subjects or ROIs in a sample. This can occur, for example, in studies of immune cold tumors such as ovarian cancer. Figure 5 (left panel) shows the empirical proportion of total cells that belong to each type of tumor-infiltrating immune cell, including B cells, CD4 T cells, CD8 T cells, and macrophages, for women in the HGSOC ovarian data with and without a BRCA genetic mutation. Note that there is a substantial right skew in the distributions of proportions for all

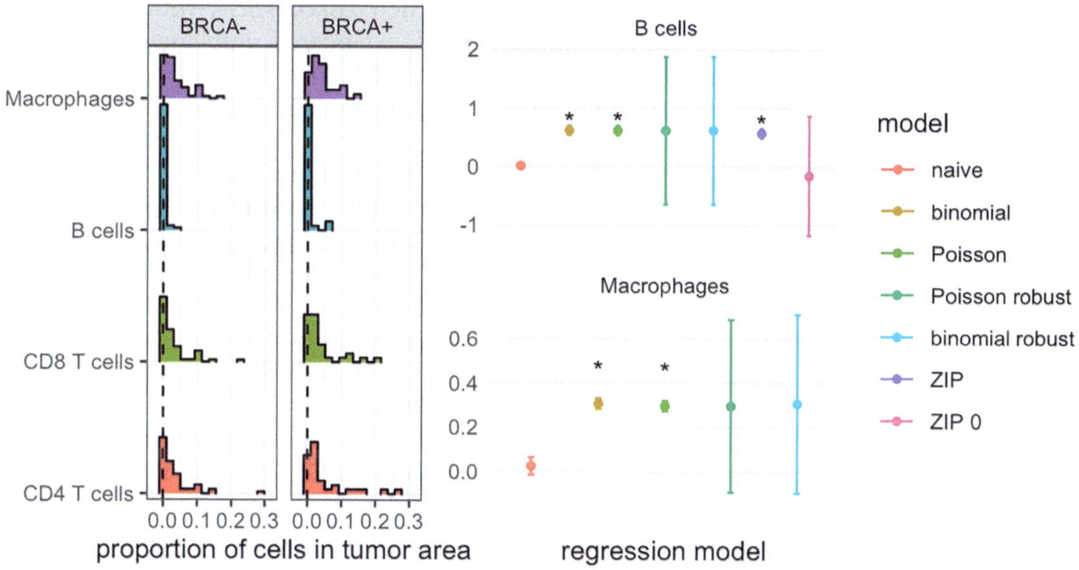

Fig. 5 Cell type composition across groups of women in the HGSOC data with (BRCA+) and without (BRCA–) a BRCA mutation. Left column shows histograms of proportions of tumor-infiltrating immune cells, including B cells, CD4 T cells, CD8 T cells, and macrophages, across subjects. Dotted vertical black line is drawn through zero. Note that there is a substantial right skew in the distributions of proportions, and many images have zero B cells. Right panel shows results from generalized linear models of the proportions of B cells (top) and macrophages (bottom), controlling for BRCA status. Estimates shown are β and 95% confidence intervals for the BRCA covariate from each model. Black stars above the estimates indicate statistical significance at the 5% α level. The "naive" model applies linear regression to empirical proportions that have been square-root transformed; "Poisson robust" and "binomial robust" indicate Poisson and binomial regression with robust standard errors, and ZIP indicates zero-inflated Poisson regression. The ZIP model was not used to model proportions of macrophages because all images contained at least one macrophage. ZIP and ZIP 0 indicate the Poisson and binomial portions of the ZIP model, respectively

immune cell subtypes, and zero inflation is present in the proportions of B cells because many images do not contain B cells. In addition, mIF data often exhibit overdispersion when modeled using common statistical methods for counts and proportions. Overdispersion occurs when the data exhibit greater variance than is dictated by the chosen model and can lead to type I error and artificially narrow *p*-values. Failing to account for non-normality, overdispersion, and zero inflation can lead to erroneous results.

3.3.1 Modeling Counts or Proportions with Overdispersion and Zero Inflation

Count data can be effectively modeled using a generalized linear modeling (GLM) framework with a data-informed choice of statistical distribution [74]. A GLM also enables straightforward modeling of how summary statistics, such as cell proportions, differ across levels of a covariate (e.g., treatment response). The binomial and Poisson distributions are common distributions for modeling proportions; however, for cell count data from mIF, these models will likely be overdispersed and underestimate the variance of model parameters, in part because they implicitly treat multiple cells within the same image as independent. One way to account for overdispersion is to use heteroskedasticity-robust standard errors, which have several standard implementations in *R* including the ***sandwich*** library [75, 76]. Another approach is to introduce an additional model parameter that explicitly captures overdispersion, and methods that account for overdispersion in counts or proportions in this way include quasi-binomial, quasi-Poisson, negative binomial, and beta-binomial regression [74].

When an excess of zero counts is present, usually due to rare cell types, zero-inflated versions of Poisson, beta-binomial, or negative binomial count models can be used. Zero-inflated models have two parts, for example, zero-inflated Poisson regression models zeros using a binomial distribution and observations greater than zero using a Poisson distribution. These models have become popular for other types of biological data where excess zeros are present, including microbiome composition [77, 78] and RNA-seq data [79–81]. Because mIF data that is zero-inflated is also likely overdispersed, zero-inflated beta-binomial and zero-inflated negative binomial are recommended over zero-inflated Poisson. For datasets with multiple tissue samples or ROIs per subject, within-subject correlation induced by repeated measures needs to be accounted for, and generalized mixed effects or estimating equation models can provide summaries of cellular composition at experiment-wide levels [82, 83]. Generalized mixed-effects models additionally provide summaries at the subject level, and both frequentist and Bayesian implementations of mixed-effect models for overdispersed and zero-inflated count data are available [84–86].

Figure 5 (right panel) shows results from generalized linear models of proportions of B cells (top) and macrophages (bottom), controlling for BRCA mutation status. Specifically, BRCA mutation status β coefficient estimates and 95% confidence intervals are shown for six different models. Black stars above the estimates indicate statistical significance at the 5% α level. The "naive" model applies linear regression to empirical proportions that have been square-root transformed; "Poisson robust" and "binomial robust" indicate Poisson and binomial regression with robust standard errors, and ZIP indicates zero-inflated Poisson regression. The naive model, though it accounts for overdispersion and provides inference about model covariates, is otherwise limited because it does not account for the total number of cells in each image and cannot be back-transformed to obtain estimates of cell counts or proportions. The binomial and Poisson models show significant differences in BRCA status for both B cells and macrophages; however, these results are likely spurious because these models do not account for overdispersion. When robust standard errors are used for the binomial and Poisson models, the proportion of B cells or macrophages is no longer significantly different across BRCA groups. Zero-inflated Poisson regression was only used to model B cell proportions because the HGSOC data did not contain excess zero counts of macrophages. While the ZIP coefficient for BRCA is significant in Fig. 5, this is also likely due to overdispersion.

As Fig. 5 demonstrates, failing to properly account for overdispersion can lead to artificially narrow confidence intervals and spurious p-values. To avoid this pitfall, we recommend using models with robust standard errors or others, such as beta-binomial, which explicitly model overdispersion. When an excess of zeros is present, it is instead more appropriate to use a model that accounts for both zero inflation and overdispersion, such as a zero-inflated beta-binomial or zero-inflated negative binomial. If one is unsure whether the data has enough zeros to constitute "excess," hypothesis tests can be used to compare zero-inflated models with their standard counterparts [87–89]. Finally, when multiple images/ROIs are available for each subject, mixed-effects versions of these models should be used to account for repeated measures.

3.3.2 Analysis of Functional Markers

Differences in the distribution and expression of functional markers across cell or patient populations are also of interest. Recent examples include whether the abundance of PD-L1 positive cells differs between responders and nonresponders to an immunotherapy in ovarian cancer [90] and whether high or low expression of major histocompatibility complex II (MHCII) in tumor cells affects overall survival in lung cancer patients [30]. To simplify the analysis of functional markers, they are often categorized into groups by thresholding continuously measured values. The choice of threshold can be arbitrary; some typical approaches include quantile-based groupings, biologically driven cutpoints, and optimal thresholds that maximize some test statistics of interest. For quantile-based groupings, a median threshold is the most common. Biologically relevant thresholds are chosen based on evidence from previous studies and are often classified at the patient level; for example, defining a patient with estrogen receptor (ER) expression on more than 1% of cells as ER+ [91], or a patient with MHCII expression on more than 5% of tumor cells as MHCII high [30]. However, biologically relevant thresholds are not always available and are sometimes inconsistently applied in the literature across studies and disease types. Finally, optimal cutpoints are chosen to maximize a test statistic of interest; for example, choosing high or low immune cell groups to assess the ovarian cancer tumor microenvironment in African American women [92]. This approach can be appropriate for data exploration and hypothesis generation, but the threshold obtained must be evaluated on external datasets to ensure its biological validity and avoid statistical bias [93, 94]. Recently, [95] proposed a threshold-free approach that classifies subjects into different categories by clustering on Jensen–Shannon distances between marker densities and may mitigate problems that arise when thresholding functional markers. All of these methods will be sensitive to upstream marker intensity normalization and phenotyping, so it is critical to carefully evaluate earlier results in the data processing pipeline.

While thresholding functional markers are common, categorizing observations from a continuous marker may reduce power and has been criticized within statistics [96]. This can be particularly problematic when the marker distribution is not bimodal and cannot be easily grouped into unexpressed and expressed cell populations. A more reproducible approach is to model marker values continuously, possibly accounting for nonlinearities with covariates using splines.

3.4 Spatial Analysis

Previous steps in the mIF data analysis pipeline are necessary for reducing bias due to technical variation, extracting interpretable image features like cell locations and cell types, and summarizing experiment-level characteristics. However, the explosion in popularity of multiplex imaging has mainly been driven by the unprecedented ability to study spatial relationships in samples with a large and growing number of markers. Many spatial analytic approaches have been adapted from ecology and spatial statistics, and new spatial analytic methods specifically for multiplex imaging data are rapidly developing [13]. Below we discuss different classes of spatial analytic tools and techniques, with a particular focus on methods grounded in point process theory. While methods for spatial transcriptomics data may be successfully leveraged for mIF data as well, these are reviewed in more detail elsewhere, for example [97].

3.4.1 Basic Tools for Data Exploration and Visualization

There are several recent software packages for high-level data exploration and visualization. Packages like Seurat and Giotto [59, 60] are multiuse toolkits that incorporate dozens of visualization methods and spatial metrics. Both of these packages were introduced for single-cell analysis of RNA sequencing data, and have evolved to accommodate spatial transcriptomics and multiplex imaging data. In contrast, the histoCAT toolbox [52] was introduced specifically to explore and visualize multiplex imaging data and provides interactive visualizations of cell phenotypes and nearest-neighbor analyses. Similarly, cytoMAP [98] is a user-friendly, comprehensive toolkit for quantitative image analysis of multiplexed tissues, and Squidpy [99] is a modular, scalable Python-based platform for spatial single-cell analysis. Each of these software platforms provides similar features, specifically, a united framework to visualize and perform basic spatial analysis of multiplexed imaging data within an interactive graphical user interface or programming environment (usually *R* or Python). A more comprehensive list of software tools for multiplex imaging data is provided in Table 1.

3.4.2 Spatial Statistical Learning Models

A few approaches that have emerged adapt statistical learning procedures to either identify spatial gene expression, which is used in spatial transcriptomics data because of the coarser spatial resolution, or to identify spatial neighborhoods [69, 100]. These approaches cluster pixels or wells while incorporating information about the spatial dependence, e.g., by smoothing or with a Markov random field. The Giotto package contains a method by [101] that integrates RNA sequencing data with mIF data using hidden Markov random fields. The BayesSpace model [102], which also uses Markov random fields, can be applied to spatial transcriptomics data to enhance resolution and cluster the image and implements a fully Bayesian approach. In [103], authors build gene or marker networks using nonnegative matrix factorization. Another approach uses spatial latent Dirichlet allocation to discover novel tissue subcompartments [104]. Many of these machine learning-based techniques for spatial 'omics are intended for spatial transcriptomics data, which have a lower spatial resolution than mIF data, but can have read counts on hundreds or thousands of genes. Because of this format, these spatial clustering methods incorporate a dimension reduction step in the marker dimension; however, when applied to mIF datasets, this dimension reduction may not be necessary, and these methods may be well adapted for handling the high spatial resolution of mIF data.

3.4.3 Spatial Metrics Based on Point Processes

The most relevant approaches for analyzing spatial relationships between cell types or quantifying the spatial co-expression of biological markers in multiplex imaging data are based on point process theory. The location of cells in mIF data assumed to follow a point process, realizations of a point process are called "point patterns," and point process models seek to understand correlations in the spatial distributions of cells. Under the assumption that the rate of a cell is constant over an entire region of interest, a point pattern will exhibit complete spatial randomness (CSR), and it is

Table 1
Software tools for data infrastructure (Store), visualization (Viz), normalization (Norm), segmentation (Seg), phenotyping (Ptype), and spatial analysis (Spatial) of multiplex imaging data

Software information			**Functionality**					
Name	***Purpose***	***Language***	***Store***	***Viz***	***Norm***	***Seg***	***Ptype***	***Spatial***
VectraPolarisData	Example mIF datasets	R (Bioc)	x					
histoCAT	Computational histology, topography, cytometry analysis toolbox	MATLAB, GUI		x	x			x
Squidpy	Tool for the analysis and visualization of spatial molecular data	Python		x		x		x
Giotto	Tools to process, analyze, and visualize single-cell spatial expression data	R, GUI, Python depends		x	x			x
imcRTools	Analysis of imaging, mass cytometry, and other highly multiplexed data	R (Bioc)	x	x				x
SVCA	Computational framework for the analysis of spatial molecular data	Python, R depends		x	x			x
SPIAT	Data processing, quality control, visualization, data analysis tools	R	x					
spatialTIME	Spatial analysis of mIF data	R	x	x				x
mxnorm	Normalization techniques for mIF data	R, Python depends	x	x	x			x
Seurat	Toolkit for QC, analysis, and exploration of single-cell omics data	R		x	x			x
QuPath	Powerful, flexible, extensible platform for bioimage analysis	GUI, Java		x	x	x	x	x
HALO	Platform for quantitative tissue analysis in digital pathology	Closed source, GUI		x		x	x	x
inForm	Visualizing, segmenting, and phenotyping images from Vectra instruments	Closed source, GUI		x		x	x	
Cytomapper	Visualization of cell-level info obtained by multiplexed imaging cytometry	R (Bioc)		x				
CyLinter	Quality control software for multiplex microscopy	Python				x		
CellProfiler Analyst	Interactive data analysis for high-throughput, image-based experiments	Python				x		

often of interest to model whether cells deviate from CSR either through clustering or repulsion. In marked point process theory points, a point pattern may carry attributes. Example attributes include cell phenotype, cell size, or expression of functional marker PDL1, and this auxiliary information is called a "mark" [105]. Marks can be qualitative (cell phenotype) or quantitative (expression of a functional marker), and marked point processes are particularly useful in the mIF data setting.

Spatial summary statistics can be calculated to quantify the clustering and co-occurrence of cells in a circular region with a particular radius r. Typically univariate (one cell type) or bivariate (two cell types) summary statistics are reported, and inference is obtained by comparing the observed spatial summary statistic to that obtained under CSR. Ripley's $K(r)$ is a popular quantity which studies the number of neighbors to a particular point within radius r and has univariate and bivariate implementations in the *spatstat* package [106, 107]. Ripley's K is characterized by clustering or repulsion depending on whether it is above or below the theoretical value under CSR of $2\pi r^2$. Modifications of Ripley's K, Besag's $L(r)$ and Marcon's $M(r)$, are interpreted as clustering or repulsion for values above or below their theoretical limits of r and 1, respectively. K and L increase with increasing radius, while M does not [108, 109]. A similar statistic called the *hypothesized interaction distribution* measures the co-localization between immune cells, and authors found that high co-clustering of certain cell types was associated with worse survival in oropharyngeal squamous cell carcinoma [110, 111]. A popular paper by [5] studying the TME of breast cancer introduced a bivariate adaptation of $K(r)$ that can be interpreted as the number of close interactions between marker-positive cells in a given region.

Other spatial summary statistics analyze the distance to a neighbor, and can be interpreted as probabilities of observing a particular cell type within a radius r. One of these metrics, $G(r)$, or the nearest neighbor distance distribution, is the cumulative distribution function of an exponential random variable [105]. Similar quantities include the empty space function $F(r)$, pair correlation function $g(r)$, and radial distribution function $J(r)$ [112, 113]. These types of spatial summaries have been used to study interactions between T-regulatory cells and tumor cells in NSCLC lung cancer [114] and to study CD68+ macrophages in human head and neck tumors [115]. An excellent and more detailed overview of these metrics, including a table with mathematical formulations, is provided in [13].

Figure 6 shows tumor-infiltrating immune cells with their respective Ripley's K functions for three images from the HGSOC dataset. The top row shows TMA cores and cell locations, with immune and other cells colored in red and gray, respectively.

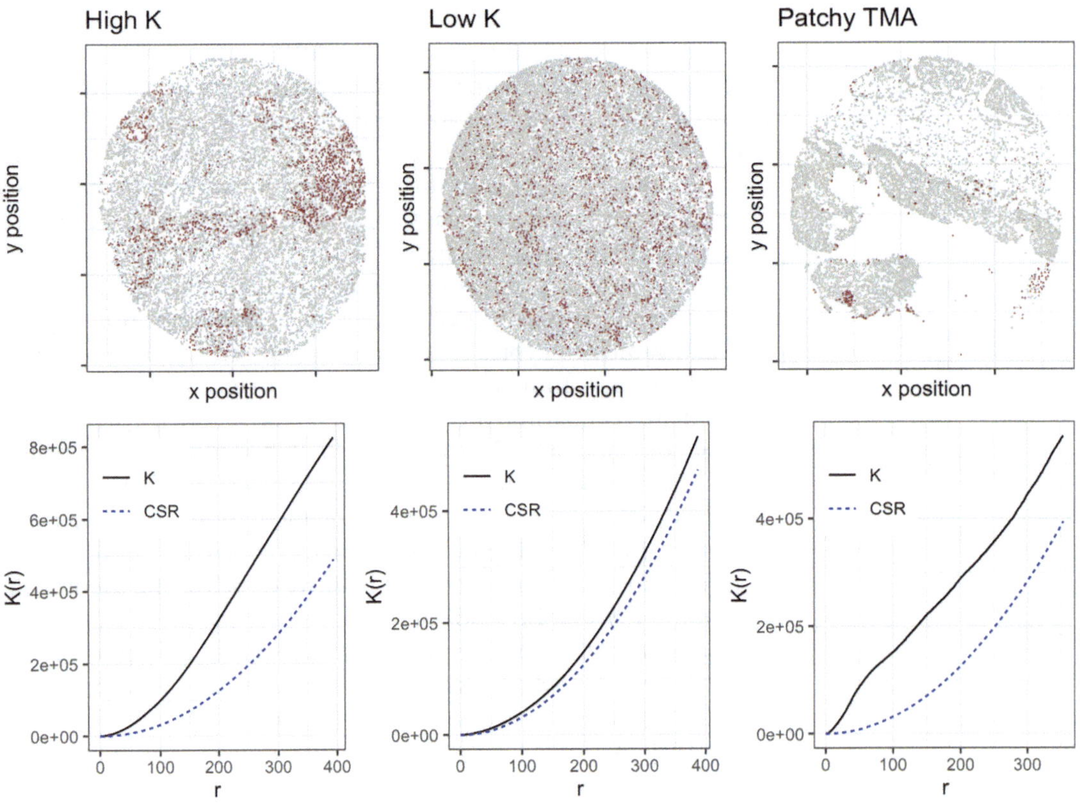

Fig. 6 Spatial summaries of tumor-infiltrating immune cells in three images from the HGSOC ovarian data. Top row shows plots of the spatial locations of all cells (gray) and tumor-infiltrating immune cells (red). Bottom row shows each estimated Ripley's *K* function *K*(*r*) for corresponding images in the top row. Estimated *K*(*r*) is shown by bold black line, and dotted blue line shows theoretical values of *K*(*r*) under complete spatial randomness. The first and second columns each show a subject with representative high or low *K* values at $r = 100$. The third column shows an image with areas of sparse cell density due to the tearing or folding of the TMA tissue

In the bottom row are Ripley's $K(r)$ for each image. The estimated $K(r)$ is indicated by the bold black line, and the dotted blue lines show theoretical values of $K(r)$ under complete spatial randomness. The subject in the first column has $K(r)$ values above the CSR line, indicating clustering of tumor-infiltrating immune cells; in contrast, the subject in the second column has a similar density of cells, but the immune cells (in red) are more randomly dispersed, leading to a $K(r)$ value close to the theoretical value under CSR.

For multiplex imaging data, it can be highly advantageous to incorporate quantitative marker intensity values with information about point locations, which can be done using second-order summary statistics. At a high level, this is typically done by incorporating measures of correlation. Moran's I is a popular global (image-level) measure of spatial autocorrelation which has bivariate and local adaptations [116, 117]. Lee's L is the spatially weighted and smoothed correlation coefficient of two markers [118]. Recently, [119] introduced a bivariate summary statistic based on the marker conditional mean. The pairwise correlation function also incorporates continuous marker information.

3.4.4 Point Process Adaptations for mIF

In mIF data, a major goal of estimating region-level spatial summaries is to perform analyses using the above spatial indices as the outcome or predictor variables [92, 120–122]. It is critical to obtain summary measures that are informative about cell-level interactions, but are not sensitive to gross tissue-level features such as tissue orientation or shape. Many of the well-established spatial summary metrics have corrections for edge effects, which arise when cells are close to the border of the image and are corrected automatically by functions in the *spatstat* package [106]. However, this is not sufficient in mIF, where estimation issues can arise when the tissue has holes due to the shape of the tissue, folds, or tears, resulting in patches of areas on the slide where no cells are present. This can bias the estimation of spatial summary statistics and is common for whole-sample mIF data as well as tumor microarrays (see Fig. 5, third column). One recent methodological development accounts for regions where no cells were present by permuting an empirical value of complete spatial randomness and then comparing observed spatial summary statistic values to those obtained by this empirical null distribution [92]. Implementations of this "patch" correction for Ripley's K and nearest neighbor-based metrics are available in the spatialTIME package [120]. In addition, many spatial summary statistics are calculated at a particular distance, r, and the selection of this distance can be arbitrary. An alternative to calculating a single spatial value for each image is to perform inference using the entire spatial summary function evaluated over a range of distances covering the image's spatial domain. Functional data analysis regression methods, which have been developed to handle functions as predictors and/or outcomes in linear models, are well suited for this. The additive functional Cox regression model [123] can be used to model spatial summary functions as covariates when the outcome is patient survival or time to disease recurrence, and this approach has been taken to study tumor-immune cross correlation functions in mIF and MIBI imaging data [122].

3.5 Software

Many software tools are in active development to assist in the analysis of bioimage, spatial transcriptomics, and spatial proteomics data. Halo and Inform are proprietary software with a graphical user interface (GUI) to allow point-and-click analysis. Most other tools are open source and primarily available in R or Python, though some open-source software are GUI based as well. Some are specialized to address one step of the mIF processing pipeline (e.g., normalization), while others can be used for visualization, normalization, phenotyping, segmentation, and/or spatial analysis. Existing software packages that can be used to analyze multiplex imaging data are highlighted in Table 1.

4 Conclusions

This chapter is a brief overview of the steps of the mIF analysis pipeline after single-cell segmentation, including normalization and batch correction, cell phenotyping, analysis of cell proportions and expression, and spatial analysis. We focused on reviewing these components of the pipeline because we believe these are areas where biostatistics can have the greatest impact. There are several areas of research that are beyond the scope of this review but are important to mention. While we focused on steps of the pipeline downstream of slide staining and image acquisition, many of the image quality issues, in particular batch effects, can be mitigated in the image acquisition process. Some work has focused on maximizing reproducibility in the image acquisition process [124]. Secondly, we assumed effective single-cell segmentation prior to many of the analyses discussed here, but estimation, bias, and variability in single-cell segmentation as well as a survey of current methodologies are, themselves, an entire literature to review. In addition, there is increasing attention to multimodal analysis, such as integrating mIF data with spatial transcriptomics, single-cell RNA-seq, and other omics data types to perform patient-level analyses that draw information across these different data types. Finally, the steps of the mIF pipeline are modular, and can each be improved upon independently or in parallel; however, it is critical to evaluate how these methodological developments influence biological interpretations in downstream analysis steps. As long as statistical methods are thoughtfully and carefully applied at each step of the analysis pipeline, there is much to be learned from multiplex data, and it will be exciting to see what discoveries will be made with these new technologies.

Acknowledgments

This work was supported by the National Institutes of Health (U2CCA233291), the Colorado Clinical and Translational Sciences Institute (UL1TR002535), and the Lupus Research Alliance (Lupus Innovation Award).

References

1. Bendall SC, Nolan GP, Roederer M, Chattopadhyay PK (2012) A deep profiler's guide to cytometry. Trends Immunol 33:323–332. https://doi.org/10.1016/j.it.2012.02.010
2. Young YK, Bolt AM, Ahn R, Mann KK (2016) Analyzing the tumor microenvironment by flow cytometry. Methods Mol Biol 1458:95–110. https://doi.org/10.1007/978-1-4939-3801-8_8
3. Hedvat CV (2010) Digital microscopy: past, present, and future. Arch Pathol Lab Med 134:1666–1670. https://doi.org/10.5858/2009-0579-RAR1.1
4. Angelo M, Bendall SC, Finck R et al (2014) Multiplexed ion beam imaging of human breast tumors. Nat Med 20:436–442. https://doi.org/10.1038/nm.3488
5. Keren L, Bosse M, Marquez D et al (2018) A structured tumor-immune microenvironment in triple negative breast cancer revealed by multiplexed ion beam imaging. Cell 174: 1373–1387.e19. https://doi.org/10.1016/j.cell.2018.08.039
6. Giesen C, Wang HAO, Schapiro D et al (2014) Highly multiplexed imaging of tumor tissues with subcellular resolution by mass cytometry. Nat Methods 11:417–422. https://doi.org/10.1038/nmeth.2869
7. Martinez-Morilla S, Villarroel-Espindola F, Wong PF et al (2021) Biomarker discovery in patients with immunotherapy-treated melanoma with imaging mass cytometry. Clin Cancer Res 27:1987–1996. https://doi.org/10.1158/1078-0432.CCR-20-3340
8. Baharlou H, Canete NP, Cunningham AL et al (2019) Mass cytometry imaging for the study of human diseases-applications and data analysis strategies. Front Immunol 10:2657. https://doi.org/10.3389/fimmu.2019.02657
9. Ehrenberg AJ, Morales DO, Piergies AMH et al (2020) A manual multiplex immunofluorescence method for investigating neurodegenerative diseases. J Neurosci Methods 339: 108708. https://doi.org/10.1016/j.jneumeth.2020.108708
10. Apaolaza PS, Petropoulou P-I, Rodriguez-Calvo T (2021) Whole-slide image analysis of human pancreas samples to elucidate the immunopathogenesis of type 1 diabetes using the QuPath software. Front Mol Biosci 8: 689799. https://doi.org/10.3389/fmolb.2021.689799
11. Kondo A, Ma S, Lee MYY et al (2021) Highly multiplexed image analysis of intestinal tissue sections in patients with inflammatory bowel disease. Gastroenterology 161:1940–1952. https://doi.org/10.1053/j.gastro.2021.08.055
12. Ramaglia V, Sheikh-Mohamed S, Legg K et al (2019) Multiplexed imaging of immune cells in staged multiple sclerosis lesions by mass cytometry. elife 8:e48051. https://doi.org/10.7554/eLife.48051
13. Wilson CM, Ospina OE, Townsend MK et al (2021) Challenges and opportunities in the statistical analysis of multiplex immunofluorescence data. Cancers (Basel) 13:3031. https://doi.org/10.3390/cancers13123031
14. McKinley ET, Shao J, Ellis ST et al (2022) MIRIAM: a machine and deep learning single-cell segmentation and quantification pipeline for multi-dimensional tissue images. Cytometry A 101:521. https://doi.org/10.1002/cyto.a.24541
15. Schapiro D, Sokolov A, Yapp C et al (2022) MCMICRO: a scalable, modular image-processing pipeline for multiplexed tissue imaging. Nat Methods 19:311–315. https://doi.org/10.1038/s41592-021-01308-y
16. Vizcarra JC, Burlingame EA, Hug CB et al (2022) A community-based approach to image analysis of cells, tissues and tumors. Comput Med Imaging Graph 95:102013
17. Bray F, Ferlay J, Soerjomataram I et al (2018) GLOBOCAN estimates of incidence and mortality worldwide for 36 cancers in 185 countries. CA Cancer J Clin 68:394–424
18. Fabbro M, Colombo P-E, Leaha CM et al (2020) Conditional probability of survival and prognostic factors in long-term survivors of high-grade serous ovarian cancer. Cancers (Basel) 12:E2184. https://doi.org/10.3390/cancers12082184
19. Siegel RL, Miller KD, Fuchs HE, Jemal A (2022) Cancer statistics, 2022. CA Cancer J Clin 72:7–33
20. Remark R, Becker C, Gomez JE et al (2015) The non—small cell lung cancer immune contexture. A major determinant of tumor characteristics and patient outcome. Am J Respir Crit Care Med 191:377–390
21. Jordan KR, Sikora MJ, Slansky JE et al (2020) The capacity of the ovarian cancer tumor microenvironment to integrate inflammation signaling conveys a shorter disease-free interval. Clin Cancer Res 26:6362–6373
22. Shakya R, Nguyen TH, Waterhouse N, Khanna R (2020) Immune contexture analysis in immuno-oncology: applications and

challenges of multiplex fluorescent immunohistochemistry. Clin Transl Immunol 9:e1183

23. Gorris MA, Halilovic A, Rabold K et al (2018) Eight-color multiplex immunohistochemistry for simultaneous detection of multiple immune checkpoint molecules within the tumor microenvironment. J Immunol 200: 347–354
24. Mezheyeuski A, Bergsland CH, Backman M et al (2018) Multispectral imaging for quantitative and compartment-specific immune infiltrates reveals distinct immune profiles that classify lung cancer patients. J Pathol 244:421–431
25. Mori H, Bolen J, Schuetter L et al (2020) Characterizing the tumor immune microenvironment with tyramide-based multiplex immunofluorescence. J Mammary Gland Biol Neoplasia 25:417–432. https://doi.org/10.1007/s10911-021-09479-2
26. Viratham Pulsawatdi A, Craig SG, Bingham V et al (2020) A robust multiplex immunofluorescence and digital pathology workflow for the characterisation of the tumour immune microenvironment. Mol Oncol 14:2384–2402
27. Wrobel J, Ghosh T (2022) VectraPolarisData: Vectra Polaris and Vectra 3 multiplex single-cell imaging data. R package version 1.0.0.
28. Righelli D, Weber LM, Crowell HL et al (2022) SpatialExperiment: infrastructure for spatially-resolved transcriptomics data in R using bioconductor. Bioinformatics 38: 3128–3131
29. Steinhart B, Jordan KR, Bapat J et al (2021) The spatial context of tumor-infiltrating immune cells associates with improved ovarian cancer survival. Mol Cancer Res 19:1973–1979
30. Johnson AM, Boland JM, Wrobel J et al (2021) Cancer cell-specific MHCII expression as a determinant of the immune infiltrate organization and function in the non-small cell lung cancer tumor microenvironment. bioRxiv. https://doi.org/10.1101/2021.02.24.432729
31. Ahmed Raza SE, Langenkämper D, Sirinukunwattana K et al (2016) Robust normalization protocols for multiplexed fluorescence bioimage analysis. BioData Min 9:11. https://doi.org/10.1186/s13040-016-0088-2
32. Berry S, Giraldo NA, Green BF et al (2021) Analysis of multispectral imaging with the AstroPath platform informs efficacy of PD-1 blockade. Science 372. https://doi.org/10.1126/science.aba2609
33. Fortin J-P, Sweeney EM, Muschelli J et al (2016) Removing inter-subject technical variability in magnetic resonance imaging studies. NeuroImage 132:198–212
34. Harris CR, McKinley ET, Roland JT et al (2022) Quantifying and correcting slide-to-slide variation in multiplexed immunofluorescence images. Bioinformatics btab877:1700. https://doi.org/10.1093/bioinformatics/btab877
35. Shinohara RT, Sweeney EM, Goldsmith J et al (2014) Statistical normalization techniques for magnetic resonance imaging. NeuroImage: Clin 6:9–19
36. Johnson WE, Li C, Rabinovic A (2007) Adjusting batch effects in microarray expression data using empirical Bayes methods. Biostatistics 8:118–127. https://doi.org/10.1093/biostatistics/kxj037
37. Graf J, Cho S, McDonough E et al (2022) FLINO: a new method for immunofluorescence bioimage normalization. Bioinformatics 38:520–526. https://doi.org/10.1093/bioinformatics/btab686
38. Fortin J-P, Parker D, Tunç B et al (2017) Harmonization of multi-site diffusion tensor imaging data. NeuroImage 161:149–170. https://doi.org/10.1016/j.neuroimage.2017.08.047
39. Pomponio R, Erus G, Habes M et al (2020) Harmonization of large MRI datasets for the analysis of brain imaging patterns throughout the lifespan. NeuroImage 208:116450. https://doi.org/10.1016/j.neuroimage.2019.116450
40. Chang YH, Chin K, Thibault G et al (2020) RESTORE: robust intEnSiTy nORmalization mEthod for multiplexed imaging. Commun Biol 3:111. https://doi.org/10.1038/s42003-020-0828-1
41. Harris C, Wrobel J, Vandekar S (2022) Mxnorm: an R package to normalize multiplexed imaging data. J Open Source Softw 7: 4180
42. Wrobel J, Martin ML, Bakshi R et al (2020) Intensity warping for multisite MRI harmonization. NeuroImage 223:117242. https://doi.org/10.1016/j.neuroimage.2020.117242
43. Chen B, Cherie'R S, McKinley ET, Simmons AJ, Marisol A, Ramirez-Solano MA, Zhu X, Markham NO, et al (2021) Differential Pre-Malignant Programs and Microenvironment Chart Distinct Paths to Malignancy in Human Colorectal Polyps. Cell 184 (26): 6262-6280.e26. https://doi.org/10.1016/j.cell.2021.11.031.

44. Rozenblatt-Rosen O, Regev A, Oberdoerffer P et al (2020) The human tumor atlas network: charting tumor transitions across space and time at single-cell resolution. Cell 181: 236–249
45. Zhang Y, Parmigiani G, Johnson WE (2020) ComBat-Seq: batch effect adjustment for RNA-Seq count data. bioRxiv. https://www.biorxiv.org/content/10.1101/2020.01.13.904730v1. Accessed 18 Aug 2020
46. Freytag S, Tian L, Lönnstedt I et al (2018) Comparison of clustering tools in R for medium-sized 10x genomics single-cell RNA-sequencing data. F1000Research 7: 1297
47. Menon V (2018) Clustering single cells: a review of approaches on high-and low-depth single-cell RNA-seq data. Brief Funct Genomics 17:240–245. https://doi.org/10.1093/bfgp/elx044
48. Rashid R, Gaglia G, Chen Y-A et al (2019) Highly multiplexed immunofluorescence images and single-cell data of immune markers in tonsil and lung cancer. Sci Data 6:323. https://doi.org/10.1038/s41597-019-0332-y
49. Dao D, Fraser AN, Hung J et al (2016) CellProfiler analyst: interactive data exploration, analysis and classification of large biological image sets. Bioinformatics 32:3210–3212
50. Jones TR, Kang IH, Wheeler DB et al (2008) CellProfiler analyst: data exploration and analysis software for complex image-based screens. BMC Bioinform 9:1–16
51. Stirling DR, Carpenter AE, Cimini BA (2021) CellProfiler analyst 3.0: accessible data exploration and machine learning for image analysis. Bioinformatics 37:3992–3994
52. Schapiro D, Jackson HW, Raghuraman S et al (2017) histoCAT: analysis of cell phenotypes and interactions in multiplex image cytometry data. Nat Methods 14:873–876. https://doi.org/10.1038/nmeth.4391
53. Baker, Gregory. (2021) CyLinter (Version 0.0.30) [Computer software]. https://github.com/labsyspharm/cylintery.
54. Piccinini F, Balassa T, Szkalisity A et al (2017) Advanced cell classifier: user-friendly machine-learning-based software for discovering phenotypes in high-content imaging data. Cell Syst 4:651–655.e5. https://doi.org/10.1016/j.cels.2017.05.012
55. Blondel VD, Guillaume J-L, Lambiotte R, Lefebvre E (2008) Fast unfolding of communities in large networks. J Stat Mech 2008: P10008. https://doi.org/10.1088/1742-5468/2008/10/P10008
56. Kiselev VY, Andrews TS, Hemberg M (2019) Challenges in unsupervised clustering of single-cell RNA-seq data. Nat Rev Genet 20: 273–282. https://doi.org/10.1038/s41576-018-0088-9
57. Levine JH, Simonds EF, Bendall SC et al (2015) Data-driven phenotypic dissection of AML reveals progenitor-like cells that correlate with prognosis. Cell 162:184–197
58. Traag VA, Waltman L, van Eck NJ (2019) From Louvain to Leiden: guaranteeing well-connected communities. Sci Rep 9:5233. https://doi.org/10.1038/s41598-019-41695-z
59. Dries R, Zhu Q, Dong R et al (2021) Giotto: a toolbox for integrative analysis and visualization of spatial expression data. Genome Biol 22:78. https://doi.org/10.1186/s13059-021-02286-2
60. Hao Y, Hao S, Andersen-Nissen E et al (2021) Integrated analysis of multimodal single-cell data. Cell 184:3573–3587.e29. https://doi.org/10.1016/j.cell.2021.04.048
61. Grün D, Lyubimova A, Kester L et al (2015) Single-cell messenger RNA sequencing reveals rare intestinal cell types. Nature 525: 251–255. https://doi.org/10.1038/nature14966
62. Bodenheimer T, Halappanavar M, Jefferys S, Gibson R, Liu S, Mucha PJ, Stanley N, Parker JS, and Selitsky SR(2020) FastPG: Fast Clustering of Millions of Single Cells. bioRxiv. https://doi.org/10.1101/2020.06.19.159749.
63. Samusik N, Good Z, Spitzer MH et al (2016) Automated mapping of phenotype space with single-cell data. Nat Methods 13:493–496. https://doi.org/10.1038/nmeth.3863
64. Stassen SV, Siu DMD, Lee KCM et al (2020) PARC: ultrafast and accurate clustering of phenotypic data of millions of single cells. Bioinformatics 36:2778–2786. https://doi.org/10.1093/bioinformatics/btaa042
65. Bortolomeazzi M, Montorsi L, Temelkovski D et al (2022) A SIMPLI (single-cell identification from MultiPLexed images) approach for spatially-resolved tissue phenotyping at single-cell resolution. Nat Commun 13:1–14
66. Seal S, Wrobel J, Johnson AM et al (2021) On clustering for cell phenotyping in multiplex immunohistochemistry (mIHC) and multiplexed ion beam imaging (MIBI) data
67. Aditya P, Doron M, and Caicedo JC. (2021) Image-Based Cell Phenotyping with Deep Learning. Current Opinion in Chemical Biology, Mechanistic Biology and Machine

Learning in Chemical Biology, 65 (December): 9–17. https://doi.org/10.1016/j.cbpa.2021.04.001.

68. Alexandrov T, Chernyavsky I, Becker M et al (2013) Analysis and interpretation of imaging mass spectrometry data by clustering mass-to-charge images according to their spatial similarity. Anal Chem 85:11189–11195. https://doi.org/10.1021/ac401420z
69. Alexandrov T, Kobarg JH (2011) Efficient spatial segmentation of large imaging mass spectrometry datasets with spatially aware clustering. Bioinformatics 27:i230–i238. https://doi.org/10.1093/bioinformatics/btr246
70. Krueger R, Beyer J, Jang W-D et al (2020) Facetto: combining unsupervised and supervised learning for hierarchical phenotype analysis in multi-channel image data. IEEE Trans Vis Comput Graph 26:227–237. https://doi.org/10.1109/TVCG.2019.2934547
71. Rajaram S, Pavie B, Wu LF, Altschuler SJ (2012) PhenoRipper: software for rapidly profiling microscopy images. Nat Methods 9: 635–637. https://doi.org/10.1038/nmeth.2097
72. Uhlmann V, Singh S, Carpenter AE (2016) CP-CHARM: segmentation-free image classification made accessible. BMC Bioinform 17: 51. https://doi.org/10.1186/s12859-016-0895-y
73. Diem K, Magaret A, Klock A et al (2015) Image analysis for accurately counting CD4+ and CD8+ T cells in human tissue. J Virol Methods 222:117–121
74. Dobson AJ, Barnett AG (2018) An introduction to generalized linear models. Chapman and Hall/CRC, Boca Raton
75. MacKinnon JG, White H (1985) Some heteroskedasticity-consistent covariance matrix estimators with improved finite sample properties. J Econ 29:305–325
76. Zeileis A (2006) Object-oriented computation of sandwich estimators. J Stat Softw 16: 1–16. https://doi.org/10.18637/jss.v016.i09
77. Hu T, Gallins P, Zhou Y-H (2018) A zero-inflated beta-binomial model for microbiome data analysis. Stat 7:e185
78. Martin BD, Witten D, Willis AD (2020) Modeling microbial abundances and dysbiosis with beta-binomial regression. Ann Appl Stat 14:94
79. Jakaitiene A, Avino M, Guarracino MR (2017) Beta-binomial model for the detection of rare mutations in pooled next-generation sequencing experiments. J Comput Biol 24: 357–367
80. Love MI, Huber W, Anders S (2014) Moderated estimation of fold change and dispersion for RNA-seq data with DESeq2. Genome Biol 15:1–21
81. Robinson MD, Smyth GK (2008) Small-sample estimation of negative binomial dispersion, with applications to SAGE data. Biostatistics 9:321–332
82. Liang K-Y, Zeger SL (1986) Longitudinal data analysis using generalized linear models. Biometrika 73:13–22
83. Zeger SL, Liang K-Y, Albert PS (1988) Models for longitudinal data: a generalized estimating equation approach. Biometrics 44: 1049–1060
84. Lee AH, Wang K, Scott JA et al (2006) Multilevel zero-inflated Poisson regression modelling of correlated count data with excess zeros. Stat Methods Med Res 15:47–61
85. Jiang S, Xiao G, Koh AY et al (2021) A Bayesian zero-inflated negative binomial regression model for the integrative analysis of microbiome data. Biostatistics 22:522–540
86. Zhang X, Yi N (2020) Fast zero-inflated negative binomial mixed modeling approach for analyzing longitudinal metagenomics data. Bioinformatics 36:2345–2351
87. Vuong QH (1989) Likelihood ratio tests for model selection and non-nested hypotheses. Econometrica 57:307–333
88. Desmarais BA, Harden JJ (2013) Testing for zero inflation in count models: bias correction for the Vuong test. Stata J 13:810–835
89. He H, Zhang H, Ye P, Tang W (2019) A test of inflated zeros for Poisson regression models. Stat Methods Med Res 28:1157–1169
90. Sanchez K, Kim I, Chun B et al (2021) Multiplex immunofluorescence to measure dynamic changes in tumor-infiltrating lymphocytes and PD-L1 in early-stage breast cancer. Breast Cancer Res 23:1–15
91. Hammond MEH, Hayes DF, Dowsett M et al (2010) American Society of Clinical Oncology/College of American Pathologists guideline recommendations for immunohistochemical testing of estrogen and progesterone receptors in breast cancer (unabridged version). Arch Pathol Lab Med 134:e48–e72
92. Wilson C, Soupir AC, Thapa R et al (2022) Tumor immune cell clustering and its association with survival in African American women with ovarian cancer. PLoS Comput Biol 18: e1009900

93. Altman DG, Lausen B, Sauerbrei W, Schumacher M (1994) Dangers of using "optimal" cutpoints in the evaluation of prognostic factors. JNCI: J Natl Cancer Inst 86:829–835
94. Mabikwa OV, Greenwood DC, Baxter PD, Fleming SJ (2017) Assessing the reporting of categorised quantitative variables in observational epidemiological studies. BMC Health Serv Res 17:1–8
95. Seal S, Vu T, Ghosh T et al (2021) DenVar: Density-based Variation analysis of multiplex imaging data. bioRxiv. https://doi.org/10.1101/2021.09.27.462056
96. Harrell FE (2015) Regression modeling strategies: with applications to linear models, logistic and ordinal regression, and survival analysis. Springer, Cham
97. Dries R, Chen J, Del Rossi N et al (2021) Advances in spatial transcriptomic data analysis. Genome Res 31:1706–1718
98. Stoltzfus CR, Filipek J, Gern BH et al (2020) CytoMAP: a spatial analysis toolbox reveals features of myeloid cell organization in lymphoid tissues. Cell Rep 31:107523
99. Palla G, Spitzer H, Klein M et al (2022) Squidpy: a scalable framework for spatial omics analysis. Nat Methods 19:171–178
100. Schürch CM, Bhate SS, Barlow GL et al (2020) Coordinated cellular neighborhoods orchestrate antitumoral immunity at the colorectal cancer invasive front. Cell 182:1341–1359.e19. https://doi.org/10.1016/j.cell.2020.07.005
101. Li M, Santpere G, Kawasawa YI et al (2018) Integrative functional genomic analysis of human brain development and neuropsychiatric risks. Science 362:eaat7615. https://doi.org/10.1126/science.aat7615
102. Zhao E, Stone MR, Ren X et al (2021) Spatial transcriptomics at subspot resolution with BayesSpace. Nat Biotechnol 39:1375–1384
103. Wu S, Joseph A, Hammonds AS et al (2016) Stability-driven nonnegative matrix factorization to interpret spatial gene expression and build local gene networks. Proc Natl Acad Sci 113:4290–4295
104. Chen Z, Soifer I, Hilton H et al (2020) Modeling multiplexed images with spatial-LDA reveals novel tissue microenvironments. J Comput Biol 27:1204–1218
105. Baddeley A, Rubak E, Turner R (2015) Spatial point patterns: methodology and applications with R. CRC Press, Boca Raton
106. Baddeley A, Turner R (2005) Spatstat: an R package for analyzing spatial point patterns. J Stat Softw 12:1–42
107. Ripley BD (1976) The second-order analysis of stationary point processes. J Appl Probab 13:255–266
108. Besag J, Diggle PJ (1977) Simple Monte Carlo tests for spatial pattern. J R Stat Soc Ser C Appl Stat 26:327–333
109. Marcon E, Puech F, Traissac S (2012) Characterizing the relative spatial structure of point patterns. Int J Ecol 2012:1
110. Tsakiroglou AM, Fergie M, Oguejiofor K et al (2020) Spatial proximity between T and PD-L1 expressing cells as a prognostic biomarker for oropharyngeal squamous cell carcinoma. Br J Cancer 122:539–544
111. Rose CJ, Naidoo K, Clay V et al (2013) A statistical framework for analyzing hypothesized interactions between cells imaged using multispectral microscopy and multiple immunohistochemical markers. J Pathol Inform 4:4
112. Baddeley A, Gill RD (1997) Kaplan-Meier estimators of distance distributions for spatial point processes. Ann Stat 25:263–292
113. Baddeley AJ, Gill RD (1994) The empty space hazard of a spatial pattern. University of Western Australia, Department of Mathematics Perth, Australia
114. Barua S, Fang P, Sharma A et al (2018) Spatial interaction of tumor cells and regulatory T cells correlates with survival in non-small cell lung cancer. Lung Cancer 117:73–79
115. Bull JA, Macklin PS, Quaiser T et al (2020) Combining multiple spatial statistics enhances the description of immune cell localisation within tumours. Sci Rep 10:1–12
116. Moran PA (1948) The interpretation of statistical maps. J R Stat Soc Ser B Methodol 10:243–251
117. Anselin L (1995) Local indicators of spatial association—LISA. Geogr Anal 27:93–115
118. Lee S-I (2001) Developing a bivariate spatial association measure: an integration of Pearson's r and Moran's I. J Geogr Syst 3:369–385
119. Chervoneva I, Peck AR, Yi M et al (2021) Quantification of spatial tumor heterogeneity in immunohistochemistry staining images. Bioinformatics 37:1452–1460. https://doi.org/10.1093/bioinformatics/btaa965
120. Creed JH, Wilson CM, Soupir AC et al (2021) spatialTIME and iTIME: R package and Shiny application for visualization and analysis of immunofluorescence data. Bioinformatics 37:4584–4586. https://doi.org/10.1093/bioinformatics/btab757
121. Maisel BA, Yi M, Peck AR et al (2022) Spatial metrics of interaction between CD163-positive macrophages and cancer cells and

progression-free survival in chemo-treated breast cancer. Cancers 14:308

122. Vu T, Wrobel J, Bitler BG et al (2021) SPF: a spatial and functional data analytic approach to cell imaging data. bioRxiv. https://doi.org/10.1101/2021.09.29.462282

123. Cui E, Crainiceanu CM, Leroux A (2021) Additive functional Cox model. J Comput Graph Stat 30:780–793

124. Laberiano-Fernández C, Hernández-Ruiz S, Rojas F, Parra ER (2021) Best practices for technical reproducibility assessment of multiplex immunofluorescence. Front Mol Biosci 8:660202

Chapter 9

Statistical Analysis in ChIP-seq-Related Applications

Mingxiang Teng

Abstract

Chromatin immunoprecipitation sequencing (ChIP-seq) has been widely performed to identify protein binding information along the genome. The sequencing protocol is quite flexible and mature to measure different types of protein binding as long as sequencing parameters are properly tailored to accommodate protein features. Two distinct types of protein binding are point-source-like binding by transcription factors and diffused-distribution binding by histone modifications. Consequently, statistical approaches have been proposed to address ChIP-seq-related questions according to different protein features. In this chapter, we briefly summarize statistical principles, approaches, and tools that are widely implemented in modeling ChIP-seq data, from raw data quality control to final result reporting. We discuss the key solutions in addressing eight routine questions in ChIP-seq applications. We also include discussion on approaches fitting unique data features in different ChIP-seq types. We hope this chapter will serve as a brief guide, especially for ChIP-seq beginners, to provide them with a high-level overview to understand and design processing plans for their ChIP-seq experiments.

Key words ChIP-seq, Statistical modeling, Data interpretation

1 Introduction

Chromatin immunoprecipitation sequencing (ChIP-seq) was adopted as an alternative but superior technology over ChIP-chip more than a decade ago to efficiently scan genome-wide protein-DNA interactions and histone modifications [1–3]. It combines chromatin immunoprecipitation with next-generation sequencing technology which allows massive de novo detection of genomic information. Several key characteristics make this technology widely applied since it was introduced amid multiple similar and improved technologies proposed recently, such as ChIP-exo [4], CUT&RUN [5], and CUT&Tag [6]. First, ChIP-seq provides easy-to-prepare protocol with ChIP antibody well-developed for hundreds of known transcription factors and histone modifications [7]. These established resources allow researchers to set up

Brooke L. Fridley and Xuefeng Wang (eds.), *Statistical Genomics*,
Methods in Molecular Biology, vol. 2629, https://doi.org/10.1007/978-1-0716-2986-4_9,

experiments quickly, especially with the goal to investigate cooperative binding across a list of protein cofactors. Second, as a targeted protocol focusing on a selection of genomic regions, ChIP-seq requires relatively low sequencing volume which makes it affordable for most laboratories [8]. Third, ChIP-seq data has been largely generated across organisms, tissues, and cell types, with numerous datasets available through public data portals and databases [9]. Researchers nowadays can compare and cross-validate their own ChIP-seq data conveniently by exploring these public resources, which significantly facilitate pilot research studies. Finally, a significant number of statistical methods and tools were well established for ChIP-seq data analysis over the last decade, ranging from data production to biological result reporting [10–14]. These tools have promoted a great environment and standards for researchers to continue working with ChIP-seq technology.

A successful ChIP-seq experiment needs both proper design of the wet experiment and elegant dry analysis with the acquired dataset. Broadly, ChIP-seq represents ChIP-based experiments in querying protein-DNA binding information that requires common operations such as cross-linking, fragment length filtering, PCR amplification, etc. These operations inspire the key strategies to model the distribution of sequencing reads around protein binding sites and detect protein binding locations [13, 15, 16]. ChIP-seq can also be categorized into several sub-protocols that characterize distinct types of protein binding and consequentially require unique designs in wet and dry experiments. For example, some proteins show discrete and point-source binding profiles on genome, while some others present diffused and broad-domain distributions [11]. This features additional protein properties beyond their binding locations. Wet experimental parameters have been tuned to fully capture both common and distinct features of different protein binding, such as by selecting proper sequencing depths and fragment lengths in library preparation. For dry analysis, statistical approaches were proposed to fit data features across different ChIP-seq datasets. In this chapter, we briefly summarize statistical analysis widely applied in ChIP-seq data processing, from data quality control to protein binding annotation. We cover the statistical approaches in main analyzing steps for a typical ChIP-seq protocol as well as some of the specific methods more suitable in sub-protocols. We also summarize strategies in multimodal data integration by incorporating ChIP-seq with other genome-wide measuring technologies.

2 Methods

In this section, we summarize the statistic principle, approaches, and tools from eight major aspects in ChIP-seq data analysis, as illustrated in Fig. 1.

2.1 Quality Control

Quality control is the initial step in sequencing data analysis to statistically evaluate whether data is good enough to proceed downstream analysis. Usually, a series of quantitative metrics and virtualization plots are generated to characterize sequencing qualities regarding the studied biology. Like most of other sequencing technologies, such as RNA and exome sequencing, an overall scanning of sequencing quality is needed in ChIP-seq to evaluate sequencing contamination, duplicate level, base quality, sequence content biases, etc. Widely used tools to estimate these metrics include FASTQC [17], fastp [18], etc., with measurements usually denoted in percentages. In brief, good sequencing quality is expected if the data has low contamination, low PCR duplication, high base quality, and balanced sequence content. Adapter sequences remained in raw sequencing data could also be detected in this step for indication of further removal.

Additionally, specific metrics are needed to fully characterize the distinct nature of ChIP-seq datasets that usually exhibit unneglectable background noises due to varied antibody quality, cross-linking specificity, etc. The ENCODE consortium has established a

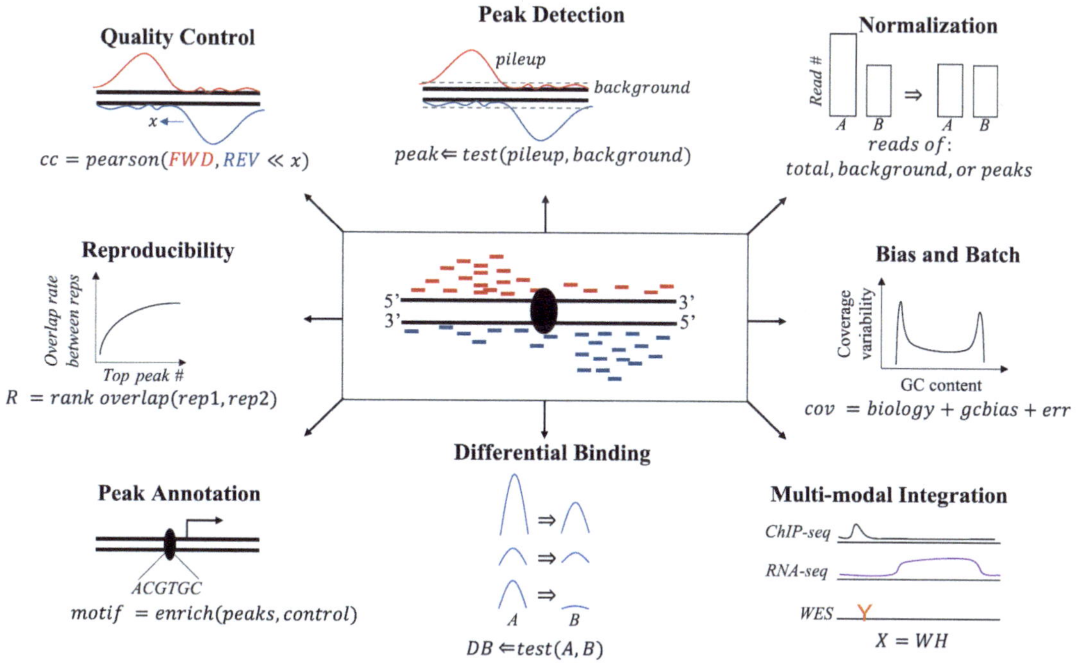

Fig. 1 Typical statistical components involved in major steps of ChIP-seq data processing

spectrum of statistical standards to help address the unique quality metrics in ChIP-seq [11]. These metrics evaluate data quality mainly based on ChIP-seq read distribution along the genome. As such, aligned sequencing reads instead of raw reads are required in calculation of these metrics. For example, ChIP-seq reads are expected to be symmetrically detected on the five primes of forward and reverse strands centering the protein binding sites, due to the randomly picking of fragment orientation during sequencing. Thus, a high cross-correlation coefficient between genome-wide forward and reverse strand ChIP-seq signals, especially with a proper shift of the signals on one strand toward the other, was proposed as a measurement of good data quality (Fig. 1). Beyond that, a detailed analysis of the signal-to-noise ratio based on mapped reads can further estimate the specificity of ChIP-seq experiments, such as those proposed by CHANCE [19]. Metrics including library complexity, sequencing depth, and fragment lengths also help better understand individual datasets. It is noted that the standards of "good" scores on these metrics might vary depending on the working data types (point-source, broad-domain, or mixed). As such, domain knowledge is highly recommended to proceed metric estimation.

Full characterization of ChIP-seq quality usually goes beyond those based on raw or mapped sequencing reads. One of the most important statistic metrics to estimate library specificity is the fraction of reads in peaks (FRiP), which is calculated after peak detection (summarized in the next section) [11]. Basically, the proportion of sequencing reads located inside the identified peaks tells researchers if sequencing coverage is spent as desired in the protein binding regions. Other metrics such as the number of identified peaks tells if the ChIP-seq library is sensitive enough to capture the majority of true binding sites. For example, CTCF is an insulator which holds thousands to tens of thousands binding sites in a human cell type [9, 20]. A low quality of CTCF ChIP-seq would be inferred if the identified number of CTCF peaks significantly differ from that. Again, the standards of good quality by these metrics vary between proteins and experiments, and domain knowledge plays important roles in fully interpreting them.

2.2 Peak Detection

The main goal of ChIP-seq study is to identify protein binding sites along the genome, usually indicated as higher ChIP-seq pileup signals (i.e., peaks) than the nonbinding sites or background. Unfortunately, the background intensity is not distributed evenly along the genome [21, 22]. It thus raises the question that to what extent or criteria the pileup signals would be nominated to represent true binding, and statistical testing is necessary here to help define the significance. Basically, the common procedure is to compare the local pileup signals with background signals [15, 16, 23–25] (Fig. 1) that are defined customized but largely based on the mean signals from genomic regions surrounding the compared pileups. The comparing significances are then evaluated by implementing statistical testing models (e.g., Poisson) in some tools (e.g., MACS2 [15]) and controlling multiple testing across genome-wide pileups. To ensure the testing procedure time- and space-efficient, some methods simplify the units of pileups based on identical sliding windows (e.g., 300 bp bins) along genome and merge all significant pileup windows after statistical testing. It is noted that ChIP-seq experiments usually enroll control samples (e.g., INPUT) to better mimic the background levels at the compared regions. In that case, normalization processing between ChIP and INPUT samples are required before significance testing [26]. Normalization options are discussed in later sections.

Beyond statistical testing with pileup enrichment, most methods implement additional criteria to help control false-positive peak calls. ChIP-seq presents unique data features, as we mentioned before, that the sequencing reads between forward and reverse strands at true protein binding sites exhibit symmetric distributions. Methods have implemented algorithms to check the presence of such pattern at individual peak candidates. Fail to meet the criteria indicates possible artifact peaks even though enriched pileup is detected over background. Another statistical operation by utilizing this symmetric characteristic is to estimate the protein binding width on DNA. Binding width is critical as another data dimension in addition to the peak center locations to be reported in ChIP-seq peak detection. As we mentioned before, different proteins present different binding types, point-source, broad-domain, or mixed. Methods usually estimate the binding width as the distance between the centers of the symmetric pileup signals of forward and reverse strands. More statistical implementations on peak detection have been detailed in these reviews [10–13, 27].

2.3 Signal Reproducibility

Similar to other high-throughput technologies, reproducibility is critical in ChIP-seq data analysis. One can assume reproducibility analysis as another round of QC check to ensure the confidence of reported ChIP-seq peaks. However, different from other studies like RNA-seq and whole exome sequencing, ChIP-seq have floating genomic features to be compared between samples and replicates. For example, in addition to the presence/absence discrepancies, some peaks shift their occupied locations along the DNA between replicates, complicating quantitative measurements with spatial characterization. A successful approach to address this challenge is to use the irreproducible discovery rate (IDR) [28] that was widely implemented by the ENCODE project. Basically, it ranks identified peaks in each of two compared replicates and uses a mixture model to provide a conservative estimate of the threshold where the two ranked lists no longer consistent. The shifted but overlapped peaks are treated as the same peaks between replicates.

Another approach is to use correspondence at the top (CAT) plot [29], which scans the replicate consistency based on top detected candidates by assuming that top candidates contain the most genuine and biological meaningful signals (Fig. 1). This approach has been adopted to evaluate high-throughput experiments including gene microarray, RNA-seq, ChIP-seq, etc. The proposed statistical metric is straightforward without sophisticated modeling. However, both IDR and CATplot were defined to evaluate reproducibility between two sequencing replicates. For experiments with three or more replicates, customized adaption of these approaches is further needed. Nevertheless, two-replicate ChIP-seq experiments are commonly designed as those characterized by the ENCODE project. It is also noted that reproducibility analysis usually merges overlapped peaks between replicates. However, for broadly and diffusely distributed ChIP-seq signals, if overlaps are consecutively found along multiple peaks, peak merging might be biased due to largely neglecting of peak spatial information.

2.4 Normalization Strategies

Normalization is needed when the goal is to compare ChIP-seq between biological conditions or between ChIP and background control samples [30]. There are no one-fit-all strategies currently for ChIP-seq normalization. It depends on the statistical hypothesis of the investigated biological settings. For instance, the most widely used strategy is to assume the same sequencing coverage (total number of sequencing reads (Fig. 1)) between compared samples for a balanced experimental design. In this case, the number of peaks identified from the compared samples is expected to be not dramatically changed. However, protein binding does alter significantly in some biological settings such as when cofactors or dependent factors of the studied proteins are knocked out or down [31]. In this case, alternative assumption could be made that only the background signals of the compared samples hold similar coverage regardless the peak intensities. This, however, sets up dependency between peak identification and normalization procedure, which could be biased if peak detection is not properly performed for the studied datasets. Some methods focus on peak signal normalization regardless of background regions, such as those implemented in differential testing models [32, 33]. Additionally, there are other special scenarios, where strong dependency can be observed between signal-to-noise ratio and peak intensities, and such dependency differ between compared samples. In this case, additional statistical correction is needed in order to make a fair justification, such as those proposed by MAnorm methods [34, 35].

Additional normalization could be added by considering INPUT samples [30]. INPUTs are able to filter nonspecific binding and background biases during peak detection as mentioned before. When accounted in comparisons between conditions, they help eliminate false-positive differential peaks due to background differences. However, INPUT samples might also introduce unwanted biases such as library preparation effects, if differed from ChIP samples, further complicating downstream comparisons. There is no easy solution to cover all covariants in ChIP-seq normalization procedure. Currently, practical solutions usually (1) detect peaks by controlling false positives using INPUTs; (2) normalize across ChIP samples without INPUTs; and (3) validate downstream picks with other approaches like low-throughout wet experiments.

2.5 Bias and Batch Correction

Different biases and batch effects were reported in ChIP-seq data beyond background noises. While not all of them can be statistically justified (e.g., antibody efficiency), some have been well modeled and greatly appreciated in improving ChIP-seq data analysis. For example, the G and C contents on DNA fragments can affect PCR amplification during sequencing library preparation [36, 37]. It is hard to distinguish if strong protein binding at GC high or low fragments is due to real biology or the GC-induced PCR bias. Fortunately, GC content can be deterministically measured as the frequencies of the two nucleotides among all four nucleotides on sequencing fragments, with effects estimated through their associations with local read coverages (Fig. 1). Methods have showed promise on justification GC bias in ChIP-seq peak detection. For batch effect justification between samples, tools developed based on other platforms can be adopted to ChIP-seq analysis by regressing batch covariants from peak signals [38, 39]. These approaches are particularly powerful with large-scale studies, where the effects of covariants can be evaluated efficiently and robustly.

2.6 Differential Binding Analysis

Another key ChIP-seq analysis is to identify differential binding regions between biological conditions. A naïve approach is to compare the common and unique peaks between compared conditions. Basically, if a peak shows in both conditions within a defined shifted distance, it is identified as non-differential. A peak showing in only one condition is identified as differentially enriched in that condition. This approach, however, is too conservative to detect small changes where peaks are detected in both conditions with altered intensities.

Quantitative testing by considering the peak intensities provides higher sensitivity. The first step is to determine the target regions that can be quantified uniformly across ChIP-seq samples (Fig. 1). Due to peak centers and boundaries differ between individual samples, three strategies are widely used to decide target regions: (1) regions of interests based on prior knowledge, such as promoter regions of annotated genes (commonly applied in studies with active histone modifications); (2) regions that are generated by merging overlapped peaks from individual ChIP-seq samples (commonly applied for transcription factors that usually bind to promoters and enhancers); and (3) peaks called based on pooled ChIP-seq data from all samples. Pros and cons of different strategies have been discussed by literatures [40]. Once a uniform list of target regions is selected, ChIP-seq reads can be quantified at these sites with differential testing followed to determine binding differences by proteins, such as that proposed by the DiffBind tool [41]. A few statistical models are available to help report differential binding sites. For example, models proposed in differential gene expression analysis with RNA-seq can be adopted here [32, 33, 42].

It is noted that, differential binding analysis with the above strategies could be biased by selection of peak detection criteria. For instance, a q-value 0.05 in peak detection might be overconservative, thus resulting in exclusion of interesting regions and generating false negatives during differential analysis. In contrast to that, some approaches were proposed to work unbiasedly by modeling differential regions based on genome-wide fixed-width bins, either sliding or nonoverlapped windows [43]. These approaches, however, need additional justification to report results after differential testing, such as removing dependencies between neighbored windows, merging consecutive windows, and tweaking merged-region boundaries.

2.7 Peak Annotation

Peak annotation based on existing knowledge or other associated datasets is critical toward understanding the identified peaks of interests, either through peak detection or differential binding analysis. The common questions are as follows: (1) where do the peaks distribute? (2) What DNA sequence patterns underlie the peaks? (3) Which genes are associated with the peaks? To address these questions, statistical summarization or modeling is required to draw conclusions with genome-wide peak signals.

The first question can be addressed by calculating the linear distances between the identified peaks and known genes/transcripts on the DNA if the studied genome is well-annotated. For example, transcription factors are expected to be mostly located at promoters and enhancers [44]. Large distances from genes indicate transcription factor binding in intergenic regions and are lack of interpretation with existing knowledge. Distance measurements provide filters to zoom into the gene-close peaks for downstream analysis, widely implemented in histone modification analysis. The second question requires extraction of short DNA sequences (i.e., motifs) that appear frequently among identified peaks, with the null hypothesis that the frequency of appearance is no larger than background frequencies that can be estimated with simulation or random sampling [45]. Once a motif is identified based on the studied protein ChIP-seq, one can compare it with known protein motifs to predict co-binding/locating relations between proteins. It is noted that motif analysis for broad ChIP-seq signals might lack of specificity due to complicated sequence contents in wider peaks.

Answers to the third question transfer the interpretation of peaks to the knowledge of genes which are usually better annotated. The common strategy defines the closest genes to the ChIP-seq peaks as the potential regulating targets of the binding proteins. Then, statistical enrichment analysis, such as pathways and ontologies, can be performed based on the peak-linked genes. However, this might introduce ambiguity when annotating peaks located at gene dense regions. Tools have proposed improved approaches to annotate peaks with the aid of additional data such as gene expression, etc. [46].

To present annotated peaks, elegant visualization strategies have been proposed such as deepTools [47]. For example, meta peaks have been defined to represent the overall profile of genome-wide peaks by pooling peaks together aligned along the peak summits. ChIP-seq signals surrounding the peak summits are aggregated across peaks with mean and variation indicated at base-pair or genomic-window level. Plotting of meta peak signals enables intuitive comparisons of ChIP-seq signals between samples. In addition, peak coverage heatmap is highly applied in scientific researches, as they highlight both overall profiles and individual intensities for genome-wide peaks. Gene-centered heatmap or meta-region plots are also frequently presented in publications, especially for proteins and histone modifications presented along gene bodies. In brief, ChIP-seq signals are plotted based on their relative locations to gene transcription start/end sites, with all genes are scaled to the same length. In addition, genome browsers are great resources to provide views of ChIP-seq signal coverage, in supporting of both data exploration and result reporting [48, 49].

2.8 Multimodal Integration with Other Genomic Information

With high-throughput sequencing adopted as routine approaches in individual labs and the emergence of novel sequencing technologies, multimodal data integration is necessary in modern genomic studies to demonstrate findings within complex genomic context. One can annotate ChIP-seq peaks using static information of different genomic dimensions as we discussed in the previous section or jointly process and interpret sequencing data from multiple genomic dimensions that are generated simultaneously.

Joint analysis applications are usually built based on two strategies. One strategy processes data separately in each dimension and interpret data dimensions together with features of interests. For example, after separately identifications of differential binding regions from ChIP-seq and differentially expressed genes from RNA-seq, results were combined to recognize genes whose differential expression can be explained by differential protein binding [50]. Some studies interpreted the ChIP-seq signal variations across population samples using the individual allele discrepancies by correlation analysis [51]. The other common strategy chooses to model and interpret multidimensional data simultaneously

[52]. For example, features of protein binding, gene expression, and DNA mutations could be simultaneously modeled using matrix factorization methods (Fig. 1), with the goal to find latent factor variables across all data dimensions. One can further infer relations between features of different dimensions by correlation analysis based on the latent factors. These methods also maximum the clustering analysis of studied samples using multidimensional datasets.

Simultaneous modeling usually requires data reformatting or transformation to fit the model design, while separate processing with joint interpretation shows much flexibility in preprocessing datasets. Particularly, recent technique advances have brought many new data structures which challenge the existing joint models, including those measuring 3D chromatin interactions [53] to identify gene regulation associated with protein binding. Separate processing is usually the easy-to-setup option when integrating them with ChIP-seq, with the hope that simultaneous modeling could cover them in near future.

3 Summary

In this chapter, we focused on the statistical components involved in the routine questions in ChIP-seq data processing with the aim to provide a high-level overview. As summarized in Subheading 2, key processing steps were covered including QC control, peak detection, differential binding analysis, peak annotation, etc. In addition, we briefly mentioned in each subsection the special requirements regarding different types of ChIP-seq data, point-source, broad-domain, and mixed. However, the involvement of domain knowledge is highly recommended toward a successful data analysis plan, due to proteins usually hold unique characteristics which can guide downstream filtering, validation, and interpretation. There are many other statistical operations we didn't enumerate here as they are either shared techniques with other sequencing technology (e.g., RNA-seq) or designed to fit customized ChIP-seq settings. For instance, peak calling tools usually equipped with a list of type-in parameters that allow users to customize processing plans. Although default parameter settings are widely appreciated, customized parameter tuning could better accommodate protein features such as library preparation differences.

It is noted that many alternative approaches are available but not discussed here for each key processing step of ChIP-seq data. Our summarization focused on the main statistical principle underlining these approaches instead of the exact tools. These tools, built under different statistical assumption, modeling, and parameters, share similar performance and sometimes complement each other for specific data types. We only listed a selection of tools here. In practice, users usually follow a generic protocol to select the right statistical methods in each step based on the following standards. (1) How is the method performance for the studied data type? (2) Is the method well maintained and easy to use? (3) How compatible the input and output data formats between methods of different analyzing steps? Our chapter here hopefully serves as a brief guide to help users set up analysis plans.

Many tools are available for data visualization, yet in-house programming is still required to generate customized figures. For example, most tools by default plot results at a genome-wide level, which could be overwhelming and sometimes blur the signal patterns in regions of interests. The regions, such as those close to genes in critical pathways, if zoomed in, provide intuitive understanding of the key genes in cells. Again, this highlights the importance of domain knowledge in ChIP-seq data analysis.

References

1. Johnson DS, Mortazavi A, Myers RM, Wold B (2007) Genome-wide mapping of in vivo protein-DNA interactions. Science 316(5830):1497–1502. https://doi.org/10.1126/science.1141319
2. Barski A, Cuddapah S, Cui K, Roh TY, Schones DE, Wang Z, Wei G, Chepelev I, Zhao K (2007) High-resolution profiling of histone methylations in the human genome. Cell 129(4):823–837. https://doi.org/10.1016/j.cell.2007.05.009
3. Mikkelsen TS, Ku M, Jaffe DB, Issac B, Lieberman E, Giannoukos G, Alvarez P, Brockman W, Kim TK, Koche RP, Lee W, Mendenhall E, O'Donovan A, Presser A, Russ C, Xie X, Meissner A, Wernig M, Jaenisch R, Nusbaum C, Lander ES, Bernstein BE (2007) Genome-wide maps of chromatin state in pluripotent and lineage-committed cells. Nature 448(7153):553–560. https://doi.org/10.1038/nature06008
4. Rhee HS, Pugh BF (2011) Comprehensive genome-wide protein-DNA interactions detected at single-nucleotide resolution. Cell 147(6):1408–1419. https://doi.org/10.1016/j.cell.2011.11.013
5. Skene PJ, Henikoff S (2017) An efficient targeted nuclease strategy for high-resolution mapping of DNA binding sites. elife 6. https://doi.org/10.7554/eLife.21856
6. Kaya-Okur HS, Wu SJ, Codomo CA, Pledger ES, Bryson TD, Henikoff JG, Ahmad K, Henikoff S (2019) CUT&Tag for efficient epigenomic profiling of small samples and single cells. Nat Commun 10(1):1930. https://doi.org/10.1038/s41467-019-09982-5
7. Consortium EP (2012) An integrated encyclopedia of DNA elements in the human genome. Nature 489(7414):57–74. https://doi.org/10.1038/nature11247
8. Jung YL, Luquette LJ, Ho JW, Ferrari F, Tolstorukov M, Minoda A, Issner R, Epstein CB, Karpen GH, Kuroda MI, Park PJ (2014) Impact of sequencing depth in ChIP-seq experiments. Nucleic Acids Res 42(9):e74. https://doi.org/10.1093/nar/gku178
9. Davis CA, Hitz BC, Sloan CA, Chan ET, Davidson JM, Gabdank I, Hilton JA, Jain K, Baymuradov UK, Narayanan AK, Onate KC, Graham K, Miyasato SR, Dreszer TR, Strattan JS, Jolanki O, Tanaka FY, Cherry JM (2018) The encyclopedia of DNA elements (ENCODE): data portal update. Nucleic Acids Res 46(D1):D794–D801. https://doi.org/10.1093/nar/gkx1081
10. Bailey T, Krajewski P, Ladunga I, Lefebvre C, Li Q, Liu T, Madrigal P, Taslim C, Zhang J (2013) Practical guidelines for the comprehensive analysis of ChIP-seq data. PLoS Comput Biol 9(11):e1003326. https://doi.org/10.1371/journal.pcbi.1003326
11. Landt SG, Marinov GK, Kundaje A, Kheradpour P, Pauli F, Batzoglou S, Bernstein BE, Bickel P, Brown JB, Cayting P, Chen Y, DeSalvo G, Epstein C, Fisher-Aylor KI, Euskirchen G, Gerstein M, Gertz J, Hartemink AJ, Hoffman MM, Iyer VR, Jung YL, Karmakar S, Kellis M, Kharchenko PV, Li Q, Liu T, Liu XS, Ma L, Milosavljevic A, Myers RM, Park PJ, Pazin MJ, Perry MD, Raha D, Reddy TE, Rozowsky J, Shoresh N, Sidow A, Slattery M, Stamatoyannopoulos JA, Tolstorukov MY, White KP, Xi S, Farnham PJ, Lieb JD, Wold BJ, Snyder M (2012) ChIP-seq guidelines and practices of the ENCODE and modENCODE consortia. Genome Res 22(9):

1813–1831. https://doi.org/10.1101/gr.136184.111

12. Shin H, Liu T, Duan X, Zhang Y, Liu XS (2013) Computational methodology for ChIP-seq analysis. Quant Biol 1(1):54–70. https://doi.org/10.1007/s40484-013-0006-2
13. Nakato R, Shirahige K (2017) Recent advances in ChIP-seq analysis: from quality management to whole-genome annotation. Brief Bioinform 18(2):279–290. https://doi.org/10.1093/bib/bbw023
14. Nakato R, Sakata T (2021) Methods for ChIP-seq analysis: a practical workflow and advanced applications. Methods 187:44–53. https://doi.org/10.1016/j.ymeth.2020.03.005
15. Zhang Y, Liu T, Meyer CA, Eeckhoute J, Johnson DS, Bernstein BE, Nusbaum C, Myers RM, Brown M, Li W, Liu XS (2008) Model-based analysis of ChIP-Seq (MACS). Genome Biol 9(9):R137. https://doi.org/10.1186/gb-2008-9-9-r137
16. Zang C, Schones DE, Zeng C, Cui K, Zhao K, Peng W (2009) A clustering approach for identification of enriched domains from histone modification ChIP-seq data. Bioinformatics 25(15):1952–1958. https://doi.org/10.1093/bioinformatics/btp340
17. Andrews S (2010) FastQC: a quality control tool for high throughput sequence data. Available online at: http://www.bioinformatics.babraham.ac.uk/projects/fastqc
18. Chen S, Zhou Y, Chen Y, Gu J (2018) fastp: an ultra-fast all-in-one FASTQ preprocessor. Bioinformatics 34(17):i884–i890. https://doi.org/10.1093/bioinformatics/bty560
19. Diaz A, Nellore A, Song JS (2012) Chance: comprehensive software for quality control and validation of ChIP-seq data. Genome Biol 13(10):R98. https://doi.org/10.1186/gb-2012-13-10-r98
20. Schmidt D, Schwalie PC, Wilson MD, Ballester B, Goncalves A, Kutter C, Brown GD, Marshall A, Flicek P, Odom DT (2012) Waves of retrotransposon expansion remodel genome organization and CTCF binding in multiple mammalian lineages. Cell 148(1–2):335–348. https://doi.org/10.1016/j.cell.2011.11.058
21. Flensburg C, Kinkel SA, Keniry A, Blewitt ME, Oshlack A (2014) A comparison of control samples for ChIP-seq of histone modifications. Front Genet 5:329. https://doi.org/10.3389/fgene.2014.00329
22. Park PJ (2009) ChIP-seq: advantages and challenges of a maturing technology. Nat Rev Genet 10(10):669–680. https://doi.org/10.1038/nrg2641
23. Ji H, Jiang H, Ma W, Johnson DS, Myers RM, Wong WH (2008) An integrated software system for analyzing ChIP-chip and ChIP-seq data. Nat Biotechnol 26(11):1293–1300. https://doi.org/10.1038/nbt.1505
24. Rozowsky J, Euskirchen G, Auerbach RK, Zhang ZD, Gibson T, Bjornson R, Carriero N, Snyder M, Gerstein MB (2009) PeakSeq enables systematic scoring of ChIP-seq experiments relative to controls. Nat Biotechnol 27(1):66–75. https://doi.org/10.1038/nbt.1518
25. Kharchenko PV, Tolstorukov MY, Park PJ (2008) Design and analysis of ChIP-seq experiments for DNA-binding proteins. Nat Biotechnol 26(12):1351–1359. https://doi.org/10.1038/nbt.1508
26. Liang K, Keles S (2012) Normalization of ChIP-seq data with control. BMC Bioinform 13:199. https://doi.org/10.1186/1471-2105-13-199
27. Wilbanks EG, Facciotti MT (2010) Evaluation of algorithm performance in ChIP-seq peak detection. PLoS One 5(7):e11471. https://doi.org/10.1371/journal.pone.0011471
28. Li Q, Brown JB, Huang H, Bickel PJ (2011) Measuring reproducibility of high-throughput experiments. Ann Appl Stat 5(3):1752–1779. https://doi.org/10.1214/11-AOAS466
29. Irizarry RA, Warren D, Spencer F, Kim IF, Biswal S, Frank BC, Gabrielson E, Garcia JG, Geoghegan J, Germino G, Griffin C, Hilmer SC, Hoffman E, Jedlicka AE, Kawasaki E, Martinez-Murillo F, Morsberger L, Lee H, Petersen D, Quackenbush J, Scott A, Wilson M, Yang Y, Ye SQ, Yu W (2005) Multiple-laboratory comparison of microarray platforms. Nat Methods 2(5):345–350. https://doi.org/10.1038/nmeth756
30. Diaz A, Park K, Lim DA, Song JS (2012) Normalization, bias correction, and peak calling for ChIP-seq. Stat Appl Genet Mol Biol 11(3): Article 9. https://doi.org/10.1515/1544-6115.1750
31. Jin H, Kasper LH, Larson JD, Wu G, Baker SJ, Zhang J, Fan Y (2020) ChIPseqSpikeInFree: a ChIP-seq normalization approach to reveal global changes in histone modifications without spike-in. Bioinformatics 36(4):1270–1272. https://doi.org/10.1093/bioinformatics/btz720
32. Robinson MD, McCarthy DJ, Smyth GK (2010) edgeR: a bioconductor package for differential expression analysis of digital gene expression data. Bioinformatics 26(1):139–140. https://doi.org/10.1093/bioinformatics/btp616

33. Love MI, Huber W, Anders S (2014) Moderated estimation of fold change and dispersion for RNA-seq data with DESeq2. Genome Biol 15(12):550. https://doi.org/10.1186/s13059-014-0550-8
34. Shao Z, Zhang Y, Yuan GC, Orkin SH, Waxman DJ (2012) MAnorm: a robust model for quantitative comparison of ChIP-seq data sets. Genome Biol 13(3):R16. https://doi.org/10.1186/gb-2012-13-3-r16
35. Tu S, Li M, Chen H, Tan F, Xu J, Waxman DJ, Zhang Y, Shao Z (2021) MAnorm2 for quantitatively comparing groups of ChIP-seq samples. Genome Res 31(1):131–145. https://doi.org/10.1101/gr.262675.120
36. Teng M, Du D, Chen D, Irizarry RA (2021) Characterizing batch effects and binding site-specific variability in ChIP-seq data. NAR Genom Bioinform 3(4):lqab098. https://doi.org/10.1093/nargab/lqab098
37. Teng M, Irizarry RA (2017) Accounting for GC-content bias reduces systematic errors and batch effects in ChIP-seq data. Genome Res 27(11):1930–1938. https://doi.org/10.1101/gr.220673.117
38. Leek JT (2014) Svaseq: removing batch effects and other unwanted noise from sequencing data. Nucleic Acids Res 42(21). https://doi.org/10.1093/nar/gku864
39. Risso D, Ngai J, Speed TP, Dudoit S (2014) Normalization of RNA-seq data using factor analysis of control genes or samples. Nat Biotechnol 32(9):896–902. https://doi.org/10.1038/nbt.2931
40. Wu DY, Bittencourt D, Stallcup MR, Siegmund KD (2015) Identifying differential transcription factor binding in ChIP-seq. Front Genet 6:169. https://doi.org/10.3389/fgene.2015.00169
41. Ross-Innes CS, Stark R, Teschendorff AE, Holmes KA, Ali HR, Dunning MJ, Brown GD, Gojis O, Ellis IO, Green AR, Ali S, Chin SF, Palmieri C, Caldas C, Carroll JS (2012) Differential oestrogen receptor binding is associated with clinical outcome in breast cancer. Nature 481(7381):389–393. https://doi.org/10.1038/nature10730
42. Ritchie ME, Phipson B, Wu D, Hu Y, Law CW, Shi W, Smyth GK (2015) Limma powers differential expression analyses for RNA-sequencing and microarray studies. Nucleic Acids Res 43(7):e47. https://doi.org/10.1093/nar/gkv007
43. Lun AT, Smyth GK (2016) Csaw: a Bioconductor package for differential binding analysis of ChIP-seq data using sliding windows. Nucleic Acids Res 44(5):e45. https://doi.org/10.1093/nar/gkv1191
44. Gates LA, Foulds CE, O'Malley BW (2017) Histone marks in the 'Driver's seat': functional roles in steering the transcription cycle. Trends Biochem Sci 42(12):977–989. https://doi.org/10.1016/j.tibs.2017.10.004
45. Machanick P, Bailey TL (2011) MEME-ChIP: motif analysis of large DNA datasets. Bioinformatics 27(12):1696–1697. https://doi.org/10.1093/bioinformatics/btr189
46. Hariprakash JM, Ferrari F (2019) Computational biology solutions to identify enhancers-target gene pairs. Comput Struct Biotechnol J 17:821–831. https://doi.org/10.1016/j.csbj.2019.06.012
47. Ramirez F, Dundar F, Diehl S, Gruning BA, Manke T (2014) deepTools: a flexible platform for exploring deep-sequencing data. Nucleic Acids Res 42(Web Server issue):W187–W191. https://doi.org/10.1093/nar/gku365
48. Kent WJ, Sugnet CW, Furey TS, Roskin KM, Pringle TH, Zahler AM, Haussler D (2002) The human genome browser at UCSC. Genome Res 12(6):996–1006. https://doi.org/10.1101/gr.229102
49. Robinson JT, Thorvaldsdottir H, Winckler W, Guttman M, Lander ES, Getz G, Mesirov JP (2011) Integrative genomics viewer. Nat Biotechnol 29(1):24–26. https://doi.org/10.1038/nbt.1754
50. Wang C, Zhang L, Ke L, Ding W, Jiang S, Li D, Narita Y, Hou I, Liang J, Li S, Xiao H, Gottwein E, Kaye KM, Teng M, Zhao B (2020) Primary effusion lymphoma enhancer connectome links super-enhancers to dependency factors. Nat Commun 11(1):6318. https://doi.org/10.1038/s41467-020-20136-w
51. Kasowski M, Grubert F, Heffelfinger C, Hariharan M, Asabere A, Waszak SM, Habegger L, Rozowsky J, Shi M, Urban AE, Hong MY, Karczewski KJ, Huber W, Weissman SM, Gerstein MB, Korbel JO, Snyder M (2010) Variation in transcription factor binding among humans. Science 328(5975):232–235. https://doi.org/10.1126/science.1183621
52. Shen R, Mo Q, Schultz N, Seshan VE, Olshen AB, Huse J, Ladanyi M, Sander C (2012) Integrative subtype discovery in glioblastoma using iCluster. PLoS One 7(4):e35236. https://doi.org/10.1371/journal.pone.0035236
53. Mumbach MR, Rubin AJ, Flynn RA, Dai C, Khavari PA, Greenleaf WJ, Chang HY (2016) HiChIP: efficient and sensitive analysis of protein-directed genome architecture. Nat Methods 13(11):919–922. https://doi.org/10.1038/nmeth.3999

Chapter 10

Bioinformatic and Statistical Analysis of Microbiome Data

Youngchul Kim

Abstract

Since advances in next-generation sequencing (NGS) technique enabled to investigate uncultured microbiota and their genomes in unbiased manner, many microbiome researches have been reporting strong evidences for close links of microbiome to human health and disease. Bioinformatic and statistical analysis of NGS-based microbiome data are essential components in those microbiome researches to explore the complex composition of microbial community and understand the functions of community members in relation to host and environment. This chapter introduces bioinformatic analysis methods that generate taxonomy and functional feature count table along with phylogenetic tree from raw NGS microbiome data and then introduce statistical methods and machine learning approaches for analyzing the outputs of the bioinformatic analysis to infer the biodiversity of a microbial community and unravel host-microbiome association. Understanding the advantages and limitations of the analysis methods will help readers use the methods correctly in microbiome data analysis and may give a new opportunity to develop new analytic techniques for microbiome research.

Key words Microbiome, Metagenomics, 16s rRNA sequencing, Phylogeny tree, Alpha diversity, Beta diversity, Microbiome-wide association

1 Introduction

Over the course of our lives, humans are colonized by largely diverse commensal microbes including bacteria, archaea, viruses, and fungi, which comprise the human microbiome. Many research works on microbiome have shown that microbial communities (microbiota) and their collective genomes (microbiome) and metabolic products influence human health and disease including cancer, obesity, metabolic diseases, and even psychological disorders in various ways [1, 2]. Understanding human microbiome composition, function, and range of variation across different human body sites and living environments will provide a promising opportunity to promote health and prevent disease through the discovery and development of biomarkers for prevention, assessment, treatment, and management of human diseases.

Brooke L. Fridley and Xuefeng Wang (eds.), *Statistical Genomics*,
Methods in Molecular Biology, vol. 2629, https://doi.org/10.1007/978-1-0716-2986-4_10,

A traditional method to investigate microbial composition in an ecosystem is to sequence DNAs of microbiological cultures at a web-lab. However, this method is very labor intensive, and therefore a large proportion of microorganisms still cannot be cultured to date. In contrast, culture-independent low-cost high-throughput next-generation sequencing (NGS) technique enables to efficiently detect even unculturable microorganisms and infer their functions by sequencing all of their DNA and RNA sequence fragments in a given sample or environment.

Two NGS approaches have been broadly used for a number of microbiome researches. One is a phylogenetic marker-gene amplicon sequencing approach that amplifies specific regions of a highly conserved gene, e.g., 16S ribosomal RNA (rRNA) gene, providing species-specific signatures and then sequences the regions to profile a taxonomic composition of a microbial community in an ecosystem. The other approach is a whole metagenome shotgun sequencing that sequences all of DNAs within a sample and catalogues the composition of microbial organisms as well as their functional capability [3]. Both approaches have their own advantages and limitations. In brief, the marker-gene sequencing method is known to be cost-effective and provides a fast identification of unknown bacteria or microorganism. However, it is still laborious approach and has a possibility of introducing a polymerase chain reactin (PCR) bias in taxa identification because species are not equally amplified in the PCR experiment process during a library preparation despite of the use of well-designed primers. Furthermore, its taxonomic resolution is limited to genera or a part of species due to the conservation of the marker gene. In contrast, the whole shotgun metagenomic sequencing is an unbiased approach and provides not only a taxonomic profile at strain-level resolution but also capture profiles of gene function entities. However, it often struggles with a host DNA contamination. That is, microbial DNA sequences remaining after a bioinformatic decontamination analysis may not provide an enough sequencing depth to profile a whole microbial community, especially rare species. Therefore, it requires relatively more sequence reads within each sample to detect unique taxonomic identifiers, and thus the cost of the metagenomic sequencing is still high. Lastly, whole shotgun metagenome sequencing data analysis is more complex and computationally intensive than marker-gene sequencing data analysis.

In human health and biomedical research on microbiome, bioinformatic and statistical analyses of massive high-dimensional microbiome data generated by either a targeted marker-gene sequencing or whole metagenome shotgun sequencing are a key component to explore and understand how human microbiome distributes and also unravels their roles and association in relation to human health and disease. In this chapter, we introduce bioinformatic analysis pipelines and statistical analysis methods of NGS-based microbiome data from a quality control of raw sequencing data to a microbiome-wide association and biomarker discovery analysis.

2 Datasets Used to Illustrate the Methods

Oropharyngeal cancer (OPC) is a disease in which malignant cancer cells form in the tissues of the oropharynx that is the middle part of throat behind the mouth. The incidence of OPC continues to increase dramatically in the USA. Smoking and human papillomavirus are known risk factors of OPC. Accumulating evidence suggests that oral microbiota (communities of microorganisms that reside in the oral cavity) may influence cancer treatment-related toxicities and that microbiome profiling may hold promise as a prognostic biomarker of treatment-related toxicities. To explore the composition and diversity of oral microbiota in OPC patients prior to treatment and evaluate the associations of oral microbiota with development of oral mucositis and candidiasis after chemoradiation or radiation treatment, 16S rRNA marker-gene sequencing data were generated from mouthwash-based oral gargles of OPC patients before receiving the treatment at H. Lee Moffitt Cancer Center and Research Institute [4]. An operational taxonomic unit (OTU) table was generated using QIIME bioinformatic tool [5], and Ribosomal Database Project (RDP) classifier was used to assign a phylogenetic taxonomy to individual OTUs [6].

3 Bioinformatic and Statistical Methods for Microbiome Data Analysis

3.1 Overview of Bioinformatic Pipeline for Raw Sequencing Data Analysis

Raw microbiome sequencing data are subject to a computationally intensive bioinformatic downstream analysis. Figure 1 depicts an overall overview of the bioinformatic analysis pipeline, and Table 1 summarizes bioinformatic tools and softwares available for each step of the pipeline. In practice, bioinformatic pipelines for analyzing marker-gene sequencing data commonly begins with a quality control analysis of the raw sequencing data and generates representative consensus sequences in terms of operational taxonomic units (OTUs) or amplicon sequencing variants (ASVs). Finally, taxonomic classification and phylogenetic tree inference are performed with the consensus sequences. QIIME and Mothur are two well-known bioinformatic suite tools for analyzing raw marker-gene sequencing data. For metagenomic sequencing data analysis, bioinformatic analysis pipeline also begins with quality control and needs a host DNA decontamination process. Finally, reference mapping-based taxonomic classification or de novo genome assembly/binning for gene prediction and functional annotation analysis is performed depending on the goal of study.

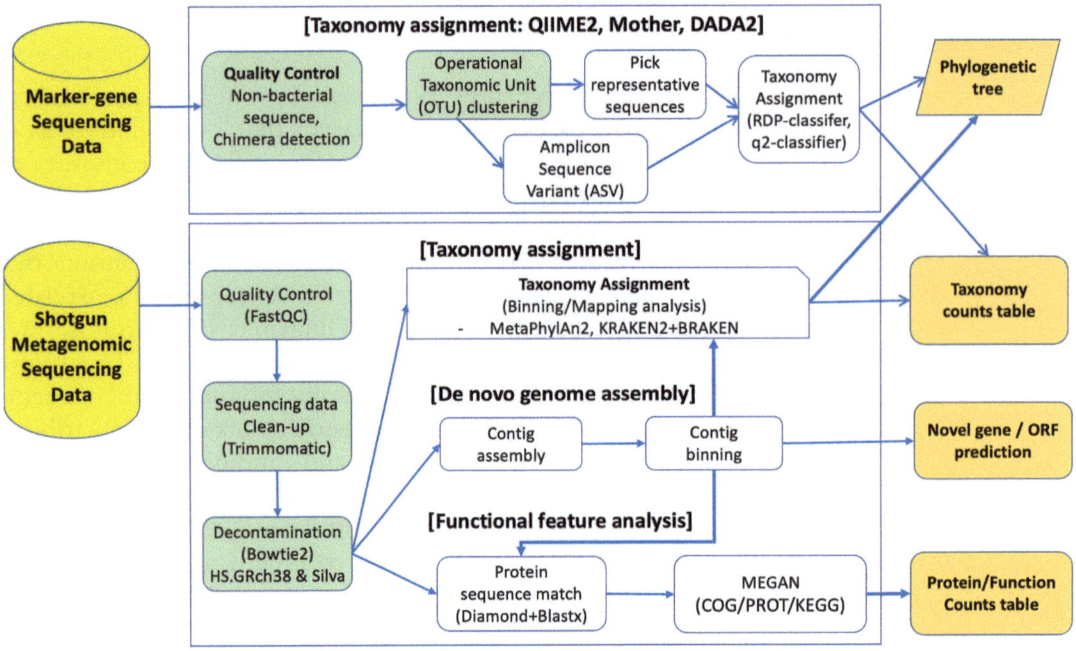

Fig. 1 Bioinformatic analysis procedures

3.2 Bioinformatic Analysis of Marker-Gene Sequencing Data

3.2.1 Sequencing Error Control and Variant Call

The first step in marker-gene amplicon sequencing data analysis is to reduce the effect of PCR amplification and sequencing errors or artifacts. A traditional and popular approach is to cluster similar sequences into single features so-called operational taxonomic units (OTUs) based on a predefined identity threshold, commonly 97%, into OTUs [7]. This clustering approach merges sequence variants including those that may occur due to sequencing errors into a single OTU for a downstream analysis. However, the OTU clustering approach may lose ecologically important sequence variations between microbes. As an alternative approach, oligotyping method distinguishes closely related but biologically distinct sequences with a substantial nucleotide variation by calculating Shannon entropy at nucleotide sites with high variation [8]. Recent advanced approach to addressing the sequencing error is to determine exact marker-gene sequence variants like single-nucleotide polymorphisms by statistically modeling sequence error profiles. DADA2, Deblur, and UNOISE are representative methods. DADA2 records the mismatches between every sequences among all samples and then probabilistically models nucleotide transition rates at which an amplicon read with one original sequence is produced from another sample sequence [9]. Deblur method computationally models a sequencing error profile per each sample as a function of the Hamming distance between sequences and then applies to the estimated distribution of the error profile a predefined upper bound that was selected from multiple validation datasets with known spikes to minimize false-positive sequences [10]. The resulting output from

Table 1
Bioinformatic analysis of raw NGS data for microbiome study

	Goal	Approach	Software and reference
Marker-gene sequencing data analysis	Sequencing error control and variant call	Read quality evaluation	FastQC
		Adapter removal/low-quality read trimming	BBTools [22]
		Clustering of sequence into OTUs	UCLUST [5] Average linkage in Mothur [87] UPARSE [7]
		Oligotyping	Oligotyping [8]
		Sequence error model	R DADA2 [9] QIIME deblur [10]
	Taxonomy assignment	Sequence alignment	BLAST [12] USEARCH [13] VSEARCH [88]
		Machine learning	RDP-classifier [6] QIIME: q2-feature classifier [14]
	Phylogenetic tree	De novo tree construction	*MAFFT* [16], *FastTree* [17] *phangorn* [18]
		Placement-based approach	PAGAN [19], pplacer [20]
Shotgun metagenomic sequencing data analysis	Quality control and host decontamination	Read quality evaluation	FastQC
		Adapter removal/low-quality read trimming	BBTools [22]
		Host genome decontamination	Trimmomatic [21]
	Reference-based analysis	Taxonomic identification	MetaPhlAn [23] Kraken [24] + Braken [25]
		Gene/function annotation	Diamond [89] +MEGAN [89] HUMAnN2 [27]
	De novo assembly analysis	Contig assembly	MetaSPAdes, MEGAHIT
		Quality assessment of assembled genomes	MetaQUAST
		Contig binning	MaxBin [90], MetaBAT [91], CONCOCT
		Evaluate quality of bins for completeness/ contamination	CheckM [92], binrefiner
		Gene prediction from assembled genomes	Glimmer, Prodigal [30]

either DADA2 or Deblur is exact-sequence variant table that consists of the count of each sequence variant observed in each sample. The sequence variants are called amplicon sequence variants (ASVs) in DADA2 and sub-OTUs in Deblur method. Hereafter, we will just call them ASVs. To note, OTU approach is typically limited to genera-level taxonomy. When a microbiome study needs a higher taxonomy resolution beyond genus, ASV approaches are recommended because they have the potential to account species- and strain-level variations of marker genes [11].

3.2.2 Taxonomic Classification

The next important step is taxonomic classification analysis that assigns taxonomic names to error-free representative DNA sequences of OTUs or ASVs. Taxonomy is typically assigned by using sequence alignment search tool or machine learning approaches for taxonomy classification. The former is represented by BLAST that compares individual sequence against a well-annotated microbial reference sequence database such as Greengenes, RDP, and Silva [12]. Similarly, USEARCH and VSEARCH tools also assign the taxonomy to a sequence by searching for local and global sequence alignments and known to be faster than BLAST [13]. However, this alignment search approach yields a high specificity but often suffers a poor sensitivity because there are many unknown taxa in the reference databases. By contrast, the latter approach trains machine-learning classifiers for taxonomy classification and use the classifiers to predict taxonomy of an input sequence. For example, Wang et al. (2007) built the machine learning classifier, the so-called *RDP-classifier*, that provides 16S rRNA sequence with taxonomic assignment from domain to genus which was trained on oligonucleotide frequencies at the genus level by using naïve Bayes algorithm and achieves around 80% accuracy in genus-level assignments [6]. In addition, the q2-feature-classifier implemented in QIIME tool uses multiple machine learning classifiers available in scikit-learn for marker-gene taxonomy classification [14].

3.2.3 Phylogenetic Tree Construction

Marker-gene sequence data can be used to construct a phylogenetic tree of taxa that depicts the lines of evolutionary descent of different species or genes from their common ancestor or root. In particular, the phylogenetic tree is crucial for interpreting evolutionary relationship among different species and understanding the biodiversity of microbial community in an ecosystem [15]. In microbiome study based on NGS technology, a phylogenetic tree can be built based on distances between DNA sequences of taxa or genes. Sequence distance-based tree-building methods can be divided into (1) de novo tree construction method that infers the phylogenetic tree using only data on sequences and (2) phylogenetic-placement or reference-based method that place the query sequence of an unknown or newly cultured organism onto a pre-existing reference tree of known species. For the de novo tree construction, QIIME tool first uses MAFFT algorithm to make a multiple sequence alignment for all taxa sequences and then masks phylogenetically uninformative or ambiguously aligned sequences [16]. Next, FastTree algorithm is used to build a large-scale phylogenetic tree by the use of a hierarchical neighbor-joining method [17]. The R phangorn package also offers ungma() function to build a de novo phylogenetic tree by the use of UPGMA method [18]. The UPGMA method assumes the same evolutionary speed for all lineages, meaning that all leaves have the same distance from the root. For the phylogenetic-placement method, PAGAN and Pplacer tools are broadly used. In brief, PAGAN uses partial-order graphs for phylogenetic alignment and placement of sequences [19]. Pplacer calculates the posterior probability of a placement on an edge to statistically quantify uncertainty on an edge-by-edge basis [20]. Of note, the de novo tree method is indeed more widely used in NGS-based microbiome studies than the phylogenetic-placement method because it does not need a reference tree and also have the potential to reveal a new taxonomic branch from unclassified DNA sequences.

3.3 Bioinformatic Analysis of Metagenome Shotgun Sequencing Data

Shotgun metagenomic sequencing approach enables investigators to comprehensively catalogue all genes in diverse microbial organisms such as bacteria, archaea, fungi, and viruses present in a sample or an environment. A series of bioinformatic analysis of raw shotgun sequencing data is essential to evaluate the microbial abundance, biodiversity, and the functional profile of genes of microbial organisms in various environments.

Depending on the goal of the microbiome study, bioinformatic analysis of raw sequencing data can be divided into reference-based analysis and de novo genome assembly-based analysis. Briefly, the reference-based analysis aims to obtain the abundance table of taxa or genes. It aligns DNA sequence reads onto reference genome databases such as NCBI Refseq and SILVA database to obtain taxonomic catalogue. Similarly, protein-transformed sequences can be queried in NCBI nonredundant protein database to detect genes of the microbial communities. The de novo assembly-based analysis is used to seek for a novel genome that is not present in the reference genome database. It first builds a metagenome assembly from decontaminated sequenced reads and use assembled genomes to profile taxonomic and function abundance or predict a novel microorganism or gene function.

Both analyses will finally result in a taxonomic abundance table as well as a gene/metabolic function table. The number of bioinformatic tools and algorithms available for metagenomic sequencing data analysis is rapidly growing. In following subsections, we explain a bioinformatic analysis workflow of shotgun metagenomic sequencing data as a step-by-step guide and introduce popular bioinformatic tools and algorithms at each step.

3.3.1 Quality Control and Decontamination

Quality control is an essential prerequisite that involves quality trimming and contamination removal from raw sequence reads. Raw sequence reads with poor quality, e.g., Phred quality score of <20, and adapter sequences can be trimmed using Trimomatic [21] and BBTools [22]. Above all, because metagenomic sequencing data contains a high proportion of host-associated sequence reads, it is necessary to remove them prior to the reference-based analysis or de novo assembly analysis. Burrows-Wheeler Aligner (BWA) and Bowtie2 software are widely used for such a decontaminate analysis by mapping quality-trimmed reads to a host reference genome. KneadData is a convenient batch tool that internally integrates FastQC, Trimmomatic, and Bowtie2 to facilitate from quality trimming of raw sequencing data to host decontamination in a single run (https://huttenhower.sph.harvard.edu/kneaddata/).

3.3.2 Reference-Based Taxonomy Identification

The assignment of sequence to a taxonomy is critical to understand microbiome composition and diversity in a sample or environment. MetaPhlAn and Kraken are two recommended tools for taxonomy identification of metagenome sequencing data. MetaPhlAn is a computational tool to profile the composition of microbial communities and generate relative abundance of microbial organisms from metagenomic shotgun sequencing data. MetaPhlAn uses its own clade-specific marker genes that were identified from over 17,000 reference genomes [23]. While MetaPhlAn generates a species-level data for a sample, its family tool, StrainPhlAn, can be further used to construct the sample-specific strains of all species present in the sample. Kraken uses a k-mer-based approach and provides a fast and accurate taxonomic classification of metagenomic DNA sequences [24]. It relies on exact *k*-mer matches between lowest-common ancestor (LCA) of every taxa. Kraken provides a pre-built-in reference database and also allows to build a custom database. Optionally, taxonomic assignment outputs of Kraken can be reanalyzed using Braken that computes a Bayesian estimates of the abundance of genus or species when a sample contains multiple near-identical species [25].

3.3.3 Reference-Based Functional Classification

Functional classification of metagenomic sequencing data is an important step for investigating the functional and metabolic roles of microbial organisms in a sample or ecosystem. DIAMOND+MEGAN and HUMAnN are popular bioinformatic pipelines to profile the abundance of gene families and pathways from metagenomic sequencing data. DIAMOND+Megan pipeline uses a pure translated search strategy. Metagenomic DNA sequence reads are aligned using DIAMOND tool against a comprehensive reference protein database such as NCBI nonredundant (nr) protein database [26]. The resulted sequence alignment output is further analyzed using MEGANIZER program that is a part of MEGAN package and extract gene functions and metabolic pathways on the basis of EggNOG, InterPro, SEED, or KEGG database. The graphical user interface of MEGAN is then used to view the output of MEGANIZER as the feature-by-sample table and also provides a variety of interactive graphical visualizations of the table [27]. HUMAnN2 uses a three-tiered search strategy to integrate taxonomic information with microbial community functional profiles [28]. The first tier is to identify a community's known species using MetaphlAn pipeline. In the second tier, sequence reads are mapped to relevant pangenomes of the identified species. Finally, it performs translated search on unclassified reads against a reference database, e.g., UniRef database [29] by default.

3.3.4 De Novo Metagenomic Assembly Analysis

De novo metagenomic assembly analysis mainly aims to explore unknown microorganisms, predict new genes and multigene biosynthetic, or reconstruct metabolic pathways. It involves a more comprehensive bioinformatic analysis than the reference-based analysis and consists of the following three steps: (1) contig assembly, (2) contig binning, and (3) gene prediction steps. The contig assembly step is to assemble the short sequence reads into longer sequences called contigs. metaSPAdes and MEGAHIT are recommended metagenome assemblers for this step. It is a good practice to check the quality of assembly results by the use of MetaQUAST tool. The contig binning step sorts and groups contigs from the same or closely related microorganisms by examining their nucleotide composition, sequence similarity, and abundance patterns across samples. Recommended binning tools are MaxBin, MetaBAT, and CONCOCT. CheckM and Binning_refiner tools are then used to evaluate the quality of the binned and assembled genomes. Taxonomic profiles and novel genome identification from the assembled genomes could be made using the tools for the reference-based analysis such as Kraken and DIAMOND tools. At the gene prediction step, Glimmer and Prodigal algorithms are popularly used, and both showed a high accuracy (>97%) in detecting known protein-coding open reading frames [30]. Ideally, the result of a de novo metagenome assembly analysis should be a set of genomes of all the species in a given sample. However, in practice, it is challenging to achieve the aim because a high microbial diversity, low coverage, and the presence of closely related strains in metagenomic samples yield fragmented assemblies and can obscure taxa from downstream analyses [31].

3.4 Statistical Analysis of Microbiome Data

The main goal of microbiome study is to characterize the composition of microbial community and understand their roles and mechanisms in host and environments. Statistical analysis of microbiome data is essential to achieve the goal by not only exploring the composition and structure of the microbial community through graphical visualizations and descriptive summaries but also testing statistical hypotheses related to research objectives. Although

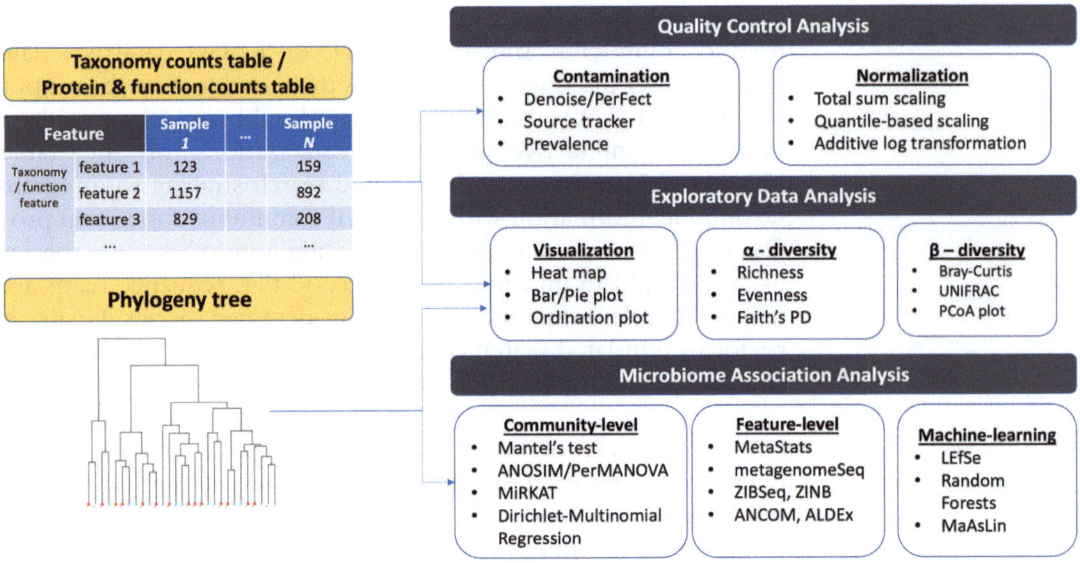

Fig. 2 Overview of statistical analysis of microbiome data

classical statistical methods for analyzing quantitative data such as t-test, analysis of variance (ANOVA), and linear regression model can be used to analyze data on the abundances of microbiota, the data often violates the assumptions underlying those methods about distribution and independence of measurements. For instance, the number of taxonomic or functional features in microbiome data is much greater than the number of samples so that a standard multivariable linear regression model cannot be fitted to relate all the features as independent variables to a host phenotype as a response variable. In addition, rare microorganisms contribute many zero counts to the taxa abundance data despite a large depth of sequencing coverage, and thus the distribution of abundances of a feature often deviates from a normal distribution that t-test and ANOVA F-test assume for a comparative analysis of the abundance. Therefore, there have been huge efforts to develop statistical methods and tools for appropriately analyzing the microbiome data. In this section, we first describe the structure and property of microbiome data and then introduce statistical analysis methods for examining the quality of microbiome data, exploring the composition and variation of microbial features, and analyzing the association of a whole microbial community and individual features with host phenotypes or environmental variables. Figure 2 shows an overview of statistical analysis of microbiome data.

3.4.1 Structure of Microbiome Data

No matter what kind of next-generation sequencing technique is used, from a statistical point of view, the microbiome data obtained from a series of bioinformatic analyses of raw sequencing data is made up of a high-dimensional "feature-by-sample" or "sample-

	S01	S02	S03	S04	S05	S06	S07	S08	S09	S10	S11
Selenomonas	0	4	0	0	0	0	0	1	0	0	0
Streptococcus.2	0	0	780	3002	0	0	0	29	55	0	1
Streptococcus.3	0	0	0	0	0	0	183	0	0	0	0
Catonella	216	0	0	13	0	336	2	37	33	114	18
Prevotella.12	176	0	0	18	0	0	484	0	0	215	1
Streptococcus.4	281	611	0	60	0	0	394	0	1	303	0
Streptococcus.5	0	0	0	0	0	1	0	0	5183	0	0
Fusobacterium.3	0	0	0	0	0	1873	1	0	0	0	0
Dialister	75	0	0	14	0	0	296	0	271	0	0
Veillonella.1	0	0	1	1925	1	0	34	0	0	0	0
Leptotrichia.2	1319	0	0	2	0	302	0	8	6	0	0
X1.68	1	0	0	0	0	0	0	0	0	1	0
Prevotella.13	0	0	0	1817	0	0	9	0	1	0	0
Bulleidia	0	0	0	26	0	36	0	0	0	0	0
X.Prevotella..2	0	0	0	0	0	0	0	0	398	0	0
Gemella.1	122	0	312	17	0	0	89	88	106	0	0
Prevotella.14	0	0	0	340	0	0	0	0	1048	0	1309
Prevotella.15	0	289	0	0	0	0	0	0	0	0	0
Veillonella.2	2538	0	0	0	0	58	3	40	14	135	0
Butyrivibrio	0	2971	4	112	0	0	13	0	0	0	19

Fig. 3 Feature-by-sample OTU count data

by-feature" contingency table. Figure 3 shows a part of the bacteria count table at genus rank in oral microbiome data from the example OPC study. The rows of this feature-by-sample table represent the taxonomic features (genus) that can be OTUs, ASVs, proteins, metabolic pathways, etc. The columns correspond to hosts, sites, or environments from which microbiome sequencing samples are collected. The cells show the count of a given taxonomic feature in a given sample. A count-based taxonomic abundance table or functional abundance table will be subject to a downstream statistical analysis to summarize the biodiversity of a microbial community or to explore correlation relationships and the difference in distribution of individual feature or a whole community in relation to phenotypes or clinical characteristics of samples. In addition to the contingency count table, a phylogenetic tree of taxonomic features is a unique feature of microbiome data and can be analyzed to visualize the evolutionary relationships and diversity of microbial organisms.

3.4.2 Property of Microbiome Data

Among microbiome data types, a feature count table typically has a larger number of taxonomic or functional features than the number of samples. When these multiple features are subjected to a statistical hypothesis test for association with environment variables or host phenotypes, multiple testing correction is necessary to reduce false-positive findings. In addition, fundamentally rare taxa lead to a high proportion of zero values in taxa count table as shown in Fig. 3 mainly because of high taxa diversity or a lack of sequencing depth. In addition, those zero values may occur in the process of filtering sequencing error. Therefore, it is hard to know whether zero abundance actually means the absence of taxon in environment or lack of sequencing depth. Next, the abundance of a taxonomic feature in a sample depends on the library size (sequencing depth) of the sample. To account for different sequencing depths among samples, the feature count table needs to be transformed using a fraction-based relative abundance or logarithmic abundance depending on assumptions of statistical analysis methods. In particular, when the relative abundance is calculated as a part of data normalization, it forms a compositional data with the constraint that the sum of feature abundances in a sample is 1. That is, relative abundances of features are dependent to each other in the sample under the constraint. This dependency motivated the development of statistical analysis methods specific to microbiome data such as ANCOM that we will introduce later in the microbiome-wide association analysis section (*see* Subheading 3.4.12). In summary, microbiome data is high-dimensional, sparse with a large proportion of zero counts, and compositional.

3.4.3 Quality Control of Microbiome Data

Sample contamination, sequencing errors, and artifacts in marker-gene and metagenomic sequencing experiment affect badly the accuracy of microbial feature detection in microbiome studies. Although a bioinformatic quality control analysis can detect and remove contaminants, misclassified taxonomic features may remain as rare features with low prevalence in the feature count table. Simply filtering features with low prevalence is widely used to improve the quality of microbiome analysis dataset and reduce the data complexity by removing low-quality features. For instance, taxonomic features observed in less than 1% of samples were removed in the Human Microbiome Project (https://hmpdacc.org/HMQCP/). Cao et al. (2021) [32] assessed how this simple and intuitive filtering procedure influences a downstream analysis by analyzing the mock community data with known taxonomic composition. It showed that the simple filtering procedure reduces the degree of sparsity of microbiome data and removes correctly contaminant features while preserving the integrity of the microbiome data in biodiversity estimation and microbiome association analysis. However, in real practice, this approach needs to define a prevalence criterion that is subjective and often hard to justify. In addition, an inadequate threshold could lead to loss of rare but truly meaningful features. Therefore, there have been huge efforts to develop statistical analysis methods and tools for detecting

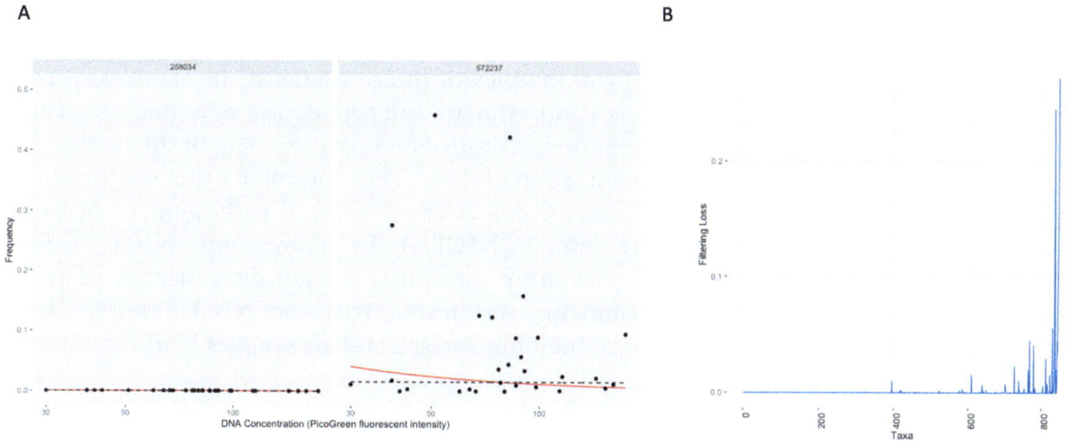

Fig. 4 Statistical quality control analysis of microbiome data

contaminants. For example, Davis et al. (2017) [33] introduced the R decontam package with two statistical classification methods that examine two well-known patterns of external contamination: (1) the frequency method assuming that contaminants appear more frequently in samples with a low sample DNA concentration and (2) the prevalence method assuming that contaminants are often observed with higher prevalence in negative control samples in comparison to true experimental samples. Figure 4a shows frequencies of a non-contaminant feature (left) and a contaminant feature (right) detected by the decontam package. Of note, the prevalence method was not applicable to the OPC data because the study experiment does not contain negative control samples.

Sminonove et al. (2019) [34] also developed a statistical test, PERFect, to decide the number of features to be filtered out before a downstream analysis. Briefly, PERFect measures the degree of contribution of individual feature to the total covariance of taxa abundance that is a basis of the biological network of taxa. Finally, it quantifies the chance of the loss of contribution of taxa due to randomness and finally filters out taxa with insignificant contribution to the total covariance [34]. Figure 4b displays the result of PERFect test with OPC example data. Taxa at x-axis are sorted by their prevalence in ascending order, and y-axis represents the amount of contribution of individual feature. The result suggests that around 400 features can be removed from OPC data and that the amount of contribution starts to increase dramatically from around 700th prevalent feature.

Lastly, the R sourcetracker package along with QIIME tool can also be used to identify the proportion of contaminant taxa in a sample of marker-gene sequencing data when additional samples can be obtained from known or suspected sources of contaminants during environmental sampling and sample processing steps such as laboratory desk surface, water, soil, and experimenter's skin samples. The sourcetracker method matches the observed taxonomic feature table from samples of interest against the samples of contaminant sources and estimate the proportion of contaminant taxa by fitting a mixture distribution of taxa onto data of the sample and the contaminant sources [35]. For example, the sourcetracker method was used to detect the source of pathogens in neonatal intensive care units (NICUs) by comparing NICU-derived sequences to previously published marker-gene studies of other indoor environments and found that the NICU samples were similar to typical building surface and air samples [36].

3.4.4 Normalization of Microbiome Data

A major goal of microbiome study from a statistical viewpoint is to infer microbial feature abundance in the environment of interest from a random sample that is just a part of the original environment and ecosystem. Especially, many microbiome researchers are interested in identifying specific taxonomic feature with significantly differential abundance between different environments or sample phenotypes such as cancer patients and healthy subjects. As previously mentioned, library sizes of microbiome sequencing data vary across samples. In general, more microbial species and functional features are detected with more sequenced reads so that an observed difference in feature abundance table may reflect not a true biological difference but just a difference in efficiency of the sequencing process. Figure 5 illustrates a toy example where two ecosystems are composed of two microbial species. The red species has equal abundance, and the blue species does not between the two ecosystems. However, when samples are collected with

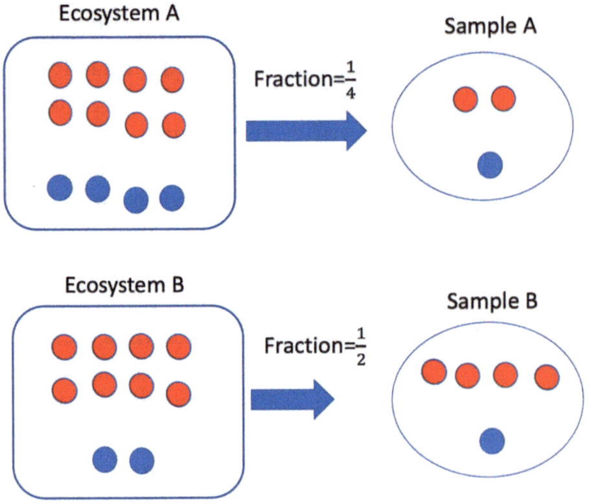

Fig. 5 Impact of sampling efficiency on differential abundance and statistical decision

Table 2
Normalization method of feature count data

Normalization method	Usage	Software	References
Rarefaction	Biodiversity estimation and comparison	R vegan QIIME, Mothur	Hurlbert et al. [93]
Total sum scaling	Differential abundance analysis of feature count data	R phyloseq R prop.table()	
Cumulative sum scaling	Differential abundance analysis of feature count data	R metagenomeSeq	Paulson et al. [38]
Additive log-ratio transformation	Differential abundance analysis of compositional data	R compositions	Aitchison et al. [39]
Centered log-ratio transformation	Differential abundance analysis of compositional data	R compositions	Aitchison et al. [39]

different sampling efficiency from the two ecosystems, the red species has a different abundance between the two samples (false-positive finding), while the blue species has an equal abundance (false negative).

To avoid those false findings, microbiome data often undergo a normalization analysis which is a process of transforming the data for an accurate comparison of statistics from different measurements by eliminating unexpected artifactual biases which are likely due to technological variation rather than from true biological difference. After an adequate normalization, data from different samples can be compared to each other. And ordination analysis such as principal component analysis can be performed with normalized data to visualize how different bacterial communities are across samples. Table 2 presents a summary of normalization methods that are commonly used for normalization of microbiome data. Note that different normalizations can be used separately to the same data depending on research questions and downstream analysis methods.

First, rarefaction is a normalization technique developed in ecology research area. It ensures that total sum of reads are the same across all samples by randomly sampling equal number of sequenced reads per sample. Rarefying samples is currently the standard method when inferring the biodiversity of microbial community. In rarefaction analysis, a rarefaction curve is constructed to assess species richness change from results of sampling at different efficiency. More detailed description about the rarefaction curve is provided in Subheading 3.4.6. Weiss et al. (2017) evaluated the impact of rarefication on differential abundance analysis and concluded that rarefying microbiome data does not increase the false discovery rates of several differential abundance test methods but potentially reduces statistical power depending on how much data is eliminated [37].

Scaling-based normalization techniques are also widely used to arrange a count table for a differential abundance analysis, and it does not require any random sampling of sequenced reads. Total sum scaling (TSS) is the most common scaling technique for normalizing various types of NGS data, especially RNA-seq platform data. It simply divides feature read counts by the total number of reads in each sample to remove a technical bias related to different sequencing depths across sample libraries. TSS has been shown to bias differential abundance estimates in RNA-seq data because a few taxa or genes are sampled preferentially as sequencing yield increases, and these measurements therefore have an unexpected influence on normalized counts. To address the disadvantage of TSS method, cumulative sum scaling (CSS) was developed [38]. CSS normalization method divides raw feature counts by the cumulative sum of feature counts in a range between the minimum feature count and a certain percentile. It assumes that feature counts are derived from the same distribution with the range. The percentile is determined by examining change points of the distributions of cumulative sum of raw counts across all samples, and it chooses the percentile before which count distribution becomes relatively stable and invariable across samples.

Additive log-ratio (ALR) normalization method was introduced to accounts for the compositional property of microbiome data [39]. It first selects an arbitrary feature as the reference and divide abundance of all other features by the reference feature abundance and then takes the natural logarithm. Suppose that microbiome relative abundance data have D features in the standard simplex (simplicial space) with sum-to-one constraint. ALR transformation reduces the dimension of features to $D - 1$ in the Euclidean space.

$$\mathrm{ALR}(x) = \left[\log\frac{x_1}{x_D}, \log\frac{x_2}{x_D}, \cdots, \log\frac{x_{D-1}}{x_D}\right].$$

where x_i is the count of the i-th feature in an environment. However, interpretation of ALR normalized data depends on the selected reference taxa. To avoid this difficulty, centered log-ratio transformation (CLR) normalization can be alternatively used. CLR normalization divides relative abundances of taxa in a community by the geometric mean of the relative abundances of all taxa as in the following formula.

$$\mathrm{CLR}(x) = \left[\log\frac{x_1}{g(x)}, \log\frac{x_2}{g(x)_D}, \cdots, \log\frac{x_{D-1}}{g(x)}, \log\frac{x_D}{g(x)} \text{ where } g(x) = \sqrt[D]{x_1 x_2 \cdots x_{D-1} x_D}\right.$$

Because these log-transformation methods are not applicable to features with zero abundance that are often over half of feature counts in microbiome data, it is used to replace zeros with a small value that is called a pseudocount [40].

3.4.5 Exploratory Analysis of Microbiome Data

Two primary research questions in microbiome study are about (1) what microorganisms reside in the host or environment and (2) what they are doing, i.e., function, in relation to the environment characteristics or host phenotypes. The first basic step to seek for a good answer to these questions will be to explore how microbial community and its members under study look like by a graphical visualization of microbiome data and examining descriptive summary statistics of data distribution. Graphical visualization of microbiome sequencing data is very useful in understanding overall structure and property of an entire microbial community as well as individual features in terms of species richness, abundance, and prevalence. Descriptive statistics are also necessary for understanding overall composition of microbiome community and controlling the quality of microbiome data samples. There are various graphs and programming tools to facilitate a graphical visualization of microbiome data. Here we introduce a few useful graphs and R packages.

First, heat map is a graphical representation of data matrix by the use of color contrast to depict quantitative values and can be used to provide a comprehensive view of high-dimensional microbiome community in multiple samples. The "heat map" function in the R ComplexHeatmap package [41] generates a heat map of microbial abundance with annotation legends on sample phenotypes and/or environmental variables. Figure 6 is a heat map of the example OPC study data at genus taxonomy rank. Rows are taxa,

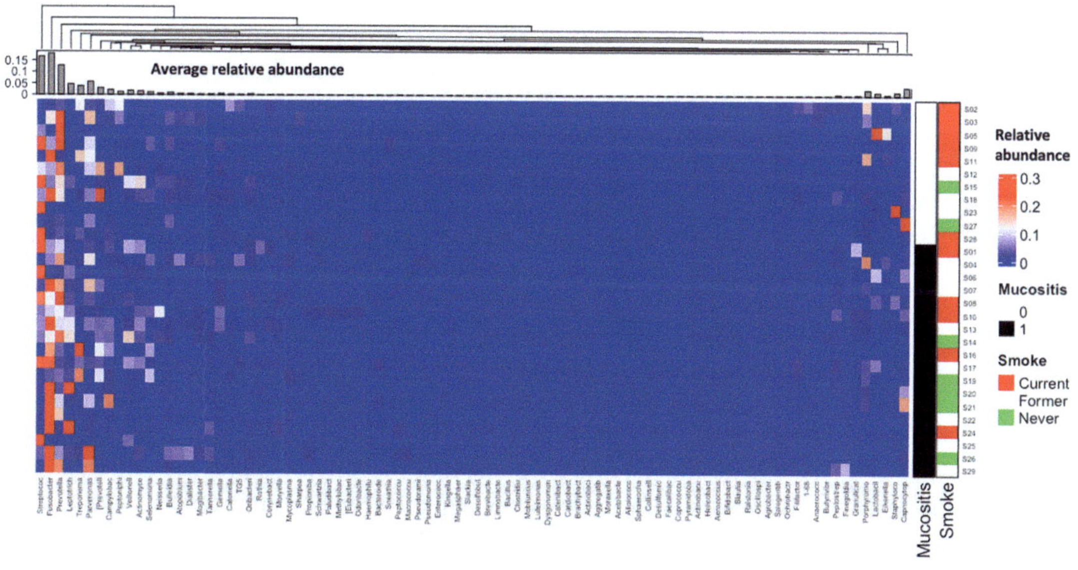

Fig. 6 Heat map of taxa abundance

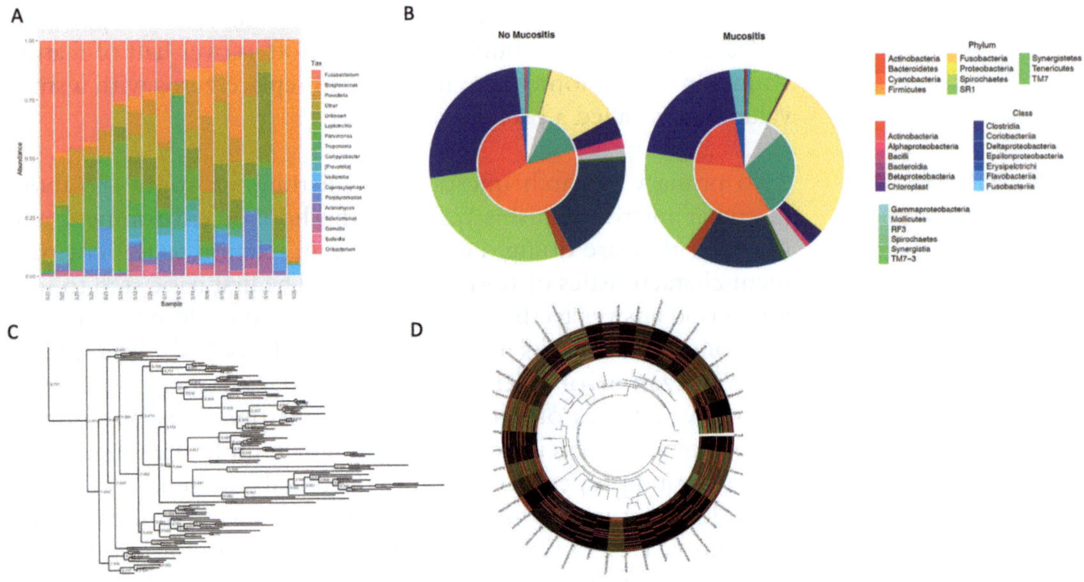

Fig. 7 Bar and pie chart and phylogeny tree

columns are samples, and a cell represents the relative abundance (proportion) of a taxonomic unit in the corresponding sample. This heat map shows a large proportion of blue cells that are zero abundance, which apparently describes a sparse property of microbiome data.

Second, bar chart and pie chart are also widely used in many microbiome studies to depict a whole microbial community composition. The "plot_bar()" function in R phyloseq package [42] for reproducible interactive analysis and graphics of microbiome data is powerful and flexible to quickly create informative summary graphics of the differences in taxa abundance across a set of samples in a microbiome study as shown in Fig. 7a. Pie chart is useful to view and compare relative abundances as a whole at higher taxonomic ranks such as phylum and class level. The "plot_pie()" function in the R psadd package draws a pie chart of relative abundance of microbiome community at phylum and class ranks at default. Figure 7b is an example of double-layered pie charts based on average relative abundance of oral microbiota samples from patients who experienced post-radiotherapy mucositis and those who did not. The inner layer at each pie chart is at phylum level, and the outer layer is at class level. We can easily see a large difference in abundance of *Fusobacteria* at phylum level as well as at class level between the two groups.

Lastly, the phyloseq package has the "plot_tree()" function that visualizes the phylogenetic tree which is distinct feature of microbiome data. Figure 7c displays the phylogenetic tree of the OPC example data. In practice, microbiome data tends to have a large number of taxa depending on taxonomic so that it may be slow to render an entire phylogenetic tree into a single plot. It will be helpful to select subsets of top abundant taxa or to divide taxa into multiple plots. In addition, the R circlize package [43] has a function to make a circular visualization of the phylogenetic tree along with taxa abundance as shown in Fig. 7d that is very useful to explore the microbiome data by presenting taxa abundance at different taxonomic ranks simultaneously on multiple tracks.

3.4.6 Alpha Diversity

Standard descriptive statistics such as mean, variance, frequency, and percentage are used to summarize counts or relative abundance of microbial features in microbiome study. Because the analysis unit of microbiome data analysis is a microorganism from a microbial ecosystem, various ecological diversity metrics have been applied to describe the biological diversity of microbial community. Here, we introduce two community-wide biodiversity measures, alpha and beta diversities, and explain how to calculate and graphically visualize them.

First, alpha diversity is so-called local biodiversity or within-sample diversity and describes the degree of species richness as well as species evenness (dominance) within a sample or an environment. Species richness is the simplest metric for assessing species diversity and is defined as the number of unique microbial species detected in a specific microbial environment or sample. Species evenness or equitability describes how equally abundant those microbial species are in the microbial ecology community of a sample and examines if there is a dominance by a few specific taxa in the microbial community. Species evenness is highest when all species in the community have the same abundance and evenness approaches zero as relative abundances vary. Table 3 lists commonly used alpha diversity metrics and software to calculate them.

Among alpha diversity indices, observed species and Chao1 infer the species richness, while Shannon's evenness index and Simpson's diversity index summarize species evenness. Briefly, the sample richness index is defined as the number of unique taxa, species, or OTUs detected per each sample and is a very intuitive richness estimator. Chao1 richness is another common estimator that was developed to consider potentially unobserved species and defined as the following equation [44].

$$S_{\text{Chao1}} = S_{\text{obs}} + \frac{f_1(f_1 - 1)}{2(f_2 + 1)}$$

where S_{obs} is the number of observed unique species (sample species index) and f_1 and f are the numbers of singleton species and doubleton species, respectively [44]. Note that these richness indices do not use any information on abundance of individual species or OTU.

Next, Shannon's diversity index (SDI) is commonly used to measures both the number of species and the inequality between species abundances, and its mathematical formula is as follows [45].

Table 3
Alpha and beta diversity indices

Diversity	Diversity statistics	Description	Use of abundance	Use of phylogeny tree	Software	References
Alpha diversity	Sample richness	The number of unique species distributed per sample	Yes	No	R phylogseq [42]	
	Chao1 species estimator	Species richness that accounts for possibly unobserved species	Yes	No	R vegan [94] QIIME Mothur	Chao, A. [44]
	Shannon's diversity	It measures both the richness and evenness	Yes	No		Shannon, C. E. [45]
	Shannon's evenness index	Estimate the evenness of species	No	No		
	Simpson's diversity	The degree of concentration of species found in the sample, which is a range from 0 to 1	Yes	No		Simpson, E. H. [95]
	Faith's phylogenetic diversity	Phylogenetic diversity (PD) distance	No	Yes	R picante package [50]	Faith, D. P. [47]
Beta diversity	Jaccard	The ratio of the number of common features to the number of all features across two samples	No	No	R Vegan QIIME	Jaccard, P. [52]
	Bray-Curtis	The ratio of the sum of lesser abundance of common features to the sum of abundances of all features across two samples	Yes	No	R Vegan QIIME	Bray et al. [53]
	Aitchison distance	Square root of difference in CLR-transformed relative abundance	Yes	No	R microViz	Aitchison et al. [39]
	Unweighted/ weighted UniFrac	The ratio of the sum of unshared branch length to the sum of all tree branch lengths	No/Yes	Yes	R GUniFrac QIIME	Lozupone et al. [55]

$$\mathrm{SDI} = -\sum_{i=1}^{S_{\mathrm{obs}}} P_i \log P_i$$

where P_i is the relative abundance of the i-th species in the sample of interest. A large value of SDI means that many species have well-balanced abundances and thus have a high diversity. The lower the value of SDI, the lower the diversity. In particular, it takes a zero value when there is a single dominant species. Shannon evenness index (SEI) was later devised as a pure diversity index, independent of species richness as below. SEI measures how evenly the microbes are distributed in a sample without considering the number of species. It ranges from 0 indicating a high dominance of a single species to 1 indicating equal abundances across all species.

$$\mathrm{SEI} = \frac{-\sum_{i=1}^{S_{\mathrm{obs}}} P_i \log P_i}{\log S_{\mathrm{obs}}}$$

Simpson's diversity index (D) was introduced to measure the degree of concentration and defined as the probability that two species taken at random from the sample of interest are the same. Its sample estimate is defined as follows:

$$D = 1 - \frac{\sum_{i=1}^{S_{\mathrm{obs}}} f_i(f_i - 1)}{N(N-1)}$$

where N is the total number of individuals in the microbial community of a given sample. SDI is known to be more sensitive to species richness, while Simpson's D index is more sensitive to species evenness [46].

However, above alpha diversity indices use only abundance data but not the phylogenetic relationship of taxa that is an important factor of microbiome data. To address it, Faith's phylogenetic diversity (PD) was developed to incorporate the phylogenetic relationships into alpha diversity [47]. For a given subset of all taxa, it first searches for the minimum spanning path at the whole phylogenetic tree that is made up of the smallest assemblage of branches connecting any two members of the subset. PD of the subset of taxa is then defined as the sum of the lengths of all branches that are members of the minimum spanning path. PD has been extended as abundance-weighted PD to account for taxon counts [48] and relative abundance [49]. The "pd()" function in the picante R package calculates Faith's phylogenetic diversity from both the multispecies abundance table and the phylogenetic tree [50].

In practice, because the total number and abundance of species increases with the sequencing depth of a sample, observed difference in species richness could just be due to differences in sequencing depth. To address this, we could try to rarefy species richness to

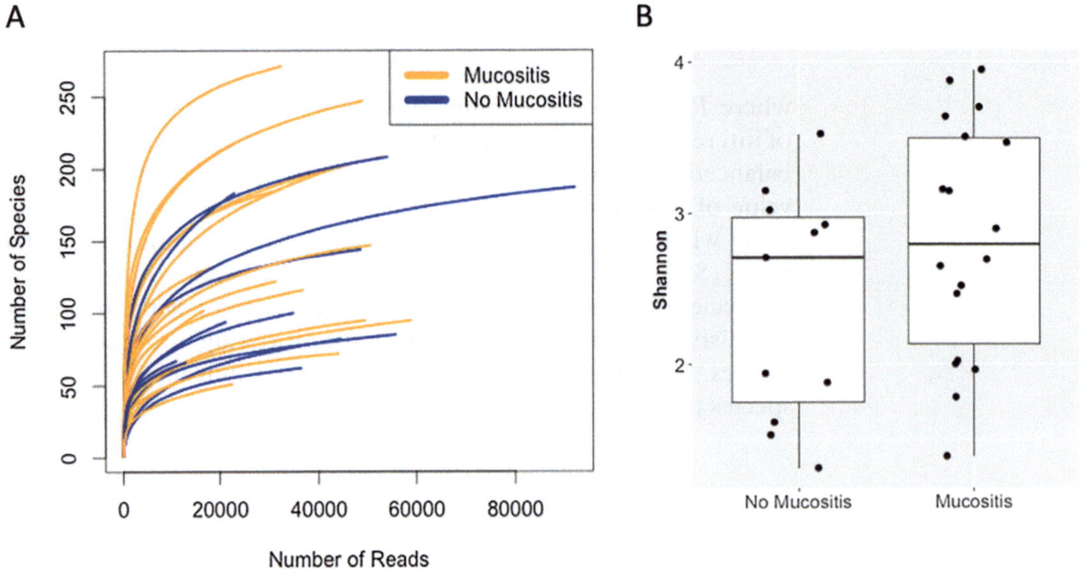

Fig. 8 Rarefaction curve and Shannon's diversity index

the same number of sequence reads. A rarefaction curve is often used to evaluate and compare species richness among samples with different sequencing depths [51]. As described in the previous section on normalization, rarefaction is a procedure of resampling a certain number of sequence reads randomly from each sample and assess the number of unique species detected in the randomly sampled sequence reads. A rarefaction curve plots the number of unique species as a function of the number of randomly selected reads by gradually increasing the amount of random sampling. Figure 8a is the rarefaction curve of the oral microbiome from OPC example data. For a comparison purpose, lines are colored according to whether they experienced mucositis after receiving radiotherapy. In individual rarefaction curve, richness increases rapidly as most common species are found and later remains at a steady level as only rare species remain to be detected. Fixing the number of reads of all samples at around 20,000 lets us know that there are three patients in the mucositis group who have more enriched number of species in comparison to others. Apart from comparison of species richness, we can also judge if the sequencing depth of a sample was sufficient to survey most of species in the sample, in other words, if each rarefaction curve reaches a plateau. Thus, the rarefaction curve informs us an adequate sequencing depth for a future microbiome study design. Figure 8b displays distributions of Shannon diversity metrics and species richness for samples of the OPC example study. It also tells us that patients who experienced mucositis after radiotherapy tend to have higher alpha diversity than those who did not.

3.4.7 Beta Diversity

While alpha diversity summarizes the richness and dominance of taxa within a single sample, beta diversity quantifies the degree of dissimilarity or the distance in taxa abundance between two samples or environmental sites. For microbiome data obtained from n environmental sites, the $n(n - 1)/2$ possible pairs are evaluable for beta diversity. Table 3 lists commonly used beta diversity metrics and currently available software.

First, Jaccard index or similarity coefficient is defined as the ratio of the number of shared taxa between two samples to the number of all features detected in the two samples [52]. The corresponding formula is:

$$J_{ij} = \frac{C}{A + B + C}$$

where A, B, and C denote the number of taxa present in sample i, j, and in both i and j samples, respectively. It is a major drawback that Jaccard index uses only information on the presence or absence of features in two comparison samples and ignores the exact abundance of features. In particular, it is uncertain to establish the presence or absence of low-abundance taxa in microbiome data so that Jaccard index is seldom used in the microbiome diversity analysis. In contrast, Bray-Curtis dissimilarity index quantifies the compositional dissimilarity between two different samples on the basis of abundances of taxa. It is calculated as the ratio of the sum of abundances of features that are not common between two samples to the sum of abundance of all features in two samples and can be formulated as below:

$$BC_{ij} = 1 - \frac{2C_{ij}}{S_i + S_j}$$

where C_{ij} is the sum of the lesser abundance for the species found in each sample i and j. S_i and S_j are the total abundance of taxa found in each sample. The BC dissimilarity is between 0 and 1, where 0 indicates that the two samples share all features and 1 with a completely lack of common features [53]. Aitchison proposed a measure of distance to take into account the compositional property of taxa abundance table because Euclidean distances does not make sense for compositional data [39]. For relative abundances of two samples i and j with D taxa, Aitchison's distance is defined as:

$$D(x_i, x_j) = \sqrt{\sum_{g=1}^{D} \left[\log \frac{x_{gi}}{g(x_i)} - \log \frac{x_{gj}}{g(x_j)} \right]^2}$$

where $g(x_i)$ is the geometric mean of the i-th sample. This distance is simply the Euclidian distance between two samples after the CLR transformation and has scale invariance, perturbation invariance, permutation invariance, and sub-compositional dominance [54].

Although a phylogenetic tree can be inferred as a cladogram in addition to feature abundance table from microbiome sequencing data, the above beta diversity index and distance metrics do not take into account the dissimilarity of microbiome communities at the phylogenetic tree. To address this, Lozupone et al. (2005) [55] developed the *UniFrac* phylogenetic distance to detect a biologically meaningful pattern of variation between samples. It searches for phylogenetic tree branches of taxa unique to each of two microbiome communities and calculates the proportion of the total lengths of the branches among the all branches at the phylogenetic tree.

$$U = \frac{\text{Sum of unshared branch length}}{\text{Sum of all tree branch lengths}} = \frac{\sum_{i=1}^{N} b_i \mid p_i^A - p_i^B \mid}{\sum_{i=1}^{N} l_i \max(A_i, B_i)}$$

where N is the number of nodes (taxa) in the tree, l_i is the branch length between node i and its parent, A_i and B_i are indicators being 1 if descendants of node i are present in microbiome sample A and B, and 0 if absent. Because this original UniFrac distance ignores the taxa abundance information, they introduced the weighted UniFrac distance that weights the branch length of a taxon with a difference in relative abundance of the taxon between the two communities. In R program, the *vegdist()* function in the vegan package offers various beta diversity measures that are summarized in Table 3. The GUniFrac R package has the GUnifrac function calculating weighted- and unweighted Unifrac distance from taxa abundance table and phylogenetic tree data.

In practice, it is highly recommended to arrange beta diversity estimates of all samples into the form of pair-wise distance matrix so that the beta diversity matrix can be visualized for exploration of data through a hierarchical clustering analysis or a multivariate ordination analysis. In particular, the multivariate ordination analysis projects high-dimensional community data of samples into a lower-dimensional space while preserving the inter-sample dissimilarity or distance. Two or three dimensional ordination plots are used to explore the pattern of samples in relation to their demographic or phenotypic characteristics. The principal component analysis (PCA), principal coordinate analysis (PCoA), nonmetric multidimensional scaling (NMDS), uniform manifold approximation, and project (UMAP) techniques can be used for the ordination analysis of beta diversity of microbiome data to explore the relationships among samples or clusters of samples. Here we explain more details of each technique.

First, the principal component analysis (PCA) is the most representative ordination method in classical multivariate statistical analysis. PCA decomposes covariance of original variables where each element can be viewed as a linear function of Euclidian distance among variables that are taxa in microbiome data [56]. The eigenvectors of covariance matrix are used as weight vectors to generate new orthogonal dimensions that are linear combinations of original variables or taxa. The linear combinations are called principal components (PCs), and the first two or three PCs with largest variances are used to visualize the original data in a reduced dimension. Notably, the total sum of variance of PCs is preserved to be the same with the total sum of variances of the original variables. However, PCA itself is often inappropriate for microbiome data where the dissimilarity among samples is not measured by Euclidian distances but nonmetric beta diversity indices.

Second, PCoA is called metric dimensional scaling (MDS) or classical scaling and was developed to obtain a Euclidean representation of multiple objects whose relationships are measured using any dissimilarity or distance metric [57]. It is often confused with PCA that focuses on preserving the total variance of original variables as described above. On the other hand, PCoA extracts new dimensions that maximally preserve the distances among samples. When Euclidean distance is used to measure the dissimilarity among samples, PCoA yields the same result as PCA of the covariance matrix of the same samples. The "ordinate()" function and "plot_ordination()" function in the R phyloseq package perform PCoA on a dissimilarity or distance matrix and produce the corresponding ordination plot.

Lastly, NMDS is a nonmetric alternative version of PCoA analysis [58]. While PCoA analyzes raw distance or dissimilarity values, NMDS converts them into the ranks and extracts new ordination axes by the use of numerical optimization technique while preserving the ranking of the distances. Its utility is in that it is robust to missing data or the presence of outliers. In the R phyloseq package, the "plot_ordination()" function can be used to generate various ordination plots to explore the microbiome data.

Figure 9a, b are heat map graphs of the matrix of Bray-Curtis dissimilarity index and the matrix of Unifrac distance, respectively, for the OPC example data. A hierarchical cluster analysis was used to group patients together that are close to one another. Figure 9c, d are PCoA and NMDS ordination plots based on the matrix of Bray-Curtis dissimilarity index to visualize beta diversity of microbial communities across samples. In this ordination plot, we can find that there is one sample in the group of patients who did not have posttreatment mucositis located far from other samples in the group. A good practice of ordination analysis is to repeat the ordination analysis with different numbers of reduced dimensions as well as other beta diversity indices that are complementary to each other and examines whether all these analyses lead to the same conclusion. Lastly, the beta diversity matrix can be further analyzed for various statistical association tests for evaluating the association of overall microbial community with environmental factors and sample phenotype. More details will be followed in the next section.

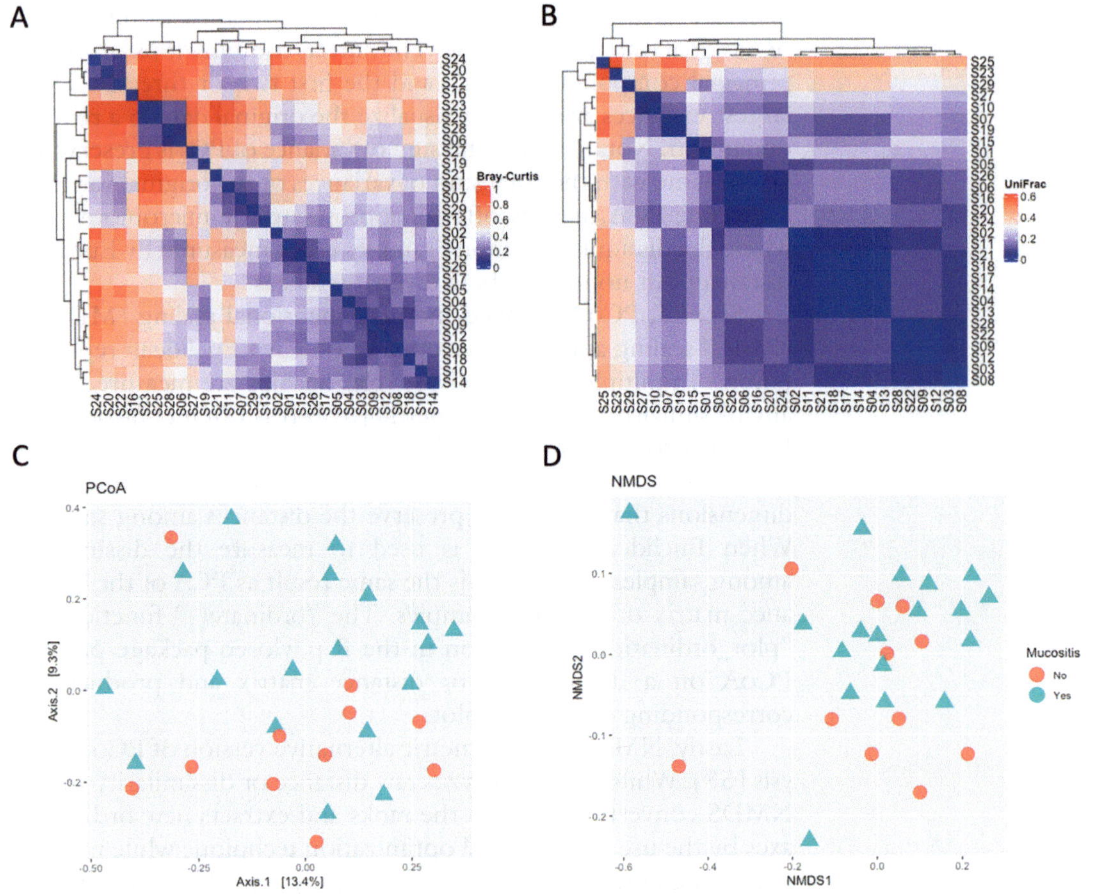

Fig. 9 Heat map and PCoA plots of beta diversity of oral microbiome in OPC study

3.4.8 Microbiome-Wide Association Analysis

As evidences for the involvement of microbiome in human health and disease have been growing, many microbiome studies focus on understanding how microbiome composition links with host phenotypes or environmental characteristics. Therefore, it is of great interest in many public health and biomedical studies to evaluate the association of dynamic microbial community to the diseases and outcomes through a comprehensive microbiome-wide association analysis and to find clinically translatable targets for precision diagnostics and personalized therapies on the basis of host-specific microbiome data. Therefore, statistical hypothesis testing and modeling of the host-microbiome association are necessary in most of those microbiome studies. Largely, there are two ways to identify the links between microbiome and host. One way is the microbial community-level association analysis focusing on the association of entire microbial community as a whole with host or environment characteristics. To examine the association of microbial community with sample characteristics or environmental variables, most common approach is to first summarize all microbiota abundances into alpha or beta diversity metrics and then compare those diversity metrics between different sample groups or examine a correlation relationship with continuous sample characteristics. There is also a diversity-free association test that models taxa abundance probabilistically with a vector of taxa probability per sample group and then tests the homogeneity of the vectors of taxa probability. The other way is feature-level association analysis that investigates the association of sample characteristic with individual microbial feature separately. Note that these twos approaches just differ in how to view microbiome data and are complementary to each other. Thus we will need both approaches to correctly understand the host-microbiome association. In the following sections, we introduce currently available statistical hypothesis tests and visualization methods for the community-level association analysis and feature-level association analysis.

3.4.9 Community-Level Association Analysis Based on Alpha Diversity

Classical statistical tests are widely used to test for difference in alpha diversity between sample groups or to relate to quantitative phenotypes. Table 4 lists parametric statistical tests and corresponding nonparametric tests that are widely used to compare alpha diversity metric in microbiome data analysis. Most of parametric tests assume a normal distribution for alpha diversity metrics. The assumption of normality can be checked graphically histogram and Q-Q plot or perform Kolmogorov-Smirnov test or Shapiro-Wilk test. Nonparametric test does not need a distributional assumption but still requires the homogenous variance between groups. When these assumptions are not satisfied, a randomization test based on permutation method can be used to

Table 4
Alpha diversity-based association analysis

Parametric test	Nonparametric test	Description of general use
Two-sample T-test	Mann Whitney test	Test for difference in mean alpha diversity between two independent groups or ecosystems
Paired T-test	Wilcoxon signed rank test	Test for difference in mean alpha diversity between two correlated or paired samples, e.g., pre- and posttreatment measurement from a patient
Analysis of variance (ANOVA)	Kruskal-Wallis analysis	Test for different in mean alpha diversity between 3 or more independent groups or ecosystems
Pearson correlation test	Spearman correlation test	Assess the degree of a linear relationship of alpha diversity to quantitative host phenotype and test for the significance of the relationship
Linear regression	Rank-based regression	Describes a linear relationship between alpha diversity and quantitative host phenotype

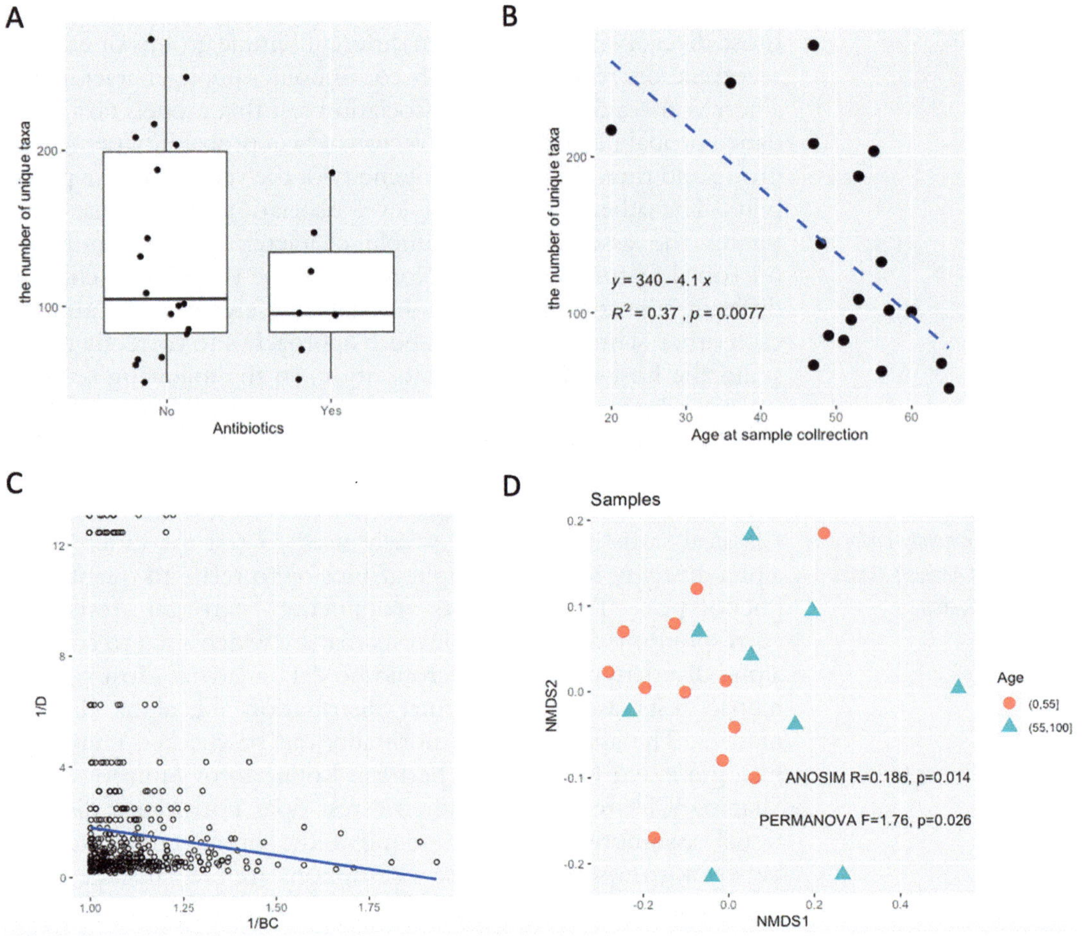

Fig. 10 Community-level association analysis

compute the sampling distribution of the test statistics under the null hypothesis of no association and calculate a *p*-value [59].

Figure 10a, b illustrates the usage of classical statistical analyses of alpha diversity metrics for examining the entire microbiome community-level association. The number of unique taxa was calculated to measure the species richness that is an aspect of alpha diversity in OPC study participants who received antibiotics and those who did not. A higher average of richness was observed in the latter group but no statistically significant difference (Fig. 10a, average richness: 134 vs. 109; $p = 0.305$). There is a linear relationship with a positive spearman correlation coefficient of 0.61 ($p < 0.01$) between age at sample collection and the richness (Fig. 10b).

Table 5
Beta diversity-based association analysis

Method	Input data	Type of covariate	Test statistics	R package	Reference
Mantel's r test	Dissimilarity matrix	Continuous	$\sum_i \sum_j X_{ij} \Upsilon_{ij}$	"'mantel.rtest" function of ade4 package	Mantel (1967)
ANOSIM	Dissimilarity matrix	Categorical	$R = (r_B - r_w)/(n(n-1)/2)$	"'anosim" function of vegan package	Clarke (1993) [62]
PERMANOVA	Dissimilarity matrix	Categorical	$F = (SS_B/a)/(SS_w/(N\text{-}a))$	PERMANOVA	Anderson (2001) [63]
MirKAT	Dissimilarity matrix	Continuous or dichotomous	Score test statistics	MiRKAT	Zhao et al. (2015) [96]

3.4.10 Community-Level Association Analysis Based on Beta Diversity

As explained above, alpha diversity is calculated as a single metric per sample so that its visualization and statistical analysis for association with host or environment characteristics are quite intuitive and straightforward. In contrast, beta diversity analysis for association requires more sophisticated ideas and specialized methods for graphical visualization and statistical tests because beta diversity is a kind of distance between pairs of sample. Table 5 is a summary of available statistical analysis methods and software for identifying a community-level microbiome association based on beta diversity.

First, Mantel et al. (1967) [60] proposed a nonparametric test, so-called Mantel's test, for detecting a temporal and spatial cluster of childhood leukemia by evaluating the sum of cross-products of the temporal closeness and the spatial closeness over all pairs of observed disease cases. Mantel's test is conceptually equal to assessing the magnitude of correlation between the two dissimilarity matrices, for example, (1) microbial beta diversity matrix of human hosts and (2) Euclidean distance of the same hosts on gene expression. In its application for microbiome association analysis, a reciprocal transformation needs to be adopted to change both the dissimilarity matrix of samples on microbiome community (beta diversity) and a distance matrix of hosts on their phenotypes (e.g., Euclidean distance) into the corresponding closeness matrices. A randomization test-based permutation method is then used to evaluate the significance of the correlation of the two closeness matrices without any distributional assumption. As an example of Mantel's test for microbiome association study, Li et al. (2017) used Mantel's test to examine a correlation between the beta diversity of gut microbiota of cyprinid fishes and the distance of the same fishes on metabolite profiles in their multi-omics data analysis and revealed a strong positive correlation [61]. In addition, Fig. 10c is an illustration of Mantel's test in the example OPC study data. X-axis represents a reciprocal value of beta diversity (1/BC) as a closeness among patients, and y-axis is a reciprocal value of Euclidean distance ($1/D$) on their standardized age and tumor stage variables. In this scatter plot, a negative correlation relationship was observed between the transformed beta diversity and phenotype distance. However, the degree of the relationship is weak and insignificant with Mantel's R statistic of 0.001 and p-value of 0.45.

Analysis of similarities (ANOSIM) is another nonparametric statistical test for the difference in dissimilarity of microbial compositions between two or more groups of subjects [62]. ANOSIM can be understood as a distribution-free version of one-way ANOVA by using the rank order of given dissimilarity value, which is invariant to a scale change. More specifically, the rank transformation assigns a rank of 1 to the lowest dissimilarity (highest similarity). If two groups of sampling units are really different in their species composition, then compositional dissimilarities, i.e., beta diversity, between the groups should be greater than those within the groups. The test statistic R of ANOSIM is defined as the difference of the average of the ranks of similarities for all pairs of samples between groups (r_B) from the average of all ranks of similarities among samples within group (r_W). Positive R value indicates that the average of the ranks of between-group dissimilarities is greater than the average rank of within-group dissimilarities. The statistical significance of observed statistics is assessed by permuting the group labels to obtain the empirical sampling distribution of the test statistic under the null hypothesis that there is no difference in dissimilarity among comparison groups. Figure 10d is the PCoA plot of UniFrac dissimilarity matrix from OPC example data. Dots were colored according to corresponding patients' age at diagnosis. Younger patients cluster closely with negative values at x-axis, and they are apart from older patients. ANOSIM was used to statistically compare UniFrac dissimilarity between the two age groups, and there was a significant difference ($R = 0.186$, $p = 0.014$). One drawback of ANOSIM appears when there are two or more grouping factors. Although ANOSIM can be applied by generating a combinatory grouping variable of the group factors or tests one factor within particular levels of other factors, it is an inefficient procedure and cannot partition the effect of individual factors on the microbial community composition.

Anderson et al. (2001) [63] developed nonparametric PERMANOVA to address the drawback of ANOSIM and other methods that are not able to partition variation across multiple environmental or experimental factors. PERMANOVA takes any measure of dissimilarity as the input response variable and partitions variation of dissimilarity directly among individual factors in a multifactorial ANOVA model. For an experimental design with a

Table 6
Summary result of PERMANOVA

	Df	Sums of squares	Mean squares	F-ratio	R_2	*P*-value
Age	1	0.288	0.288	1.750	0.084	0.044
Gender	1	0.147	0.147	0.894	0.043	0.584
Residuals	18	2.966	0.164		0.871	
Total	20	3.401			1	

single experimental factor A with k levels, total sum of squares (SST) is the sum of squared distances between every pair of all observations divided by the total sample size (N). Within-group sum of squares (SSW) is the sum of squared distances between pair of observations within each group divided by the number of samples per group. Between-group sum of squares (SSA) attributed to the experimental factor A is calculated by subtracting SSW from SST. To test the null hypothesis that there is no difference in dissimilarity between all levels of factor A, a pseudo F-statistic is calculated as $F = [\mathrm{SSA}/(k - 1)]/[\mathrm{SSW}/(N - k)]$, where k is the number of groups. The F-statistics is a multivariate analogue to Fisher's F-ratio or signal-to-noise ratio. *P*-values are calculated using a permutation method that randomly shuffles group labels and generates the distribution of the pseudo F-statistic under a true null hypothesis. *P*-value is then defined the proportion of random permutation samples of which F-statistic is equal to or larger than the observed pseudo F-statistic. Table 6 summarizes the result of PERMANOVA for evaluating the association of host age and gender with oral microbiome composition being summarized as UniFrac beta diversity metric. The result indicates that beta diversity of oral microbiome community significantly differs by age at diagnosis of patients while adjusting for gender ($F = 1.75$ and $p = 0.044$ for age).

One major drawback of ANOSIM and PERMANOVA is that these methods cannot analyze the association of beta diversity with continuous covariates or experimental factors. MiRKAT is an alternative method to address the drawback. MiRKAT stands for microbiome regression-based kernel association test, and it is a kernel machine regression approach to statistically test the association between overall microbiome composition and a continuous or binary outcome or sample characteristics [64]. For a continuous outcome Y, it uses the following linear kernel machine model.

$$y_i = \beta_0 + \beta' X_i + f(Z_i) + \epsilon_i$$

Y_i is a response variable, $X_i$is a covariates, and Z_i is a vector of the abundances of all microbial features for the i-th sample. $f(Z_i) = \sum_{i'=1}^{n} \alpha_{i'} K(Z_i, Z_{i'})$ where K is a kernel function that measures the inter-sample similarity. MiRKAT uses Chen & Li's algorithm to construct a kernel matrix from a pairwise beta diversity matrix and construct the score statistics to test the null hypothesis that is equal to H_0: $f(Z) = 0$ [65]. Depending on the types of similarity measure, the kernel function captures a linear or nonlinear effect of taxa on the outcome. KRV statistic and R-squared are used as measures of effect size. One of the major advantages of MiRKAT over other microbiome community-level association tests is that MiRKAT is able to test the microbiome association with the outcome of interest while adjusting for any of continuous and discrete covariates. In addition, MiRKAT is more computationally efficient for large samples. If a sample size is large (>200), moment-based approximation method can be used to compute p-value. However, it is difficult to determine an appropriate distance matrix and kernel for MiRKAT because it has highest power when the form of association between the microbiota and the outcome assumed by the kernel matches the true form of association. Poor choice of kernel will lead to reduced power. To address it, R MiRKAT package provides an omnibus test that takes into account multiple kernels simultaneously and has intermediate substantial power gain compared to when an improper kernel is used and has little power loss compared to when the best kernel is used. MiRKAT has been further extended to accommodate survival outcome (MiRKAT-S [66]) and multivariate outcomes (M-MiRKAT [67]) and to analyze repeated microbiome data under generalized linear mixed model framework (GLMMMiRKAT [68]). In the OPC example study, MiRKAT was used to test for the association of oral microbiota community with age while adjusting for gender and stage and resulted in R^2 of 0.08 and p-value of 0.023, which is consistent with the above PERMANOVA result. However, the small R^2 value indicates that those three host characteristics are still limited to explaining only a small fraction of beta diversity of oral microbiome.

3.4.11 Biodiversity-Free Test of Microbiome Community Association

In general, the multinomial distribution can be used to probabilistically model taxa counts by a vector of taxa probabilities. However, the multinomial distribution is appropriate only when the true frequency of each taxon is consistent across all samples. Because there exists variability across samples in microbiome data, it is limited to directly use the multinomial distribution to model microbiome taxa count data. La Rosa et al. (2012) [69] introduced a new parametric analysis approach that models taxa count table of microbiome data by the use of Dirichlet-multinomial distribution per subpopulation or sample group. It assumes that taxa probabilities follow a Dirichlet distribution that is the conjugate prior of the multinomial distribution. In particular, they adapted the Dirichlet-multinomial distribution with an overdispersion parameter that can capture a larger data variability than the expected variability, so-called overdispersion problem, under the multinomial distribution [70]. They used a maximum likelihood estimation of the distribution parameters and proposed a likelihood ratio test (LRT) for the null hypothesis of the same microbial community structure that the vectors of taxa probabilities are the same across groups against the alternative hypothesis that the vectors are different. In particular, the difference in probability vectors between groups can be summarized using a modified Cramer's criterion that ranges from 0, for the same taxa probabilities between groups, to 1, for maximally different taxa probabilities. Using the Cramer's criterion as an effect size of LRT enables a power/sample size calculation for designing a microbiome study. In practice, when there are rare taxa across samples, the asymptotic Chi-square distribution of LRT statistics may lead to inaccurate inference result. It thus recommends pooling rare taxa prior to performing an LRT test. This approach will also be helpful to avoid uncertain rare taxa that arise due to sequencing errors.

3.4.12 Univariate Feature-Wise Associated Analysis Methods

Univariate feature-wise association analysis is to determine whether individual taxonomic feature is associated with a sample outcome or an environmental variable. Most common approaches are either comparing the abundances of individual taxon between groups or assessing the correlation relationship of the abundance with a quantitative outcome variable. Although classical statistical test methods such as t-test or ANOVA F-test can be used for identifying differentially abundant taxa between sample groups, microbiome data distribution often deviates from statistical assumptions underlying those statistical tests. Various statistical analysis methods have been developed specifically for univariate microbiome association analysis. Table 7 lists a summary of popular analysis methods and

Table 7
Feature-level association analysis methods

Method	Distribution of abundance	Composition	Sparsity	Covariate	Software	References
MetaStats	Normal or hypergeometric	No	Yes	No	R EDDA/Mothur	White et al. (2009) [71]
metagenomeSeq	Zero-inflated log-normal	No	Yes	Yes	R metagenomeSeq	Paulson et al. (2013) [38]
ZIBSeq	Zero-inflated beta distribution	No	Yes	Yes	R ZIBSeq	Peng et al. (2016) [75]
ZINB	Zero-inflated negative binomial	No	No	Yes	pscl and glmmTMB	Greene et al. (1994) [97]
ALDEx/ ALDEx2	Multinomial-Dirichlet	Yes	Yes	No/Yes	R ALDEx2	Fernandes et al. (2013) [80]
ANCOM	Nonparametric	Yes	No	Yes	QIIME	Mandal et al. (2015) [40]

software. Here we explain more details of those methods that will help use the methods correctly.

Among the methods, Metastats is the first statistical analysis method developed for a comparison of the abundances of individual taxonomic feature while considering the sparsity of microbiome data [71]. Metastats is a hybrid method that uses either a two-sample t-test or Fisher's exact test according to the degree sparsity of a taxon. In the t-test that compares relative abundances of a taxon between two groups, the Welch's two-sample t-test statistics with heterogeneous variances is calculated for each of taxa separately. Permutation-based *p*-values are then calculated using the Storey and Tibishirani's permutation method where group memberships of samples are randomly permuted to keep the original correlation structure among taxa. For sparsely sampled taxon, Metastats forms a 2×2 contingency table by calculating the sum of counts of the sparse taxon and the sum of counts across all other taxa in each of the two comparison groups. Fisher's exact test using the hypergeometric distribution for the count data is then used to compare proportions of the rare taxon between the two groups. Fisher's exact test is applied only if the total count of the taxon is less the number of subjects in each group. Otherwise, the above permutation-based t-test is used. For multiple testing correction, q-values are calculated from *p*-values of all taxa to control a false discovery rate (FDR). We explained Metastats for two-group comparison so far, but note that it can be used to experiment with more than two sample groups by substituting the t-test with a one-way ANOVA F-test as well as performing Fisher's exact test [72]. Of note, Metastats has a limitation that it does not allow us to adjust the effects of continuous covariates or confounding factor on the abundance of individual feature.

metagenomeSeq was developed not only to address the limitation of Metastats but also to take into account the sparsity of taxonomic count data due to undersampling with a lack of sequencing depth [38]. It first normalizes taxa count data by the use of cumulative-sum scaling (CSS) that divides raw counts by the cumulative sum of counts up to a data-driven percentile. After the CSS normalization, counts are modeled with a zero-inflated Gaussian (ZIG) mixture distribution that has two distribution components: (1) normally distributed log-transformed feature count abundances and (2) a spike mass at zero corresponding to a detection distribution that depends on each sample's sequencing depth. Next, the expectation-maximization (EM) algorithm is used for fold-change estimation. In the E-step of the EM algorithm, the mean model is expressed by the linear modeling that has three components: (1) intercept, (2) a parameter of fold change of interest, and (3) a transformed CSS normalization factor. Finally, assessing significance for the fold change of a feature is completed through LIMMA approach that was originally developed for differential gene expression analysis of microarray data that can control for covariates and technical confounding factors [73]. In a simulation study, the ZIG method outperforms Metastats, DESeq, and the traditional Kruskal-Wallis test [74].

Besides zero-inflated counts, microbiome data are represented as compositions (proportions) with a large number of zeros and skewed distribution. To account for both the sparse nature of metagenomics data and the compositional data property, Pen et al. (2016) [75] developed a zero-inflated beta regression analysis (ZIBSeq) for identifying differentially abundant features between multiple sample groups. Briefly, beta regression is an extension of the generalized linear model (GLM) approach with the assumption that proportional dependent variable can be characterized by the beta distribution that is an appropriate distribution for a continuous variable taking values in a known range, that is, from 0 to 1 for relative abundances of taxa in microbiome data. However, the beta distribution does not cover 0 and 1, while most microbiome data have many taxa with zero relative abundances and no sample with only one detected feature. Therefore, ZIBSeq employs a more general class of beta distribution, zero-inflated beta distribution, that is, a mixture distribution of a beta distribution with a probability mass at 0. To relate the relative abundance to independent variables, ZIBSeq fits a GLM model with the logit link function for the mean of relative abundances and computes the maximum likelihood estimates of model parameters, and the Wald test is used to calculate *p*-values. Lastly, to consider multiple hypothesis testing correction, q-values are calculated to control FDR.

Under the GLM framework, the Poisson regression model is a longstanding approach to analyzing the association between count response variable and a set of covariates. However, the Poisson distributional assumption that the mean is equal to the variance is frequently violated in next-generation sequencing (NGS) data, which is called overdispersion problem. To address it, negative binomial (NB) regression model was introduced as an alternative to Poisson regression model and becomes popular model for analyzing NGS data. Nevertheless, NB regression is not appropriate to model the microbiome count data because of its excessively many zeros. Fortunately, zero-inflated negative binomial (ZINB) regression model is available to account for many zeros and the overdispersion [76]. In addition, it is noteworthy that ZINB and the above regression models can adjust for the effects of covariates as well as different library sizes as an offset variable. Campbell et al. [77] showed that ZINB adequately controlled the type I error rate and showed a reasonably high statistical power in detecting differentially abundant features in microbiome association analysis. Furthermore, there have been large efforts to extend ZINB to accommodate repeated measurement data and their correlations by random effects [78].

Likewise ZIBSeq, ANCOM (analysis of composition of microbiomes) was developed to account for the compositional property of microbiome data [40]. It addresses the weakness of the methods using relative abundance (RA) of features because RA can vary severely by a small change of abundance of one feature. The key idea of ANCOM is to test all possible sub-hypotheses about log ratios of a given feature to each of other features. That is, when there are G features and K comparison groups, all pairwise $G(G-1)/2$ p-values are calculated by a nonparametric test procedure such as Wilcoxon rank sum test (when $K = 2$), Kruskal-Wallis test ($K \geq 3$), and Friedman test for repeated measurement samples. In case of adjusting for covariates through parametric modeling procedure, statistical linear models such as standard analysis of covariance model can be used. For example, when log ratio of the relative abundance of the i-th feature to the relative abundance of the r-th feature as the reference taxon are compared between K different groups while adjusting for a continuous covariate Z, the following linear model can be fitted to infer the significance.

$$\log \frac{\gamma_{ij}^{(k)}}{\gamma_{rj}^{(k)}} = \alpha_{ir} + \beta_{irk} + \eta_{ir} Z_{jk} + \varepsilon_{irk}$$

where β_{irk} is the effect of the k-th group, η_{ir} is the effect of the covariate Z, ε_{irk} is identically and independently distributed error terms across samples $j = 1,2,\ldots,n_k$, and groups $k = 1, 2, .., K$. For testing for significance of the i-th feature, the number of rejected sub-hypotheses (W_i), can be computed. ANCOM rejects the null hypothesis if $W_i = G - 1$. However, this cutoff is too stringent and makes a very conservative decision. Alternative to the very conservative cutoff, the authors suggested to examine the empirical cumulative distribution F_w of $[W_1, W_2, \cdots, W_G]$ that typically has two increasing regions, one on the left side for non- differentially abundant feature and the other on the right side for differentially abundance feature, and a long flat region between the two regions. They choose a cutoff w_0 as a W value close to the starting point of the flat region and reject the null hypothesis of no difference between groups if $W_i > w_0$. They examined the performance of ANCOM in comparison to ZIG and t-test in simulation studies and showed that ANCOM yielded a higher power and a lower FDR than ZIG method regardless of the proportion of differentially abundant features. The major advantages of ANCOM are that it can use a linear model framework that enables to adjust for the effects of covariates and to model a repeated microbiome data.

ALDEx (ANOVA-like differential expression analysis) was initially developed for a differential expression analysis of meta-transcriptomics data on gene expression levels from target microorganisms in a sample [79]. Likewise metagenomics sequencing data, meta-transcriptomics data has many observed zero counts due partly to a sparse sampling. Their relative abundances are also zero through TSS normalization, but this normalization does not account for the sparse sampling. To address this, ALDEx assumes that counts of features in a sample follows a Dirichlet distribution as a prior distribution and used a Bayesian approach to obtain a marginal posterior distribution of relative abundances of features. Next a large number of random samples are generated from the posterior distribution, and their averages are taken as sparsity-adjusted relative abundance estimates that are greater than zero even though the observed count is zero while preserving the compositional property. In addition, one proportional value depends on other values in compositional data, and this dependency makes it difficult to use a standard statistical comparison analysis. To transform the relative abundances to independent components, they used CLR transformation procedure that Aitchison developed. That is, ALDEx defines the reference as the geometric mean of relative abundances across all features and calculate a single log ratio of the relative abundance of a feature to the reference, while ANCOM method uses the relative abundance of each of all other features as the reference. To identify differentially abundant features, ALDEx uses an "approximate ANOVA"-like procedure for the analysis of experiments with a small sample size. In brief, for a given feature, ALDEx generates multiple Monte Carlo samples on relative abundance from the Dirichlet-multinomial posterior distribution and computes a CLR-transformed relative abundance and calculates absolute differences in the transformed relative abundance between two conditions (Δ_A) and the between-sample difference within each condition (Δ_W). A relative effect size (Δ_R) is calculated as the ratio of the median fold change to the maximum of between-sample differences. Finally, the median of the relative effect sizes (Δ_R^{50}) and the minimum fold change (Δ_A^0) are computed. Using simulation and real dataset analyses, the authors suggested a practical threshold that requires both $| \Delta_R^{50} | \geq 1.5$ to identify differentially expressed features and avoid an inflation of false-positive identifications and also recommend $| \Delta_R^{50} | \geq 2$ when a conservative estimate is needed. ALDEx2 was further developed to enable an analysis of CLR-transformed relative abundance by the use of standard statistical analysis methods including t-test, Kruskal-Wallis test, and general linear model analysis [80].

3.4.13 Visualization of Univariate Association Analysis

Volcano plots are widely used to visualize the results of univariate feature association analysis. It is a scatterplot that displays statistical significance in terms of a function of *p*-value, e.g., −log10 (*p*-value), as well as the magnitude of differential abundance of a feature as an effect size. The volcano plot allows us to quickly identify statistically significant features with large difference. Figure 11a is volcano plot based on the result of metagenomeSeq

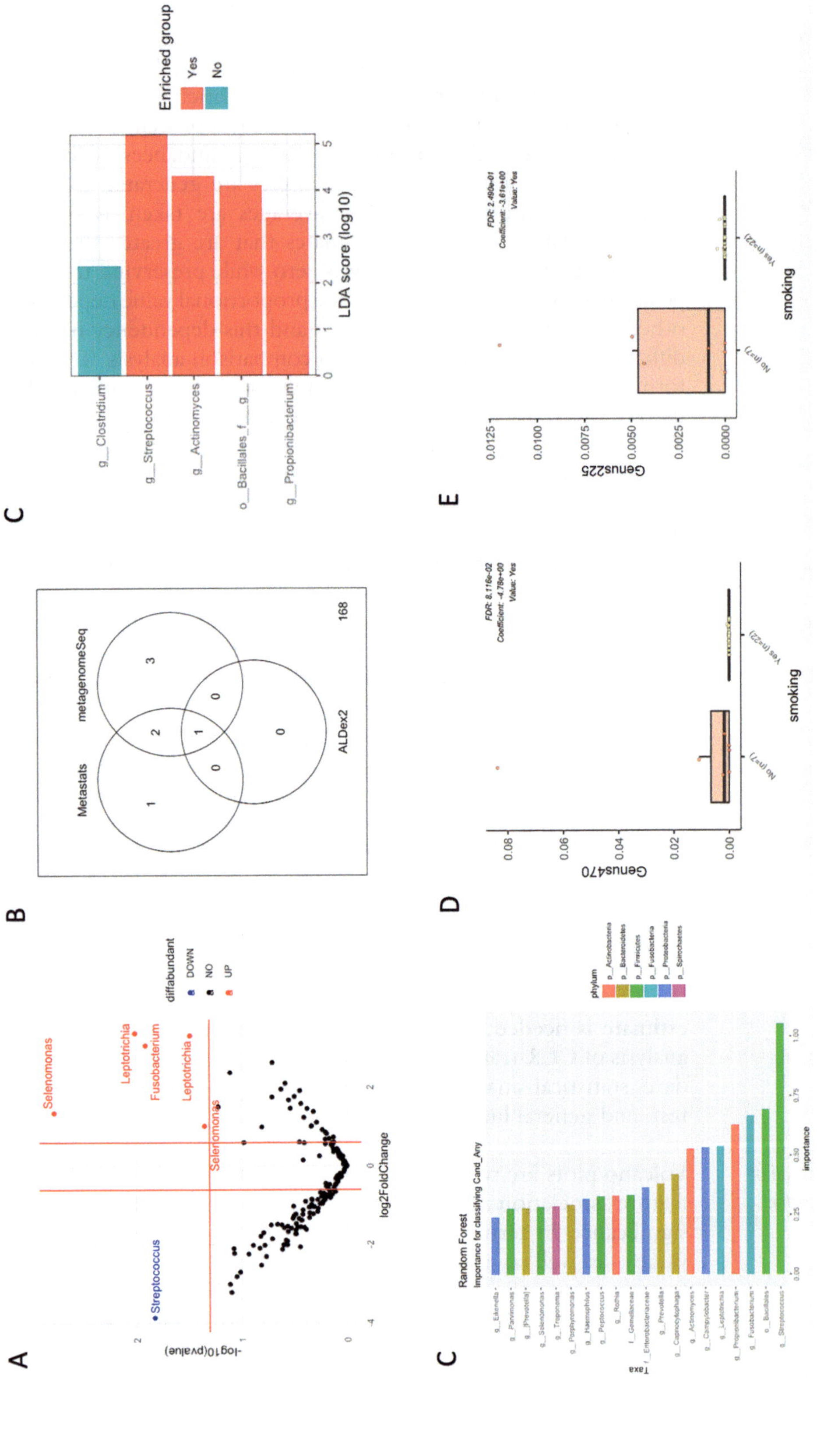

Fig. 11 Visualization of feature-level association analysis and biomarker discovery results

analysis of the OPC example data to identify taxa associated with post-radiotherapy mucositis. The most enriched taxa in patients who experienced mucositis are located with positive log2 fold-change values on the right side of the plot. In contrast, the less abundant taxa are located on the left side with negative log2 fold-change values. Therefore, taxa on the far left or right side could be biologically important genes. In addition, the most statistically significant taxa are toward the top in the volcano plot. In summary, there are six statistically significant taxa with absolute fold change of >1.5. Among them, only *Streptococcus* is less abundant in the patients who underwent mucositis. In addition, as we introduced in the above section, there are multiple analysis methods to discover microbial features that associated with host phenotypes. Because these methods work best when the analysis data follows their underlying distribution assumptions, they may lead to different conclusions for the same features. It is thus recommended to apply all plausible methods to the same data and compare the results. A Venn diagram is then a common way of visualizing the overlap between the results from the different analysis methods. The VennDiagram() function in the R limma package is useful to easily draw a Venn diagram. Figure 11b is the Venn diagram comparing the results of three association analyses of the OPC example data. This Venn diagram indicates that Metastats and metagenomeSeq methods found three common taxa with differential abundance between patients with mucositis and those without. ALDex2 method detects the most significant taxa among the three taxa.

3.4.14 Machine Learning Methods for Microbial Biomarker Discovery

With the growing importance of microbiome research, many microbiome studies have increased the availability of human microbiome data, and these data become the essential resource to explore host-microbiome associations and their relation to the development and progression of various human diseases. In particular, there have been large efforts to develop machine learning approaches that discover microbial biomarkers for classifying or predicting host health traits by examining the variation in microbial communities across hosts. For instance, Yang et al. (2019) developed a colorectal cancer diagnostic biomarker model through gut microbiome analysis [81]. Also, in the OPC example study, it would be highly desirable to develop early noninvasive microbial biomarkers that can predict a serious adverse event or treatment outcomes before treating patients with a radiation therapy. Machine learning (ML) algorithms can be largely classified into two groups: (1) supervised learning method and (2) unsupervised learning method. In microbiome study, the supervised ML algorithms aim to build a classifier or prediction model by explaining the relationship of predictive microbial features with sample phenotype or outcomes. In contrast, the unsupervised ML focuses on discovery unknown patterns or clusters of microbial features and samples by exploring inter-microorganisms and inter-sample distance or relationships. In particular, the supervised methods are more broadly used to provide a new insight into microbiome-based biomarker discovery for predicting host phenotypes or disease. Here, we introduce and illustrate three representative microbial biomarker discovery methods, 1) LEfSE, 2) Random Forest, and 3) MaASLin, based on the supervised ML algorithms.

First, LEfSe (linear discriminant analysis effect size) is one of most widely used methods for a microbiome biomarker discovery. It was developed to identify multiple taxa or functional features with differential abundance between two or more hierarchically structured biological groups and to build a microbial signature of classifying the groups [82]. Suppose that metagenomics samples are obtained from (1) mucosal tissues from oral cavity and gut and non-mucosal tissues from skin and nasal cavity, to identify functional features with differential abundance between mucosal and non-mucosal tissue. We can first select features with differential abundance between all mucosal tissues and all non-mucosal tissues. It will then be biologically relevant to further examine if the differential abundance pattern of the selected features retains in comparisons of each of the two body sites in mucosal group to those in non-mucosal groups, which is a key idea of LEfSe that prevents biologically irrelevant positive discoveries. LEfSe is a three-step procedure: (1) It first performs Kruskal-Wallis test to screen differentially abundant features across the groups; (2) subsequently performs Wilcoxon rank sum test to further confirm if both signs and significance of differential abundance of the initially screened features are retained in pairwise comparisons between subgroups; (3) and finally builds Fisher's linear discriminant analysis (LDA) to rank the remaining features. Although LEfSe does not rigorously account for sparsity of microbiome data and multiple testing problem, it is very popularly used in many microbiome studies because of biological consistency and easy interpretation of its results. LEfSe analysis can be performed by using Galaxy web platform (https://huttenhower.sph.harvard.edu/galaxy/) or the R microbiomeMarker package (https://github.com/yiluheihei/microbiomeMarker). Figure 11c displays the LDA scores of four differentially enriched features between OPC patients who had posttreatment candidiasis and those who did not. Four taxa are found to be significantly enriched in patients with candidiasis, and *Streptococcus* genus is the largest LDA score. Only *Clostridium* genus is significantly enriched in patients without candidiasis.

Second, the random forest is also a popular machine learning method used in microbiome association studies to select important species that contribute to host phenotype and to build a multivariate microbial signature that predicts or classifies the host phenotype. For example, Loomba et al. (2017) used the random forest method to build a gut microbiome-based metagenomic signature for noninvasive detection of advanced fibrosis in human nonalcoholic fatty liver disease [83]. The random forest algorithm adopts an ensemble approach that builds a number of tree classifiers from bootstrap samples and aggregates them by voting strategy to improve robustness and precision of classification outputs [84]. While LEfSe relies on statistical abundance tests for its microbial biomarker discovery and uses a linear combination of features for classification, the random forest seeks for a set of randomly chosen features and their ideal split rules that yield the best predicted outcome. To measure the importance of a microbial feature in a random forest analysis of microbiome data, the abundance values of the microbial feature are randomly permuted to nullify its contribution to the host phenotype in each training data, and the prediction error is computed on the permuted data. The importance score for the feature is determined by averaging differences in prediction error before and after the permutation. Features with large importance scores are considered important in classifying the host phenotype. Figure 11d depicts the important values of taxa from a random forest analysis of the OPC example data that classifies patients with candidiasis from those without after radiotherapy. In line with the result of LEfSe, *Streptococcus* genera and *Bacillales* order showed largest values of importance followed by *Fusobacterium* that was not detected in LEfSe. In fact, *Fusobacterium* is more abundant in patients with candidiasis, but Wilcoxon rank sum test yields an insignificant difference. Random forest results imply that even a feature with less significant differential abundance can be an important factor to improve the performance of multivariable classifier.

Lastly, MaAsLin (Microbiome Multivariable Associations with Linear Models) is a multivariate statistical framework that facilitates exploration of associations between high-dimensional microbiome data and clinical metadata [85]. It uses boosted additive general linear models that take a group of host phenotypes as explanatory variables and another group of microbial feature relative abundance as the response. In brief, relative abundances of the features are first normalized using a variance-stabilizing arcsine square root transformation. The following multivariable linear model of the transformed relative abundance as a response variable is fitted onto multiple phenotypes of interest as independent variables.

$$\arcsin(\sqrt{r_i}) = \beta_0 + \sum_p \beta_p X_{i,p} + \varepsilon_i \text{ for subject } i = 1, \ldots, n \text{ and host phenotype } p = 1, .., P$$

Considering the sparsity of microbiome data, a model selection is performed using a boosting algorithm. To test for significance of the association of individual microbial feature with the host phenotypes, multiple comparisons are adjusted using Bonferroni correction method. In addition, multiple hypothesis tests across all taxonomic clades and phenotypes are adjusted using Benjamine and Horchberg FDR method. Recently, a subsequent version, MaAsLin2, was developed to accommodate various epidemiological study designs including cross-sectional as well as longitudinal study designs. MaAsLin2 uses a log-transformed linear model on TSS-normalized data and fits general linear models or mixed effects models according to the study design but also supports other statistical count models including zero-inflated negative binomial and zero-adjusted models [86]. Figure 11e, f show outputs of MaAsLin for identifying genus associated with smoking status in OPC example study. Two genera are significantly enriched in non-smokers in comparison to smokers.

4 Conclusions

Microbiome research is currently one of the most dynamic fields in public health and life sciences. It is highly relevant to precision medicine, microbial ecology, and other areas. Bioinformatic and statistical analyses are crucial for understanding the composition and structure of microbial community in their hosts and the mechanism by which individual features are linked to the host phenotype. In this chapter, we introduced a series of computational bioinformatic methods as well as various statistical analysis methods addressing the challenges of NGS-based microbiome data. More rigorous and efficient analysis methods are continuously emerging. Understanding the concept and strategy of the analyses will be helpful to use them correctly in microbiome data analysis and further advance our understanding of the important role of microbiome in health and disease.

References

1. Fan Y, Pedersen O (2021) Gut microbiota in human metabolic health and disease. Nat Rev Microbiol 19(1):55–71. https://doi.org/10.1038/s41579-020-0433-9
2. Gilbert JA, Blaser MJ, Caporaso JG, Jansson JK, Lynch SV, Knight R (2018) Current understanding of the human microbiome. Nat Med 24(4):392–400. https://doi.org/10.1038/nm.4517
3. Peterson D, Bonham KS, Rowland S, Pattanayak CW, Consortium R, Klepac-Ceraj V (2021) Comparative analysis of 16S rRNA gene and metagenome sequencing in pediatric gut microbiomes. Front Microbiol 12:670336. https://doi.org/10.3389/fmicb.2021.670336
4. Pierce CM, Hogue S, Paul S, Hong BY, da Silva WV, Gomez MF, Giuliano AR, Caudell JJ, Weinstock GM (2019) Mucositis, candidiasis, and associations with the oral microbiome in treatment naive patients with oropharyngeal cancer. Cancer Res 79(13):3326. https://doi.org/10.1158/1538-7445.Am2019-3326
5. Caporaso JG, Kuczynski J, Stombaugh J, Bittinger K, Bushman FD, Costello EK, Fierer N, Pena AG, Goodrich JK, Gordon JI, Huttley GA, Kelley ST, Knights D, Koenig JE, Ley RE, Lozupone CA, McDonald D, Muegge BD, Pirrung M, Reeder J, Sevinsky JR, Turnbaugh PJ, Walters WA, Widmann J, Yatsunenko T, Zaneveld J, Knight R (2010) QIIME allows analysis of high-throughput community sequencing data. Nat Methods

7(5):335–336. https://doi.org/10.1038/nmeth.f.303
6. Wang Q, Garrity GM, Tiedje JM, Cole JR (2007) Naive Bayesian classifier for rapid assignment of rRNA sequences into the new bacterial taxonomy. Appl Environ Microb 73(16):5261–5267. https://doi.org/10.1128/Aem.00062-07
7. Edgar RC (2013) UPARSE: highly accurate OTU sequences from microbial amplicon reads. Nat Methods 10(10):996–998. https://doi.org/10.1038/nmeth.2604
8. Eren AM, Borisy GG, Huse SM, Mark Welch JL (2014) Oligotyping analysis of the human oral microbiome. Proc Natl Acad Sci U S A 111(28):E2875–E2884. https://doi.org/10.1073/pnas.1409644111
9. Callahan BJ, McMurdie PJ, Rosen MJ, Han AW, Johnson AJA, Holmes SP (2016) DADA2: high-resolution sample inference from Illumina amplicon data. Nat Methods 13(7):581.-+. https://doi.org/10.1038/Nmeth.3869
10. Amir A, McDonald D, Navas-Molina JA, Kopylova E, Morton JT, Xu ZZ, Kightley EP, Thompson LR, Hyde ER, Gonzalez A, Knight R (2017) Deblur rapidly resolves single-nucleotide community sequence patterns. Msystems 2(2):e00191-16. https://doi.org/10.1128/mSystems.00191-16
11. Johnson JS, Spakowicz DJ, Hong BY, Petersen LM, Demkowicz P, Chen L, Leopold SR, Hanson BM, Agresta HO, Gerstein M, Sodergren E, Weinstock GM (2019) Evaluation of 16S rRNA gene sequencing for species and strain-level microbiome analysis. Nat Commun 10:5029. https://doi.org/10.1038/s41467-019-13036-1
12. Altschul SF, Gish W, Miller W, Myers EW, Lipman DJ (1990) Basic local alignment search tool. J Mol Biol 215(3):403–410. https://doi.org/10.1016/S0022-2836(05)80360-2
13. Edgar RC (2010) Search and clustering orders of magnitude faster than BLAST. Bioinformatics 26(19):2460–2461. https://doi.org/10.1093/bioinformatics/btq461
14. Bokulich NA, Kaehler BD, Rideout JR, Dillon M, Bolyen E, Knight R, Huttley GA, Gregory Caporaso J (2018) Optimizing taxonomic classification of marker-gene amplicon sequences with QIIME 2's q2-feature-classifier plugin. Microbiome 6(1):90. https://doi.org/10.1186/s40168-018-0470-z
15. Washburne AD, Morton JT, Sanders J, McDonald D, Zhu Q, Oliverio AM, Knight R (2018) Methods for phylogenetic analysis of microbiome data. Nat Microbiol 3(6):652–661. https://doi.org/10.1038/s41564-018-0156-0
16. Katoh K, Misawa K, Kuma K, Miyata T (2002) MAFFT: a novel method for rapid multiple sequence alignment based on fast Fourier transform. Nucleic Acids Res 30(14):3059–3066. https://doi.org/10.1093/nar/gkf436
17. Price MN, Dehal PS, Arkin AP (2009) FastTree: computing large minimum evolution trees with profiles instead of a distance matrix. Mol Biol Evol 26(7):1641–1650. https://doi.org/10.1093/molbev/msp077
18. Schliep K, Potts AJ, Morrison DA, Grimm GW (2017) Intertwining phylogenetic trees and networks. Methods Ecol Evol 8(10):1212–1220. https://doi.org/10.1111/2041-210x.12760
19. Loytynoja A, Vilella AJ, Goldman N (2012) Accurate extension of multiple sequence alignments using a phylogeny-aware graph algorithm. Bioinformatics 28(13):1684–1691. https://doi.org/10.1093/bioinformatics/bts198
20. Matsen FA, Kodner RB, Armbrust EV (2010) pplacer: linear time maximum-likelihood and Bayesian phylogenetic placement of sequences onto a fixed reference tree. Bmc Bioinformatics 11:538. https://doi.org/10.1186/1471-2105-11-538
21. Bolger AM, Lohse M, Usadel B (2014) Trimmomatic: a flexible trimmer for Illumina sequence data. Bioinformatics 30(15):2114–2120. https://doi.org/10.1093/bioinformatics/btu170
22. Bushnell B, Rood J, Singer E (2017) BBMerge - accurate paired shotgun read merging via overlap. PLoS One 12(10):e0185056. https://doi.org/10.1371/journal.pone.0185056
23. Beghini F, McIver LJ, Blanco-Miguez A, Dubois L, Asnicar F, Maharjan S, Mailyan A, Manghi P, Scholz M, Thomas AM, Valles-Colomer M, Weingart G, Zhang YC, Zolfo M, Huttenhower C, Franzosa EA, Segata N (2021) Integrating taxonomic, functional, and strain-level profiling of diverse microbial communities with bioBakery 3. elife 10:e65088. https://doi.org/10.7554/eLife.65088
24. Wood DE, Salzberg SL (2014) Kraken: ultrafast metagenomic sequence classification using exact alignments. Genome Biol 15(3):R46. https://doi.org/10.1186/gb-2014-15-3-r46
25. Lu J, Breitwieser FP, Thielen P, Salzberg SL (2017) Bracken: estimating species abundance

in metagenomics data. Peerj Comput Sci 3: e104. https://doi.org/10.7717/peerj-cs.104
26. Buchfink B, Xie C, Huson DH (2015) Fast and sensitive protein alignment using DIAMOND. Nat Methods 12(1):59–60. https://doi.org/10.1038/nmeth.3176
27. Huson DH, Auch AF, Qi J, Schuster SC (2007) MEGAN analysis of metagenomic data. Genome Res 17(3):377–386. https://doi.org/10.1101/gr.5969107
28. Franzosa EA, McIver LJ, Rahnavard G, Thompson LR, Schirmer M, Weingart G, Lipson KS, Knight R, Caporaso JG, Segata N, Huttenhower C (2018) Species-level functional profiling of metagenomes and metatranscriptomes. Nat Methods 15(11):962–968. https://doi.org/10.1038/s41592-018-0176-y
29. Suzek BE, Wang Y, Huang H, McGarvey PB, Wu CH, UniProt C (2015) UniRef clusters: a comprehensive and scalable alternative for improving sequence similarity searches. Bioinformatics 31(6):926–932. https://doi.org/10.1093/bioinformatics/btu739
30. Hyatt D, Chen GL, Locascio PF, Land ML, Larimer FW, Hauser LJ (2010) Prodigal: prokaryotic gene recognition and translation initiation site identification. Bmc Bioinformatics 11:119. https://doi.org/10.1186/1471-2105-11-119
31. Lapidus AL, Korobeynikov AI (2021) Metagenomic data assembly - the way of decoding unknown microorganisms. Front Microbiol 12:613791. https://doi.org/10.3389/fmicb.2021.613791
32. Cao Q, Sun X, Rajesh K, Chalasani N, Gelow K, Katz B, Shah VH, Sanyal AJ, Smirnova E (2021) Effects of rare microbiome taxa filtering on statistical analysis. Front Microbiol 11. https://doi.org/10.3389/fmicb.2020.607325
33. Davis NM, Proctor DM, Holmes SP, Relman DA, Callahan BJ (2018) Simple statistical identification and removal of contaminant sequences in marker-gene and metagenomics data. Microbiome 6(1):226. https://doi.org/10.1186/s40168-018-0605-2
34. Smirnova E, Huzurbazar S, Jafari F (2019) PERFect: PERmutation filtering test for microbiome data. Biostatistics 20(4): 615–631. https://doi.org/10.1093/biostatistics/kxy020
35. Knights D, Kuczynski J, Charlson ES, Zaneveld J, Mozer MC, Collman RG, Bushman FD, Knight R, Kelley ST (2011) Bayesian community-wide culture-independent microbial source tracking. Nat Methods 8(9): 761–763. https://doi.org/10.1038/nmeth.1650
36. Hewitt KM, Mannino FL, Gonzalez A, Chase JH, Caporaso JG, Knight R, Kelley ST (2013) Bacterial diversity in two Neonatal Intensive Care Units (NICUs). PLoS One 8(1): e54703. https://doi.org/10.1371/journal.pone.0054703
37. Weiss S, Xu ZZ, Peddada S, Amir A, Bittinger K, Gonzalez A, Lozupone C, Zaneveld JR, Vazquez-Baeza Y, Birmingham A, Hyde ER, Knight R (2017) Normalization and microbial differential abundance strategies depend upon data characteristics. Microbiome 5(1):27. https://doi.org/10.1186/s40168-017-0237-y
38. Paulson JN, Stine OC, Bravo HC, Pop M (2013) Differential abundance analysis for microbial marker-gene surveys. Nat Methods 10(12):1200–1202. https://doi.org/10.1038/nmeth.2658
39. Aitchison J (1982) The statistical-analysis of compositional data. J Roy Stat Soc B Met 44(2):139–177
40. Mandal S, Van Treuren W, White RA, Eggesbo M, Knight R, Peddada SD (2015) Analysis of composition of microbiomes: a novel method for studying microbial composition. Microb Ecol Health Dis 26:27663. https://doi.org/10.3402/mehd.v26.27663
41. Gu Z, Eils R, Schlesner M (2016) Complex heatmaps reveal patterns and correlations in multidimensional genomic data. Bioinformatics 32(18):2847–2849. https://doi.org/10.1093/bioinformatics/btw313
42. McMurdie PJ, Holmes S (2013) Phyloseq: an R package for reproducible interactive analysis and graphics of microbiome census data. PLoS One 8(4):e61217. https://doi.org/10.1371/journal.pone.0061217
43. Gu Z, Gu L, Eils R, Schlesner M, Brors B (2014) Circlize implements and enhances circular visualization in R. Bioinformatics 30(19): 2811–2812. https://doi.org/10.1093/bioinformatics/btu393
44. Chao A (1987) Estimating the population-size for capture recapture data with unequal catchability. Biometrics 43(4):783–791. https://doi.org/10.2307/2531532
45. Shannon CE (1948) A mathematical theory of communication. Bell Syst Tech J 27(3): 379–423. https://doi.org/10.1002/j.1538-7305.1948.tb01338.x
46. DeJong TM (1975) A comparison of three diversity indices based on their components of richness and evenness. Oikos 26(2):222–227. https://doi.org/10.2307/3543712

47. Faith DP (1992) Conservation evaluation and phylogenetic diversity. Biol Conserv 61(1): 1–10. https://doi.org/10.1016/0006-3207(92)91201-3
48. Barber NA, Jones HP, Duvall MR, Wysocki WP, Hansen MJ, Gibson DJ (2017) Phylogenetic diversity is maintained despite richness losses over time in restored tallgrass prairie plant communities. J Appl Ecol 54(1): 137–144. https://doi.org/10.1111/1365-2664.12639
49. Mccoy CO, Matsen FA (2013) Abundance-weighted phylogenetic diversity measures distinguish microbial community states and are robust to sampling depth. Peerj 1:e157. https://doi.org/10.7717/peerj.157
50. Kembel SW, Cowan PD, Helmus MR, Cornwell WK, Morlon H, Ackerly DD, Blomberg SP, Webb CO (2010) Picante: R tools for integrating phylogenies and ecology. Bioinformatics 26(11):1463–1464. https://doi.org/10.1093/bioinformatics/btq166
51. Willis AD (2019) Rarefaction, alpha diversity, and statistics. Front Microbiol 10:2407. https://doi.org/10.3389/fmicb.2019.02407
52. Jaccard P (1912) The distribution of the flora in the alpine zone. New Phytol 11(2):37–50
53. Bray JR, Curtis JT (1957) An ordination of the upland forest communities of southern Wisconsin. Ecol Monogr 27(4):326–349. https://doi.org/10.2307/1942268
54. Quinn TP, Erb I, Richardson MF, Crowley TM (2018) Understanding sequencing data as compositions: an outlook and review. Bioinformatics 34(16):2870–2878. https://doi.org/10.1093/bioinformatics/bty175
55. Lozupone CA, Knight R (2015) The unifrac significance test is sensitive to tree topology. Bmc Bioinformatics 16:211. https://doi.org/10.1186/s12859-015-0640-y
56. Hotelling H (1933) Analysis of a complex of statistical variables into principal components. J Educ Psychol 24:417–441. https://doi.org/10.1037/h0071325
57. Kruskal JB (1964) Multidimensional-scaling by optimizing goodness of fit to a nonmetric hypothesis. Psychometrika 29(1):1–27. https://doi.org/10.1007/Bf02289565
58. Kruskal JB (1964) Nonmetric multidimensional-scaling - a numerical-method. Psychometrika 29(2):115–129. https://doi.org/10.1007/Bf02289694
59. Storey JD, Tibshirani R (2003) Statistical significance for genomewide studies. Proc Natl Acad Sci U S A 100(16):9440–9445. https://doi.org/10.1073/pnas.1530509100
60. Mantel N (1967) The detection of disease clustering and a generalized regression approach. Cancer Res 27(2):209–220
61. Li T, Long M, Li H, Gatesoupe FJ, Zhang X, Zhang Q, Feng D, Li A (2017) Multi-omics analysis reveals a correlation between the host phylogeny, gut microbiota and metabolite profiles in cyprinid fishes. Front Microbiol 8:454. https://doi.org/10.3389/fmicb.2017.00454
62. Clarke KR (1993) Non-parametric multivariate analyses of changes in community structure. Aust J Ecol 18(1):117–143. https://doi.org/10.1111/j.1442-9993.1993.tb00438.x
63. Anderson MJ (2001) A new method for non-parametric multivariate analysis of variance. Austral Ecol 26(1):32–46. https://doi.org/10.1111/j.1442-9993.2001.01070.pp.x
64. Wilson N, Zhao N, Zhan X, Koh H, Fu W, Chen J, Li H, Wu MC, Plantinga AM (2021) MiRKAT: kernel machine regression-based global association tests for the microbiome. Bioinformatics 37(11):1595–1597. https://doi.org/10.1093/bioinformatics/btaa951
65. Chen J, Li H (2013) Kernel methods for regression analysis of microbiome compositional data. In: Hu M, Liu Y, Lin J (eds) Topics in applied statistics. Springer New York, New York, pp 191–201
66. Plantinga A, Zhan X, Zhao N, Chen J, Jenq RR, Wu MC (2017) MiRKAT-S: a community-level test of association between the microbiota and survival times. Microbiome 5(1):17. https://doi.org/10.1186/s40168-017-0239-9
67. Zhan X, Tong X, Zhao N, Maity A, Wu MC, Chen J (2017) A small-sample multivariate kernel machine test for microbiome association studies. Genet Epidemiol 41(3):210–220. https://doi.org/10.1002/gepi.22030
68. Koh H, Li Y, Zhan X, Chen J, Zhao N (2019) A distance-based kernel association test based on the generalized linear mixed model for correlated microbiome studies. Front Genet 10: 458. https://doi.org/10.3389/fgene.2019.00458
69. La Rosa PS, Brooks JP, Deych E, Boone EL, Edwards DJ, Wang Q, Sodergren E, Weinstock G, Shannon WD (2012) Hypothesis testing and power calculations for taxonomic-based human microbiome data. PLoS One 7(12):e52078. https://doi.org/10.1371/journal.pone.0052078
70. Tvedebrink T (2010) Overdispersion in allelic counts and theta-correction in forensic genetics. Theor Popul Biol 78(3):200–210. https://doi.org/10.1016/j.tpb.2010.07.002

71. White JR, Nagarajan N, Pop M (2009) Statistical methods for detecting differentially abundant features in clinical metagenomic samples. PLoS Comput Biol 5(4):e1000352. https://doi.org/10.1371/journal.pcbi.1000352
72. Mehta CR, Patel NR (1983) A network algorithm for performing fisher exact test in R X C contingency-tables. J Am Stat Assoc 78(382): 427–434. https://doi.org/10.2307/2288652
73. Ritchie ME, Phipson B, Wu D, Hu Y, Law CW, Shi W, Smyth GK (2015) limma powers differential expression analyses for RNA-sequencing and microarray studies. Nucleic Acids Res 43 (7):e47–e47. https://doi.org/10.1093/nar/gkv007
74. Ritchie ME, Phipson B, Wu D, Hu YF, Law CW, Shi W, Smyth GK (2015) Limma powers differential expression analyses for RNA-sequencing and microarray studies. Nucleic Acids Res 43(7):ARTN e47. https://doi.org/10.1093/nar/gkv007
75. Peng XL, Li G, Liu ZQ (2016) Zero-inflated beta regression for differential abundance analysis with metagenomics data. J Comput Biol 23(2):102–110. https://doi.org/10.1089/cmb.2015.0157
76. Xu L, Paterson AD, Turpin W, Xu W (2015) Assessment and selection of competing models for zero-inflated microbiome data. PLoS One 10(7):e0129606. https://doi.org/10.1371/journal.pone.0129606
77. Campbell H, O'Hara RB (2021) The consequences of checking for zero-inflation and overdispersion in the analysis of count data. Methods Ecol Evol 12(4):665–680. https://doi.org/10.1111/2041-210X.13559
78. Zhang X, Mallick H, Tang Z, Zhang L, Cui X, Benson AK, Yi N (2017) Negative binomial mixed models for analyzing microbiome count data. BMC Bioinformatics 18(1):4. https://doi.org/10.1186/s12859-016-1441-7
79. Fernandes AD, Macklaim JM, Linn TG, Reid G, Gloor GB (2013) ANOVA-like differential expression (ALDEx) analysis for mixed population RNA-Seq. PLoS One 8(7):e67019. https://doi.org/10.1371/journal.pone.0067019
80. Fernandes AD, Reid JN, Macklaim JM, McMurrough TA, Edgell DR, Gloor GB (2014) Unifying the analysis of high-throughput sequencing datasets: characterizing RNA-seq, 16S rRNA gene sequencing and selective growth experiments by compositional data analysis. Microbiome 2:15. https://doi.org/10.1186/2049-2618-2-15
81. Yang J, McDowell A, Kim EK, Seo H, Lee WH, Moon CM, Kym SM, Lee DH, Park YS, Jee YK, Kim YK (2019) Development of a colorectal cancer diagnostic model and dietary risk assessment through gut microbiome analysis. Exp Mol Med 51:117. https://doi.org/10.1038/s12276-019-0313-4
82. Segata N, Izard J, Waldron L, Gevers D, Miropolsky L, Garrett WS, Huttenhower C (2011) Metagenomic biomarker discovery and explanation. Genome Biol 12(6):R60. https://doi.org/10.1186/gb-2011-12-6-r60
83. Loomba R, Seguritan V, Li W, Long T, Klitgord N, Bhatt A, Dulai PS, Caussy C, Bettencourt R, Highlander SK, Jones MB, Sirlin CB, Schnabl B, Brinkac L, Schork N, Chen CH, Brenner DA, Biggs W, Yooseph S, Venter JC, Nelson KE (2017) Gut microbiome-based metagenomic signature for non-invasive detection of advanced fibrosis in human nonalcoholic fatty liver disease. Cell Metab 25(5): 1054–1062. e1055. https://doi.org/10.1016/j.cmet.2017.04.001
84. Breiman L (2001) Random forests. Mach Learn 45(1):5–32. https://doi.org/10.1023/A:1010933404324
85. Morgan XC, Tickle TL, Sokol H, Gevers D, Devaney KL, Ward DV, Reyes JA, Shah SA, LeLeiko N, Snapper SB, Bousvaros A, Korzenik J, Sands BE, Xavier RJ, Huttenhower C (2012) Dysfunction of the intestinal microbiome in inflammatory bowel disease and treatment. Genome Biol 13(9):R79. https://doi.org/10.1186/gb-2012-13-9-r79
86. Mallick H, Rahnavard A, McIver LJ, Ma S, Zhang Y, Nguyen LH, Tickle TL, Weingart G, Ren B, Schwager EH, Chatterjee S, Thompson KN, Wilkinson JE, Subramanian A, Lu Y, Waldron L, Paulson JN, Franzosa EA, Bravo HC, Huttenhower C (2021) Multivariable association discovery in population-scale meta-omics studies. PLoS Comput Biol 17(11):e1009442. https://doi.org/10.1371/journal.pcbi.1009442
87. Schloss PD, Gevers D, Westcott SL (2011) Reducing the effects of PCR amplification and sequencing artifacts on 16S rRNA-based studies. PLoS One 6(12):e27310. https://doi.org/10.1371/journal.pone.0027310

88. Rognes T, Flouri T, Nichols B, Quince C, Mahe F (2016) VSEARCH: a versatile open source tool for metagenomics. PeerJ 4:e2584. https://doi.org/10.7717/peerj.2584
89. Bagci C, Patz S, Huson DH (2021) DIAMOND+MEGAN: fast and easy taxonomic and functional analysis of short and Long microbiome sequences. Curr Protoc 1(3):e59. https://doi.org/10.1002/cpz1.59
90. Wu YW, Tang YH, Tringe SG, Simmons BA, Singer SW (2014) MaxBin: an automated binning method to recover individual genomes from metagenomes using an expectation-maximization algorithm. Microbiome 2:26. https://doi.org/10.1186/2049-2618-2-26
91. Kang DD, Froula J, Egan R, Wang Z (2015) MetaBAT, an efficient tool for accurately reconstructing single genomes from complex microbial communities. PeerJ 3:e1165. https://doi.org/10.7717/peerj.1165
92. Parks DH, Imelfort M, Skennerton CT, Hugenholtz P, Tyson GW (2015) CheckM: assessing the quality of microbial genomes recovered from isolates, single cells, and metagenomes. Genome Res 25(7):1043–1055. https://doi.org/10.1101/gr.186072.114
93. Hurlbert SH (1980) Citation classic - the non-concept of species-diversity - a critique and alternative parameters. Cc/Agr Biol Environ 23:12–12
94. Dixon P (2003) VEGAN, a package of R functions for community ecology. J Veg Sci 14(6): 927–930. https://doi.org/10.1658/1100-9233(2003)014[0927:Vaporf]2.0.Co;2
95. Simpson EH (1949) Measurement of diversity. Nature 163(4148):688–688. https://doi.org/10.1038/163688a0
96. Zhao N, Chen J, Carroll IM, Ringel-Kulka T, Epstein MP, Zhou H, Zhou JJ, Ringel Y, Li HZ, Wu MC (2015) Testing in microbiome-profiling studies with MiRKAT, the microbiome regression-based kernel association test. Am J Hum Genet 96(5):797–807. https://doi.org/10.1016/j.ajhg.2015.04.003
97. William HG (1994) Accounting for excess zeros and sample selection in Poisson and negative binomial regression models. New York University, Leonard N. Stern School of Business, Department of Economics

Chapter 11

Statistical and Computational Methods for Microbial Strain Analysis

Siyuan Ma and Hongzhe Li

Abstract

Microbial strains are interpreted as a lineage derived from a recent ancestor that have not experienced "too many" recombination events and can be successfully retrieved with culture-independent techniques using metagenomic sequencing. Such a strain variability has been increasingly shown to display additional phenotypic heterogeneities that affect host health, such as virulence, transmissibility, and antibiotics resistance. New statistical and computational methods have recently been developed to track the strains in samples based on shotgun metagenomics data either based on reference genome sequences or Metagenome-assembled genomes (MAGs). In this paper, we review some recent statistical methods for strain identifications based on frequency counts at a set of single nucleotide variants (SNVs) within a set of single-copy marker genes. These methods differ in terms of whether reference genome sequences are needed, how SNVs are called, what methods of deconvolution are used and whether the methods can be applied to multiple samples. We conclude our review with areas that require further research.

Key words Genetic polymorphisms, Genome assembly, Haplotype, Deconvolution, Strain typing

1 Introduction

The human-associated microbiota is known to have implications on host health and disease risk, with well-established links between various complex health conditions versus elements of the microbiome, including member clades, their functional capacities, transcriptional activities, and metabolic products [1, 2]. Beyond association findings at the species (or higher) taxonomic levels, studies have been increasingly interested in sub-species markers, i.e., microbial strain-level variations [3]. Within the same species, microbial strains can display additional phenotypic heterogeneities that affect host health, such as virulence, transmissibility, and

---

Short title: Resolving Microbial Strain Mixtures

This research is supported by NIH grants GM123056 and P30DK050306.

Brooke L. Fridley and Xuefeng Wang (eds.), *Statistical Genomics*,
Methods in Molecular Biology, vol. 2629, https://doi.org/10.1007/978-1-0716-2986-4_11,

antibiotics resistance [4]. Identifying and tracking such strain-level variations is an active focus of modern microbiome epidemiology and bioinformatics research [5–7], which have already yielded promising findings such as body site niche adaptation in the healthy human microbiome [8], engraftment signatures of fecal microbiome transplant [9], and host-specific evolution and transmission dynamics of skin strain mixtures [10].

To discuss these recent research, it is important that we first clarify the definition of microbial strains as reported in culture-independent, metagenomic community surveys. In culture-based context, microbial strains comprise clonal or near-clonal descendants of the same ancestor [11]. Characterizing such perfect genomic identity becomes impractical for culture-independent community-level epidemiology, and studies often opt for an effective phylogenetic identity threshold that is more stringent than the typical species-level variation (e.g., >99% identical genomes) [12]. With this definition, microbial strains as characterized in metagenomic studies are best interpreted as a lineage derived from a recent ancestor that have not experienced "too many" recombination events and can be successfully retrieved with culture-independent techniques [13].

To this end, there have been a series of computation methods for tracking microbial strains in metagenomic surveys in the past decade. Compared to standard taxonomic profiling [14, 15], these newer tools utilize more precise computational techniques to identify sub-species genomic variants and combine them as joint evidence to profile microbial strains. For 16S rRNA amplicon sequencing, newer pipelines [16] employ quality information to model sequencing errors statistically and can reconstruct exact amplicon sequences instead of traditional operational taxonomic units as ambiguous sequence clusters. These exact amplicon sequences variants (ASVs) are shown to represent sub-species strain identities. Different microbial strains can nevertheless be identical in their 16S rRNA amplicon region [17], and it is indeed far more common to track strain variations with metagenomic shotgun sequencing.

The typical approach in strain identification with metagenomic shotgun sequencing first identifies high-confidence genetic variants in a sample, e.g., single nucleotide variants (SNVs), either on sequenced species genomes (for reference-based approaches [18, 19]) or assembled and binned contig (for de novo assembly-based approaches [5, 20]). Earlier approaches simply concatenate these genetic variants to profile the dominant strain per-species [18]. More recent consensus is that sub-species genotypes are more truthfully represented by mixtures of multiple strains [5]. Thus, newer methods aim to resolve such mixtures by deconvoluting SNV profiles computationally, at the cost of increased coverage requirement such that non-dominant strains can be captured.

Given the overall framework, there are many important considerations for metagenomic strain profiling, including the choice between reference-based and assembly-based approaches, whole-genome versus single-copy universal gene variant calling, utilizing both strain-specific SNVs and structural variations, detection limits related to sequencing depth, and community complexities. Fortunately, multiple recent, high-quality reviews on these topics [3, 13, 21–23] provide detailed explanation, commentaries, and recommendations regarding different methods and their advantages. However, the modern issue of how individual microbial strains are deconvoluted from the mixture of their genetic variants has not been covered, likely because the problem itself only became reliably feasible in recent years with advances in sequencing technology and computational methods.

In this review, we defer to existing strain-profiling reviews for the reader interested in broad overviews of the topic and focus on the active research problem of resolving strain mixtures. We first provide clear definitions of the computation task, with identified genetic variants as observed input data, and individual strains haplotypes and their relative abundances as the unobserved parameter of interests. We also summarize a basic solution framework that is nevertheless ubiquitous as the core of most existing approaches in the field. We then dedicate the most of this review to various important technical considerations as extensions to the basic framework, and highlight corresponding methods. We last conclude with discussions for future research opportunities and challenges.

2 Resolving Strain Mixtures from Metagenomic Shotgun Sequencing Data

2.1 Strain Mixtures and Observed Frequency Matrix

For the majority of methods that resolve microbial strain mixtures from metagenomic shotgun samples (*see* Table 1), the first step is to generate an SNV profile $\mathbf{X}^{ij} = (\mathbf{X}_1^{ij}, \mathbf{X}_2^{ij}, \ldots, \mathbf{X}_{V^{ij}}^{ij})$ for sample i and species j. Here V^{ij} is the total number of identified SNV loci. As many methods infer microbial strains on a per-sample/species basis, we often omit the superscripts i and j to simplify notations, where it is implicitly assumed that the discussion concerns a specific sample and species pair. At the v-th SNV, $\mathbf{X}_v = (x_{1v}, x_{2v}, x_{3v}, x_{4v})$ is a length-four vector of allele frequencies for the four nucleotides. To produce this, in a reference-based approach (e.g., MetaPhlAn [18]), metagenomic shotgun reads are mapped to (parts of) a known species reference genome. Existing SNV calling methods are then used to identify the V SNVs and count the total number of reads for each allele to provide $\mathbf{X}_v$. For reference-free approaches (e.g., DESMAN [24]), assembled and clustered contig bins representing individual species (i.e., metagenome-assembled genomes, or MAGs [25]) are used in place of a reference genome. With either

Table 1
Computation methods to resolve microbial strain mixtures from metagenomic shotgun sequencing samples

Tool	Reference- or assembly-based?	How are SNVs called?	Brief description of deconvolution method	Additional comment
ConStrains [33]	Reference	Reads are mapped to species-specific marker genes for identifying SNVs.	Per-sample strains are inferred with a greedy algorithm. These are then clustered across samples to infer cohort-wise consensus strains. Strain relative abundances are then estimated with likelihood inference.	Number of strains is selected statistically with AIC.
DESMAN [24]	Assembly	Single-copy core genes (present in all strains) are selected from MAGs. Reads are mapped to these to identify SNVs.	MCMC inference of the joint likelihood for strain haplotypes and relative abundances across all samples.	Sequencing errors are considered in the phasing model. Number of strains is selected with Bayesian posterior mean deviance. Given estimated core gene haplotypes and relative abundances, strain-specific accessory genes can be inferred.
StrainFinder [9]	Reference	Phylogenetic markers are identified from reference genomes. Metagenomic reads are mapped to the marker genes to identify SNVs.	EM optimization of the joint likelihood for strain haplotypes and relative abundances across all samples	Sequencing errors are considered in the phasing model. Number of strains is selected with AIC. Strain haplotypes are relaxed to be "fuzzy" (continuous between zero and one) and L2-penalized to approximate discreteness.
BHap [20]	Assembly	SNVs are identified with branching contigs in the assembly de Bruijn graph	Greedy iterative approach. Coverage of SNV-specific contigs is decomposed into Poisson mixtures, with the smallest Poisson mean component assumed to represent an individual strain. Haplotype for the strain is then inferred given the coverage estimates. The process is repeated until contig coverages are exhausted.	Strain haplotypes are inferred per-sample. Different k-mer lengths are used to construct de Bruijn graphs and obtain consensus haplotypes.

PStrain [28]	Reference	Reads are mapped to species-specific marker genes for identifying SNVs.	Objective function is the L1 difference between observed, bias-corrected SNV frequencies, versus the theoretical frequencies given strain relative abundances and haplotypes. Inference is through iterative optimization over strain relative abundances and haplotypes.	Has bias correction on the observed SNV frequencies from prior knowledge. Incorporates "second-order" SNV frequencies as the genotypes of two adjacent loci. Selects the number of strains with an "elbow point" in the objective function.
STRONG [5]	Assembly	Assembly de Bruijn subgraphs are used to represent core genes. Read counts are tabulated against vertices in the subgraphs as opposed to SNV loci.	Likelihood is specified for observed per-vertex/sample read counts and hidden parameters including strain haplotypes and relative abundances. Inference is performed with variational approximation.	Operates directly on assembly graphs as opposed to called SNVs across reference genomes or contigs. Number of strains is selected with cross-validation of log predictive posterior likelihood.
StrainFacts [7]	Reference	Reads are mapped against reference genomes to report allele counts at each SNV site.	Likelihood is specified for observed allele counts given strain haplotypes and relative abundances along with hyperparameters. Maximum a posteriori estimates are inferred with a prior annealing approach.	Adopts "fuzzy" (between zero and one) genotypes with continuous gradients to enable fast computation. Discreteness is encouraged with shifted-scaled Dirichlet priors.

approach, the joint vector **X** thus represents the SNV profile for a given sample and species and summarizes the within-species genetic variants that are used as the input to profile the different strains present in the sample.

Now assume there are a total of S strains with varying genotypes and abundances. Notation wise, we use the length-V vector $\boldsymbol{g}_s = (g_{s1}, g_{s2}, \ldots, g_{sV})$ to represent the genotype profile for the s-th strain, i.e., its haplotype. Here $g_{sv} \in \{1, 2, 3, 4\}$ represents the nucleotide for strain s at SNV loci v. We use π_s to represent the relative proportion of strain s. $\sum_{s=1}^{S} \pi_s = 1$ by definition. The total number of strains S, their haplotypes $\mathbf{G} = (\boldsymbol{g}_1, \boldsymbol{g}_2, \ldots, \boldsymbol{g}_S)$, and relative abundances $\boldsymbol{\pi} = (\pi_1, \pi_2, \ldots, \pi_S)$ are all unobserved; providing their estimates, $(S, \mathbf{G}, \boldsymbol{\pi})$ from the input data, the allele frequency vector **X**, is the key problem of resolving strain mixtures from metagenomic shotgun data.

Given observed allele frequencies, there can be different possible strain mixtures that explain the data. This ambiguity is related to haplotype phasing problems in human genetics (*see* Fig. 1a) [26]. The human genome is diploid, and the same copy number of the paternal and maternal alleles renders different allele combinations across SNV loci equally likely. It is thus impossible to phase two parental haplotypes using SNV frequencies alone, and additional information such as population linkage is required. For microbial strain mixtures, it is highly unlikely that different strains of the same species will be of exactly even abundance. In fact, ecological dynamics suggests the configuration of a dominant strain in each community along with other subdominant ones [19]. While at the first glance the possibility of various microbial strain presence at different proportions is a complication from human diploidy, this unevenness actually provides crucial co-abundance information for resolving strain mixtures using allele frequency data (**X**) alone.

Intuitively, at each SNV loci, the most abundant allele is more likely to represent the dominant strain (*see* Fig. 1b). This is in fact the philosophy adopted in the simplistic strain mixture resolving methods described above [18, 27]. That is, without loss of generality, let the first strain be the dominant one ($\pi_1 > \max\{\pi_2, \pi_3, \ldots, \pi_S\}$). These approaches only aim to estimate this dominant strain's haplotype $\boldsymbol{g}_1$, by setting its genotype at each locus to the most frequent one, i.e., $g_{1v} = \text{argmax}_k x_{kv}$, where $k \in \{1, 2, 3, 4\}$. For samples with extreme unevenness, this approach provides good-quality representatives for strain-level variations, and often requires only moderate read coverage ($\sim 1\times$) to perform properly [21]. More recent methodology expands on this co-abundance approach, by leveraging deeper read coverage ($\geq 10\times$ [21]) and more precise statistical models, to reveal the haplotypes of less abundant strains in the mixture (*see* Fig. 1b–c). This is realized by computing the most "likely" strain mixture

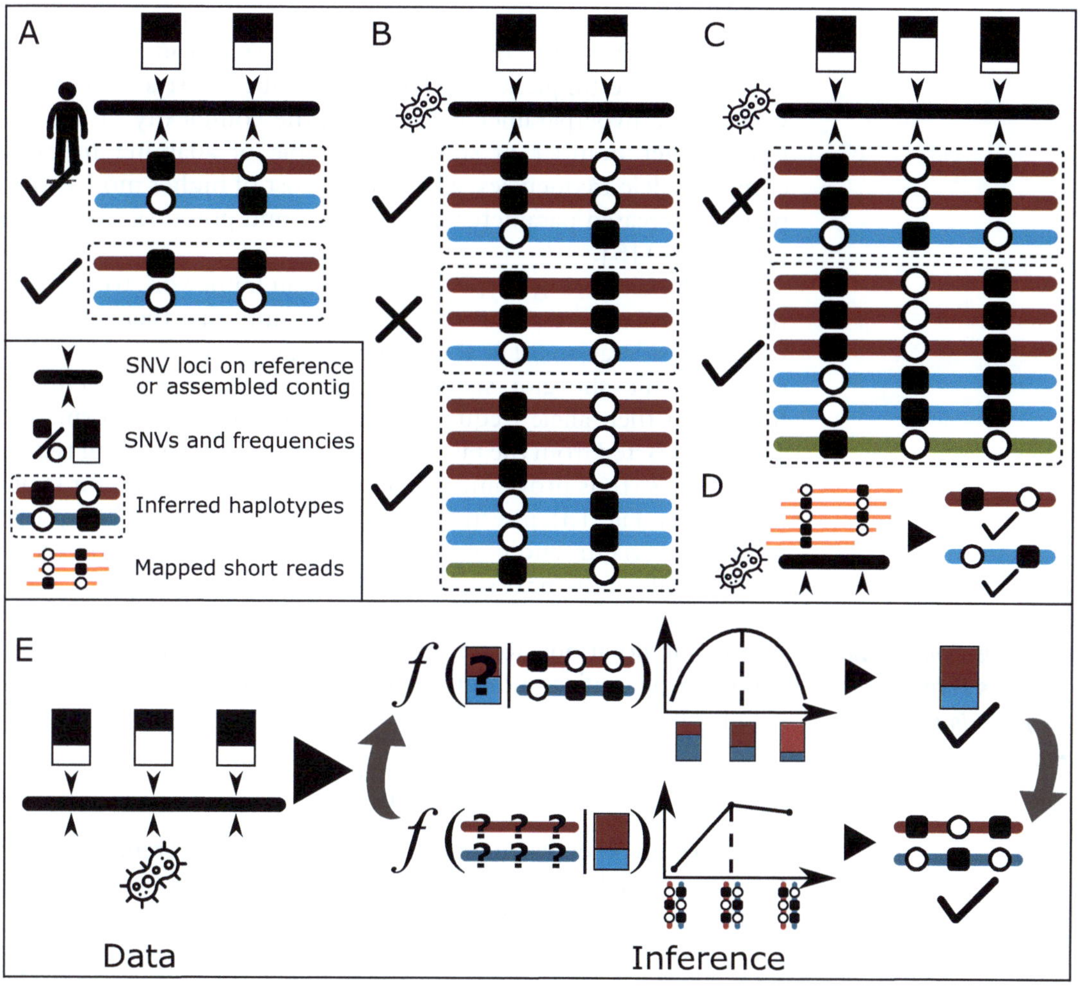

Fig. 1 Statistical phasing resolves strain-level haplotypes from sequencing allele frequencies. We assume SNV loci have been called on a reference genome or Metagenome-Assembled Genome (MAG), with allele frequencies estimated from reads mapped to each SNV site. (**a**)–(**d**): haplotyping from allele frequencies in human statistical genetics (**a**) compared to microbial species (**b**)–(**d**). A: the human genome has exactly two allele copies at each SNV site. Phasing is thus impossible from allele frequencies alone, and must incorporate additional information such as population references and recombination rate. This is different from microbial species as a mixture of strains (**b**)–(**d**), which are unlikely to be of equal abundance. (**b**): the uneven composition is leveraged to resolve haplotypes quantitatively based on co-abundance, that is, alleles of the same haplotype should have similar abundance. Different potential strain numbers add ambiguity, in which case methods often opt for simpler solutions. (**c**): more complicated allele frequency profiles can be explained by additional strain mixtures at the cost of model complexity. (**d**): in addition to quantitative phasing based on co-abundance, for SNV loci that are located closely (i.e., less than read length), physical linkage can directly resolve short-range haplotypes. (**e**): a typical algorithm for phasing microbial strain haplotypes based on species allele frequencies. Given the input data of SNVs' allele frequencies, the two parameters of interest, strain haplotypes and their composition, are inferred by iteratively optimizing an objective function (often a statistical likelihood) given the other parameter until convergence

profiles, including number of strains, their haplotypes, and relative abundances $(\hat{S}, \hat{\mathbf{G}}, \hat{\boldsymbol{\pi}})$, that provides the best explanation for the observed allele frequencies. Here the number of strains S can be viewed as a hyperparameter describing the complexity of the system. Larger number of strains is capable of adapting to more complex allele frequencies, albeit at the cost of higher model complexity, increased parameter search space, and potential overfitting (*see* Fig. 1c) [5].

Before discussing a typical computation framework to calculate the strain mixture estimates $(\hat{S}, \hat{\mathbf{G}}, \hat{\boldsymbol{\pi}})$, we note that there are additional considerations. The framework we detailed so far only utilize metagenomic shotgun sequencing data through the identified SNVs and their allele frequencies $\mathbf{X}$. However, other information can be extracted from the raw data, including reads spanning two or more SNVs (thus providing physical linkage information for strain haplotypes, Fig. 1d), insertion/deletions in addition to SNVs, large structural variants, and *k*-mer abundance profiles. This is covered in more details in the next section.

2.2 Resolving Strain Mixtures Formulated as an Optimization Problem

Given our notations above, the problem of resolving strain mixtures can be unified and formulated quantitatively as an optimization problem:

$$(\hat{S}, \ \mathbf{G}\ , \ \boldsymbol{\pi}) = \underset{(S, \mathbf{G}, \boldsymbol{\pi})}{\operatorname{argmax}} f(\mathbf{X}; S, \mathbf{G}, \boldsymbol{\pi}) \qquad (1)$$

where f is an objective function, quantifying the difference between the observed allele frequencies $\mathbf{X}$, and the theoretical ones provided by the underlying strain mixture configuration (S, $\mathbf{G}$, $\boldsymbol{\pi}$). Specifically, given (S, $\mathbf{G}$, $\boldsymbol{\pi}$), the theoretical relative frequency of allele k ($k \in \{1, 2, 3, 4\}$) at the v-th SNV is given by $\sum_{s=1}^{S} \pi_{sv} I\{g_{sv} = k\}$, which should correlate well with the observed allele frequency X_{vk}. f thus quantifies the level of this agreement, and in practice is often set as a statistical likelihood function of observing $\mathbf{X}$ given the theoretical frequencies, while accounting for the randomness in shotgun reads generation and potential sequencing errors [9, 24, 28]. As mentioned above, S is usually treated as a hyperparameter for controlling model complexity. Given a fixed S, optimization for $\mathbf{G}$ and $\boldsymbol{\pi}$ can be achieved by iteratively fixing one and solving for the other (*see* Fig. 1e): fixing the underlying strain haplotypes $\mathbf{G}$, the theoretical relative allele frequencies are linear combinations of strain relative abundances $\boldsymbol{\pi}$, which intuitively can be solved by matching against the observed allele frequencies $\mathbf{X}$. Similar arguments hold for the opposite (fixing $\boldsymbol{\pi}$ and estimating $\mathbf{G}$). A typical computation framework to solve the optimization problem (1) can thus be summarized as:

1. Determine a possible range of number of strains in the mixture $\{1, 2, \ldots, S_{\max}\}$.

2. For S in $\{1, 2, \ldots, S_{\max}\}$:
   - Optimize (1) given S, and find solutions $(\mathbf{G}^{(S)}, \boldsymbol{\pi}^{(S)})$.
   - This is often realized by optimizing $\mathbf{G}$ and $\boldsymbol{\pi}$ iteratively, using computation algorithms such as expectation-maximization [29] or Markov chain Monte Carlo [30].
3. Compare the fitted results against the different total number of strains S, and identify the optimal one S that provides the best balance between model complexity and goodness-of-fit to the data. $(S, \mathbf{G}^{(S)}, \boldsymbol{\pi}^{(S)})$ are the estimated strain mixture profiles and their relative proportions.

In this formulation, there are elements that are intentionally set to be general, such as the exact form of the objective function f, and a quantitative definition for model complexity in selecting optimal strain number S. This is so that we can encompass most existing methods. The choice of these model elements can be nuanced but important, and we also discuss these technical considerations in the following section.

3 Technical Considerations for Resolving Microbial Strain Mixtures

We first provide a systematic overview of recent methods for resolving strain mixture profiles from metagenomic shotgun sequencing samples in Table 1. Brief descriptions of each method's design and deconvolution method are provided. We then discuss specific technical considerations of these methods, organized by topics.

3.1 Reference-Based Versus Assembly-Based Approaches

As discussed above, the problem of deconvoluting SNV profiles to resolve strain mixture is somewhat independent of the requirement of a reference. With both approaches, SNVs are called with standard methodology on either a known species reference versus on MAG contig bins. Consequently, the coverage requirement for resolving multiple strains is similar between reference- versus assembly-based methods, as a $10\times$ read depth translates to similar coverage of SNVs on both the reference genome and a MAG bin.

For either method, it is important to note that not all SNVs are suitable as input to resolve strain mixtures. Specifically, only SNVs located in the species core genome are appropriate as input, in that such genes (a) are universal to all strains within the species and (b) have the same single copies for all strains. The second point is especially important, as strains with structural variants can have multiple copies of an accessory gene. This affects the formulation of (1), where it implicitly assumes that strain relative proportions are consistent across SNV loci, as multiple gene copies at a particular locus can yield overrepresented abundance for the strain. To avoid this complication, many reference-based and assembly-based approaches focus on universal, single-copy "marker" genes of a given species [24, 28]. For assembly-based approaches where reference genomes are not available, methods have proposed to utilize universal single-copy genes that are present in all bacteria and archaea. Once the strain relative abundances are determined, however, such strain-specific coverage can be utilized to identify accessory genes that are specific for a certain strain, by correlating their coverage versus that of the inferred strains. By adopting this approach, assembly-free methods such as DESMAN [24] have been successful to reconstruct the accessory genome of microbial strains entirely de novo.

3.2 Assembly-Graph-Based Approaches

As discussed earlier, the SNV frequencies do not fully represent all the information contained in metagenomic shotgun sequencing reads. To fully utilize this information, some modern assembly-based methods such as BHap [20] and STRONG [5] choose to operate computations directly on assembly graphs. That is, the observed data $\mathbf{X}$ in (1) is replaced with the constructed de Bruijn graph along with the coverage of each unitig. There are at least two benefits with this type of new approaches. First, de Bruijn graphs cover all the physical linkage information in the reads [31], not just SNVs frequencies. As such, short-range haplotypes where reads span two or more SNV loci are immediately resolved (*see* Fig. 1d). Other long-range information such as structural variation can also be potentially incorporated. Second, operating directly on the coverage of unitigs removes the intermediate step of read mapping for calling SNVs and calculating their frequencies, during which additional mapping ambiguities, biases, and errors might occur [5]. This approach does increase the input data load (unitig coverages versus SNV frequencies), and computation techniques such as greedy heuristics (BHap) or variational inference (STRONG) were adopted for feasible model inference.

It is worthwhile to discuss, realistically, how often short reads can span two or more SNVs on bacteria genomes. We assume a 1% sub-species phylogenetic divergence on a bacteria gene. That is, SNVs are located on average 100 nucleotides apart [18]. Then, for an SNV with two other SNVs flanking its either side at 100 nucleotides distance, a short read of 150 nucleotide length that cover this SNV will cover at least one of the other SNVs 2/3 of the time. While this probability will drop quickly with longer distance between SNVs, the approximate calculation does suggest that physical linkage from next generation sequencing reads can aid bacterial haplotyping, in addition to SNV frequencies. Besides the assembly-graph-based methods that directly employ such information, PStrain [28], a reference-based approach, incorporates physical linkages by including bi-SNV frequencies of adjacent loci pairs in the model objective in (1) and provides evidence for improved strain profiling. Alternatively, the introduction of long reads and corresponding methods can vastly expand the possibility of physical linkage compared to shotgun sequencing [32].

3.3 Tracking Strains Across Multiple Samples

Our discussion so far is limited to resolving strain mixtures on a per-metagenomic-sample basis. Each host subject, affected by differences in geographical regions, individualized diets, and physiology, tends to be heterogeneous in terms of both strain acquisition and internal evolutionary pressure for its microbial community [10]. It is thus often assumed that different hosts typically do not share the same bacteria strains. This, however, can be generalized in two different ways. First, in a longitudinal study or during a pandemic outbreak, certain strains are preserved within or transmitted across subjects. In this case, methods such as DESMAN [24] and STRONG [5] combine multiple samples assumed to share common strains and perform sample-aggregated inference. This enables a principled way for epidemiologically tracking strain profiles over time or across the population. An additional computational benefit of this setup is that read coverage can be considered as shared across samples, where the detection limit of a less abundant strain in at least one of samples is determined by the joint read coverage, as opposed to coverage in a single sample.

Second, by relaxing the phylogenetic divergence threshold of defining individual strains (but still stringent enough, i.e., at the sub-species level), methods such as ConStrains [33] further cluster different strains across samples into what is most accurately described as "sub-species clades." These clades no longer achieve the resolution of individual strains but rather represent population-level, within-species genetic variations that can nevertheless still characterize meaningful phenotypical and clinical associations.

3.4 Selecting the Optimal Number of Strains in the Mixture

Across different methods, determining the optimal number of strains S is a common problem. As discussed before, this effectively concerns the trade-off between model complexity versus goodness-of-fit. To this end, some methods utilize metrics in statistical literature such as the Akaike information criterion [34], by adding an additional term to the objective function f that penalizes higher model complexities and overfitting (i.e., larger S). Others adopt the cross-validation technique, by splitting the data into random training and validation sets to tune the optimal S value. This approach has been shown to achieve good empirical performance, albeit at extra computation burden (model fitting is required once per CV fold) [5]. Certain intuition-based selection, such as the elbow-point method by examining the objective function values [28], or early stopping in greedy search algorithms, where possible strain haplotypes are called one at a time [20], are also available.

4 Computation Considerations

The optimization problem (1) is non-trivial. First, the objective function f is often non-convex in its parameter space, and computation techniques such as the EM-algorithm and MCMC can only identify local optimums. In practice, multiple random initializations are required for a method to identify a most optimal strain mixture. Second, at each loci v, there are a total of 4^S possible allele configurations for the genotypes of the S strains $(g_{1v}, g_{v2}, \ldots, g_{vS})$. Optimization of this vector thus becomes exponentially expensive when the number of possible strains S becomes even moderately large. To bypass this problem, methods such as StrainFinder [9] and StrainFacts [7] have adopted "fuzzy" genotypes, where each strain genotype g_{sv} is allowed to be continuous rather than discrete, thus enabling smooth gradients for faster optimization computation. These methods also utilize specialized Bayesian priors for the continuous genotype parameters to encourage the fitted values toward one of the available alleles.

5 Discussion and Future Directions

We have reviewed some key recent statistical methods for identifying different strains in given metagenomic samples, primarily from coverage of single nucleotide variants in single-copy core genes. After such strains are identified along with their relative abundances, commonly used statistical methods can then be applied for downstream analysis, though the data will have higher sparsities compared to, e.g., species-level abundances [35].

One technical complication for downstream strain-level analysis is the reproducibility across computation methods. Because different algorithms track strains using SNVs in different genes, the resulting strains from different algorithms are not directly comparable, which should be considered in interpretation of strain-level association analysis. This problem already exists for traditional microbial taxonomic analysis [36] and will only be exacerbated for strain-level data. To deal with rare strains or rare haplotypes in association analysis, the association tests developed in human genetics such as burden tests [37] or kernel-based tests [38] can be applied.

Although the methods we reviewed can be applied to MAGs, there are inherent limitations to MAGs assembled from short-read data of shotgun sequencing. Short-read MAGs are often fragmented into hundreds or even thousands of short contigs because short-read assemblers cannot correctly deconvolute the locations of repeated genomic elements as reads are too short to span the elements and their flanking regions. These difficult-to-assemble elements include mobile genetic elements, which are commonly repeated within and across bacterial genomes [39]. These mobile genetic elements have functional potential and can vary across strains of the same species. Long-read sequencing technologies will certainly help to improve the contiguity of MAGs [39], which will also help to understand strain variations due to these mobile elements. Such long read data can potentially lead to improvements on strain-resolved metagenome assembly in environments of various complexities [40].

The strains identified and the SNVs used can also be used for strain-level metagenomic estimation of growth rate (SMEG) for inferring growth rates of bacterial sub-species, or strains, from complex metagenomic samples [41]. Emiola et al. [41] developed an approach using differential coverage patterns of SNVs in universally conserved or core genes to estimate strain-specific growth. One key step of their approach is to conduct core genome alignment and subsequently cluster strains based on their phylogenetic distance. The rationale behind clustering the strains is to reduce the SNVs dimensionality of the genome database and increase robustness of strain assignment by identifying the most likely consensus SNVs unique to a cluster. The underlying biological assumption is that strains constituting a cluster have high phylogenetic relatedness and will have similar growth rate in a sample [41].

Acknowledgements

The authors would like to thank Andrew Ghazi, Phillip Münch, Di Chen, Jordan Jensen, Yancong Zhang, Curtis Huttenhower, and Christopher Quince for their helpful input and discussions on this review.

References

1. Lloyd-Price J, Arze C, Ananthakrishnan AN et al (2019) Multi-omics of the gut microbial ecosystem in inflammatory bowel diseases. Nature 569(7758):655–662
2. Zhou W, Sailani MR, Contrepois K et al (2019) Longitudinal multi-omics of host–microbe dynamics in prediabetes. Nature 569(7758):663–671
3. Van Rossum T, Ferretti P, Maistrenko OM, Bork P (2020) Diversity within species: interpreting strains in microbiomes. Nat Rev Microbiol 18(9):491–506
4. Fournier PE, Dubourg G, Raoult D (2014) Clinical detection and characterization of bacterial pathogens in the genomics era. Genome Med 6(11):1–15
5. Quince C, Nurk S, Raguideau S et al (2021) Strong: metagenomics strain resolution on assembly graphs. Genome Biol 22(1):1–34
6. van Dijk LR, Walker BJ, Straub TJ et al (2022) StrainGE: A toolkit to track and characterize low-abundance strains in complex microbial communities. Genome Biol 23(1):1–27
7. Smith BJ, Li X, Abate A et al (2022) Scalable microbial strain inference in metagenomic data using StrainFacts. bioRxiv
8. Lloyd-Price J, Mahurkar A, Rahnavard G et al (2017) Strains, functions and dynamics in the expanded human microbiome project. Nature 550(7674):61–66
9. Smillie CS, Sauk J, Gevers D et al (2018) Strain tracking reveals the determinants of bacterial

engraftment in the human gut following fecal microbiota transplantation. Cell Host Microbe 23(2):229–240
10. Zhou W, Spoto M, Hardy R et al (2020) Host-specific evolutionary and transmission dynamics shape the functional diversification of staphylococcus epidermidis in human skin. Cell 180(3):454–470
11. Garrity GM, Parker CT, Tindall BJ (2015) International code of nomenclature of prokaryotes. Int J Syst Evol Microbiol 90(6)
12. Jain C, Rodriguez-R LM, Phillippy AM et al (2018) High throughput ani analysis of 90k prokaryotic genomes reveals clear species boundaries. Nat Commun 9(1):1–8
13. Yan Y, Nguyen LH, Franzosa EA, Huttenhower C (2020) Strain-level epidemiology of microbial communities and the human microbiome. Genome Med 12(1):1–16
14. Segata N, Waldron L, Ballarini A et al (2012) Metagenomic microbial community profiling using unique clade-specific marker genes. Nat Methods 9(8):811–814
15. Wood DE, Salzberg SL (2014) Kraken: ultrafast metagenomic sequence classification using exact alignments. Genome Biol 15(3):1–12
16. Callahan BJ, McMurdie PJ, Rosen MJ et al (2016) Dada2: High-resolution sample inference from Illumina amplicon data. Nat Methods 13(7):581–583
17. Brenner DJ, Fanning G, Steigerwalt A et al (1972) Polynucleotide sequence relatedness among three groups of pathogenic Escherichia coli strains. Infect Immun 6(3):308–315
18. Truong DT, Tett A, Pasolli E et al (2017) Microbial strain-level population structure and genetic diversity from metagenomes. Genome Res 27(4):626–638
19. Albanese D, Donati C (2017) Strain profiling and epidemiology of bacterial species from metagenomic sequencing. Nat Commun 8(1):1–14
20. Li X, Saadat S, Hu H, Li X (2019) BHap: a novel approach for bacterial haplotype reconstruction. Bioinformatics 35(22):4624–4631
21. Anyansi C, Straub TJ, Manson AL et al (2020) Computational methods for strain-level microbial detection in colony and metagenome sequencing data. Front Microbiol 11:1925
22. Garg S (2021) Computational methods for chromosome-scale haplotype reconstruction. Genome Biol 22(1):1–24
23. Ghazi AR, Münch PC, Chen D et al (2022) Strain identification and quantitative analysis in microbial communities. J Mol Biol, 167582
24. Quince C, Delmont TO, Raguideau S et al (2017) Desman: a new tool for de novo extraction of strains from metagenomes. Genome Biol 18(1):1–22
25. Pasolli E, Asnicar F, Manara S et al (2019) Extensive unexplored human microbiome diversity revealed by over 150,000 genomes from metagenomes spanning age, geography, and lifestyle. Cell 176(3):649–662
26. Loh PR, Danecek P, Palamara PF et al (2016) Reference-based phasing using the haplotype reference consortium panel. Nat Genet 48(11):1443–1448
27. Scholz M, Ward DV, Pasolli E et al (2016) Strain-level microbial epidemiology and population genomics from shotgun metagenomics. Nat Methods 13(5):435–438
28. Wang S, Jiang Y, Li S (2021) Pstrain: an iterative microbial strains profiling algorithm for shotgun metagenomic sequencing data. Bioinformatics 36(22–23):5499–5506
29. Moon TK (1996) The expectation-maximization algorithm. IEEE Signal Process Mag 13(6):47–60
30. Cowles MK, Carlin BP (1996) Markov chain Monte Carlo convergence diagnostics: a comparative review. J Am Stat Assoc 91(434):883–904
31. Zerbino DR, Birney E (2008) Velvet: algorithms for de novo short read assembly using de Bruijn graphs. Genome Res 18(5):821–829
32. Dilthey AT, Jain C, Koren S, Phillippy AM (2019) Strain-level metagenomic assignment and compositional estimation for long reads with MetaMaps. Nat Commun 10(1):1–12
33. Luo C, Knight R, Siljander H et al (2015) Constrains identifies microbial strains in metagenomic datasets. Nat Biotechnol 33(10):1045–1052
34. Sakamoto Y, Ishiguro M, Kitagawa G (1986) Akaike information criterion statistics. Dordrecht, The Netherlands: D Reidel 81(10.5555):26853
35. Li H (2015) Microbiome, metagenomics, and high-dimensional compositional data analysis. Annual Review of Statistics and Its Application 2:73–94. https://doi.org/10.1146/annurevstatistics010814020351
36. Sinha R, Abu-Ali G, Vogtmann E et al (2017) Assessment of variation in microbial community amplicon sequencing by the microbiome quality control (MBQC) project consortium. Nat Biotechnol 35(11):1077–1086
37. Neale B, Rivas M, Voight B et al (2011) Testing for an unusual distribution of rare variants. PLoS Genet 7(3):e1001322

38. Wu M, Lee S, Cai T et al (2011) Rare-variant association testing for sequencing data with the sequence kernel association test. Am J Hum Genet 89:82–93
39. Moss EL, Maghini DG, Bhatt AS (2020) Complete, closed bacterial genomes from microbiomes using nanopore sequencing. Nat Biotechnol 38:701–707
40. Vicedomini R, Quince C, Darling AE, Chikhi R (2021) Strainberry: automated strain separation in low-complexity metagenomes using long reads. Nat Commun 12:4485
41. Emiola A, Zhou W, Oh J (2020) Metagenomic growth rate inferences of strains in situ. Sci Adv 6(17):eaaz2299

Chapter 12

Statistics and Machine Learning in Mass Spectrometry-Based Metabolomics Analysis

Sili Fan, Christopher M. Wilson, Brooke L. Fridley, and Qian Li

Abstract

In this chapter, we review the cutting-edge statistical and machine learning methods for missing value imputation, normalization, and downstream analyses in mass spectrometry metabolomics studies, with illustration by example datasets. The missing peak recovery includes simple imputation by zero or limit of detection, regression-based or distribution-based imputation, and prediction by random forest. The batch effect can be removed by data-driven methods, internal standard-based, and quality control sample-based normalization. We also summarize different types of statistical analysis for metabolomics and clinical outcomes, such as inference on metabolic biomarkers, clustering of metabolomic profiles, metabolite module building, and integrative analysis with transcriptome.

Key words Mass spectrometry, Metabolomics, Statistical and machine learning, Imputation, Normalization, Integrative analysis

1 Introduction

Mass spectrometry (MS) metabolomics have been employed in numerous biomedical studies in the past decade to discover metabolic biomarkers for phenotypes and exposures, as well as to understand the etiology in disease onset and prognosis. The preprocessing of metabolomics data includes peak calling, alignment, intensity quantification, backfilling and imputation of missing values or peaks, and normalization of batch effect. Downstream analysis for the normalized metabolomics data includes but not limited to metabolic biomarker identification, clustering of metabolome samples, metabolic network or module building, and integrative analysis of metabolome and other omics, such as transcriptome. The aforementioned preprocessing and analysis procedures are usually performed by various statistical and machine learning methods.

Brooke L. Fridley and Xuefeng Wang (eds.), *Statistical Genomics*,
Methods in Molecular Biology, vol. 2629, https://doi.org/10.1007/978-1-0716-2986-4_12,

The datasets produced by MS in metabolomics usually contain a significant component of missing values, usually between 10% and 40%, that can be attributed to technological and biological hazards. A variety of reasons can yield missing values: (1) Compound intensities are too low to be detected since they are lower than the threshold for detection; (2) Compound values were not detected in the chromatogram; (3) Deconvolution technique produced false negatives due to signal separation and compound values overlapped; (4) Errors that can be attributed to computations, measurements, or gross mistakes. Missing values in metabolomics datasets can originate from numerous sources, however, they are most often classified as (a) missing completely at random (MCAR), (b) missing at random (MAR), and (c) missing not at random (MNAR). Specifically, values that are MCAR if their missing mechanisms are not related to any observed feature or biological information, e.g., random errors that are caused by peak picking software. If the mechanisms are systematically related to biological information but not to observed features, the missing values are MAR. For example, missing values that occur only in control but not in any treatment groups due to sampling difficulties are MAR. A missing value that is MNAR if the missingness is related to any observed feature. For example, compound intensities that are too low to be detected are MNAR.

Untargeted metabolomics is widely used in clinical, epidemiological, and genetics studies, which can often involve hundreds to thousands of samples. Often, experiments are conducted in batches, for example, to allow for instrument maintenance, to exchange columns and solvents, or due to instrument availability. It is not uncommon for data acquisition to span several weeks or months, resulting in systematic errors, such as temporal drift (e.g., due to a decrease in instrument sensitivity), batch effects (e.g., due to changes in tuning parameters or due to maintenance work), or due to smaller technical issues, such as slightly different pH values in the solvent. The lack of proper treatment of unwanted variances (i.e., technical errors) can significantly reduce the power of detecting metabolites associated with phenotypes of interest. For a case-control study, a 5% increase in standard deviation for metabolites with a small effect size (Cohen's d = 0.2) would require 41 more samples to achieve 80% statistical power.

Detecting differentially abundant metabolites is an essential aim in clinical studies that utilize metabolomics profiling. The parametric and non-parametric tests can be used to compare metabolite abundance between groups. Regression-based models with metabolites as outcome and the clinical variable(s) of interest as covariate or vice versa is another powerful strategy to test the association with exposure groups or phenotypes. For an observational study with a matching design, the match set indicator should be incorporated in the association analyses. Another type of analysis in clinical study is the identification of subgroups or subtypes of disease by unsupervised clustering of metabolomes.

Network-based module building is a popular analysis used in many transcriptomic, metabolomic, and proteomic studies. This analysis allows for the visualization and identification of modules and influential elements within each module and is often used as dimension-reduction technique in the interpretation of known pathways or other biological features. Network-based modules also provide insights for multi-omics data integration, which is a challenge due to high dimensionality in each data type and distinct data structures caused by different molecular sources and high-throughput technology. Several machine learning methods, such as multiple kernel learning, random forest, and integrative clustering, have been developed to leverage information from multiple data types. Therefore, in this chapter we will review the recent and commonly used methods for imputation, normalization, and different types of downstream analyses in mass spectrometry metabolomics studies.

2 ADNI Dataset

Alzheimer Disease Neuroimaging Initiative, ADNI, is one the largest cohort studies on Alzheimer's disease, aiming to gain new insight into the disease etiology through the use of neuroimaging. Barupal et al. [1] reported an untargeted lipidomics of 806 serum specimens from ADNI1 participants (188 patients with AD, 392 with mild cognitive impairment, and 226 cognitively normal subjects), as well as 83 quality control (QC) samples. Using an ultra-high-performance liquid chromatography quadruple/time-of-flight mass spectrometer (UHPLC-QTOF MS) instrument, lipids were detected and measured in both positive and negative electrospray ionization modes. A total of 630 lipid species are included in the dataset. The dataset also includes commercial internal standards, which were used for the retention time recalibration. In the dataset, there are 40% of compounds containing at least one missing value, and 13% of the compounds have more than 20% of missing samples. The relative standard deviation (RSD) of the measured lipids in the QC samples was calculated to demonstrate the reproducibility of the lipidomics assay. It is expected that lipid levels in quality control samples should not change significantly, providing evidence that the data processing techniques are robust and reproducible. Before normalization, the average RSD of the QC samples is 20.4%, and 45% of the compounds has a QC RSD less than 20%, which is a typically accepted threshold for lipidomics assay. We will illustrate the performance of SERRF and other normalization methods using this dataset in Subheading 4.3.2. This dataset is available at the ADNI repository at http://adni.loni.usc.edu/.

3 Missing Values Imputation Approaches

Missing values in metabolomics data are handled differently depending on the category of missingness. MCAR may be corrected through re-processing and re-analysis of the bio-samples for data acquisition. In instances where a baseline has been corrected but non-eluting compounds of the compound with missing values do not exist nearby, it has been suggested that peaks be filled simply by extracting and replacing with raw or baseline signals. To retrieve real signals of certain compounds, people can adjust or apply flexible parameter settings such as signal to noise ratio for peak picking at the data processing stage. The "80% rules" state that compounds with observed values less than 80% of the observed values in each biological group should be excluded from the data before filling in missing values. Various imputation strategies have been proposed to deal with missing values that persist after taking the above-mentioned steps. Next, we will give an overview of the common imputation approaches for missing values in metabolomics.

In the following context, we use X to denote a metabolomics dataset, with rows indicating p compounds and columns indicating n samples. Since missing value imputation methods can be categorized as sample-wise and compound-wise, it is necessary to clearly distinguish samples (i.e., columns) and compounds (i.e., rows). We use $x_{j,}$ to denote the jth compound and $x_{,i}$ to denote the ith sample. The first compound, $x_{1,}$, and the first sample, $x_{,1}$, are often used as an illustration example. Without loss of generality, we suppose the first s_1 compounds in the first sample, i.e., $x_{1,1}, \cdots, x_{s_1,1}$, are missing, whereas the rest $x_{s_1+1,1}, \cdots, x_{p,1}$ are observed and non-missing:

$$x_{,1} = (\underbrace{x_{1,1}, \cdots, x_{s_1,1}}_{\text{missing}}, \underbrace{x_{s_1+1,1}, \cdots, x_{p,1}}_{\text{non-missing}})'.$$

Note, we use $()'$ to indicate that $x_{,1}$ as a column vector. Similarly, suppose first t_1 samples for the first compound are missing:

$$x_{1,} = (\underbrace{x_{1,1}, \cdots, x_{1,t_1}}_{\text{missing}}, \underbrace{x_{1,t_1+1}, \cdots, x_{1,n}}_{\text{non-missing}}).$$

We give a summary in Table 1 for commonly used missing value imputation methods.

3.1 Simple Imputation

In simple imputation, missing data are filled by different values such as zero, mean, median, and half-minimum of the non-missing values for each compound, depending on the assumption of the mechanism of missingness. This compound-wise procedure can be summarized using the following simple equation

Table 1
A summary table of missing value approaches mentioned

Missing value Imputation approach	Description	R package
Mean/Median/-Half Minimum	Replaces missing elements with an mean/median average value or half-minimum of non-missing elements	metabox
Zero/LOD	Replaces all missing elements with zero/LOD (i.e., limit of detection)	metabox
kNN	Select k-nearest samples using Euclidean metric or Pearson correlation, and imputes missing elements by averaging those non-missing elements of its neighbors	impute
NS-kNN	A modified k-nearest neighbors approach to take low-intensity biasedness into account	yaImpute
kNN-TN	Utilize truncated maximum Gaussian likelihood estimator for scaling in kNN model	N/A
SVD/BPCA	A linear combination of the most significant eigen-variables or right-singular vector of sample covariance to predict the missing values	pcaMethods
TS-lasso	Use LASSO to select and re-select predictors to increase the accuracy of the missing value prediction	GMSimpute
GSimp	Estimate the Normal distribution of the data and draw samples to impute the missing values	GitHub—WandeRum
BayesMetab	Uses a Bayesian model to first discriminate between random and not random missing data, and then sample values from estimated truncated distribution	N/A
Random Forest	Ensemble non-parametric tree models and randomly select features to increase the accuracy and robustness of the prediction	missForest

$$\hat{x}_{1,i} = f(x_{1,t_1+1}, \cdots, x_{1,n}),\ \forall 1 \le i \le t_1,$$

where f can be the formula of mean, median, half-minimum, and constant zero, and the *hat* sign, $\hat{x}$, indicates the value is predicted or modeled. Mean and median imputation are often used for MCAR/MAR missing values. When the distribution is skewed, which is often observed in metabolomics data, median imputation is more robust and closer to the truth than mean imputation. Zero- or half-minimum imputation is based on the assumption that the missing values are lower than the non-missing values. In general, it is suited where the missing values are lower than the limit of detection (LOD), such as low noise levels from instrument settings, preprocessing noise level, or compound low abundance. It is recommended to impute missing values using LOD or half-LOD when the LOD value is known. In cases where exact LOD information is not available for each compound, estimating the LOD from half the non-missing values is an appropriate choice. In most studies, simple imputation methods are preferred because of efficiency. However, the use of simple imputation methods may yield biased or inaccurate values in high-dimensional data, distort the distribution of metabolites abundance, and amplify variances that affect downstream multivariate analyses.

3.2 kNN-Based Methods

The k-nearest neighbor (kNN) method is a machine learning algorithm used in classification or regression. The kNN imputation is sample-wise, namely imputing missing values sample by sample. The assumption is that biologically similar samples will have similar measured values across the vast majority of their metabolites [2]. To impute a missing value for a target sample, the k most *similar* samples are found based on a defined distance metric calculated using the values of metabolites that are present in both the target sample and candidate neighbor samples [3, 4]. Specifically, suppose our objective is to impute values for the first sample whose first s_1 compounds are missing. Suppose in the dataset X, we have $c_1 - 1$ samples $x_{,2}, \cdots, x_{,c_1}$ having observed non-missing values for compounds $s_1 + 1$ to p. Then for each of $c_1 - 1$ samples, we calculate its distance to $x_{,1}$, i.e., $d(x_{,1}, x_{,i}) > 0, \forall i = 2, \cdots, c_1$. Without loss of generality, assume $x_{,2}, \cdots, x_{,k+1}$ are the k samples whose distances are closest to $x_{,1}$. Then the missing value can be imputed using the k selected samples by weighted average:

$$\hat{x}_{j,1} = \frac{1}{k_j} \sum_{i=1}^{k_j} w_i x_{j,i}, \ \forall j \in 1, \cdots, m,$$

where k_j is the number of non-missing values for the jth compound among the chosen k samples, and w_i's are positive weights can be defined as either 1 or the absolute value of $d(x_{,1}, x_{,i})^{-1} / \sum_{i=1}^{k_j} d(x_{,1}, x_{,i})^{-1}$, i.e., the longer the distance, the less the weight. Furthermore, it is worth noting that standardization may be necessary for kNN-based methods since metabolites are often of different magnitudes. In the absence of standardization, distance may simply be dominated by a few metabolites that have high intensities. The formula of standardization for the jth compound is given by $x_{j,}^s := \left(x_{j,} - \hat{\mu}_{j,} \right) / \hat{\sigma}_{j,}$, where $\hat{\mu}_{j,} := \frac{1}{n_j} \sum_{i=1}^{n_j} x_{i,j}$, $\hat{\sigma}_{j,} := \frac{1}{n_j} \sum_{i=1}^{n_j} (x_{i,j} - \mu_{j,})^2$, $n_j := n - t_j$ is the number of non-missing values, and x^s means it is standardized.

For certain compounds with a large portion of missing values, e.g., m/p is close to 1, the distance measure may be less robust because the number of non-missing values, which are used to build a kNN model, is too small. This phenomenon is common in metabolomics, especially for low-intensity compounds whose values are close to LOD. Lee [5] proposed No Skip-kNN (NS-kNN) to detect such low-intensity compounds and replace the missing values with the minimum value of the observed non-missing metabolite intensity before building a kNN model.

Depending on different distance metric measures, variants of kNN have been developed to impute the missing values. The most frequently used distance metrics are the Euclidean distance metric and the Pearson correlation metric. Continuing with the previous example, the Euclidean distance between the target sample with missing value, i.e., $x_{,1}$, and $x_{,i}$ is defined as

$$d_E(x_{,1}, x_{,i}) := \sqrt{\sum_{j=s_1+1}^{p} (x_{j,1} - x_{j,i})^2},$$

for the ith sample. And the Pearson correlation between two samples is defined as

$$r(x_{,1}, x_{,i}) := \frac{\frac{1}{n}\sum_{j=s_1+1}^{p} x_{j,1}x_{j,i} - \hat{\mu}_{,1}\hat{\mu}_{,i}}{\hat{\sigma}_{,1}\hat{\sigma}_{,i}},$$

where $\hat{\mu}_{,1} := \frac{1}{p-s_1-1}\sum_{j=s_1+1}^{p} x_{j,1}$ and $\hat{\sigma}_{,1} := \frac{1}{p-s_1-1}\sum_{j=s_1+1}^{p} (x_{j,1} - \hat{\mu}_{,1})^2$ are the estimated sample mean and standard deviation of the target sample $x_{,1}$. Notations for neighbor samples $x_{,i}$'s, i.e., $\hat{\mu}_{,i}$ and $\sigma_{,i}$, are similar. The distance metric based on Pearson correlation is usually defined as

$$d_P(x_{,1}, x_{,i}) := 1 - |r(x_{,1}, x_{,i})| > 0,$$

for the ith sample. The absolute operator $|\cdot|$ is added to incorporate both positive and negative correlations.

One limitation of the Euclidean distance and Pearson correlation-based distance is that the estimation of the $\hat{\mu}_{,i}$, $\hat{\mu}_{j,}$, $\hat{\sigma}_{,i}$, $\hat{\sigma}_j$ does not account for the truncation at the minimum value or the limit of detection. All the missing values of ith sample and jth compound are ignored when estimating the true value of the mean $\mu_{,i}$ $\mu_{j,}$ and standard deviation $\sigma_{,i}$, $\sigma_{j,}$, thus the estimations are biased. To overcome this problem, Shah [6] proposed kNN-TN (kNN-truncation) to use truncated maximum Gaussian likelihood estimator to estimate true parameters $\mu_{i,}$, $\mu_{j,}$ and $\sigma_{i,}$, $\sigma_{j,}$. Other kNN methods include sequential K-nearest neighbor and iterative K-nearest neighbor. The basic idea is reuse imputed data to increase imputation efficiency. These two kNN variants are commonly adopted microarray analysis and less common is metabolomics.

3.3 Regression-Based Methods

Regression-based methods are one of the preferred statistical techniques for handling missing values in metabolomics. The overall regression process consists of two phases. The first step uses all the complete observations to build a regression model, and in the second step, imputes missing data based on the built regression model. A key difference compared with kNN is that regression-based methods are compound-wise, i.e., imputing missing values compound by compound. In detail, for the imputation of compound $x_{1,}$, given a set of compound variables $x_{2,}, \cdots, x_{d_1,}$, a regression model, using samples with no missing values in $x_{1,}$, i.e., $t_1 + 1, \cdots, n$, is conducted as follows:

$$\grave{x}_{1,} = \beta_1 + \beta_2 \grave{x}_{2,} + \cdots + \beta_{d_1} \grave{x}_{d_1,} + \varepsilon_1, \quad (1)$$

where $\grave{x}$ means only complete samples are used, β_j's are coefficients, and ϵ_1 is the residual. The selected d_1 compounds are those that are complete in the non-missing value of $x_{1,}$, i.e., $x_{j,i}$ are not missing for all $j \in 1, \cdots, d_1$ and $i \in 1, \cdots, t_1$. A common procure to estimate β_j, $j = 1, \cdots, d_1$, is least square technique, whose goal is to find $\hat{\beta}_j$'s to minimize the regression residual error

$$L(\beta) = \sum_{i=t_1+1}^{n} \left[\grave{x}_{1,i} - (\beta_1 + \beta_2 \grave{x}_{2,i} + \cdots + \beta_{d_1} \grave{x}_{d_1,i})\right]^2. \quad (2)$$

The least square estimator of the solution of (2) is

$$\hat{\beta} := (\grave{X}\grave{X}\prime)^{-1} \grave{X}\grave{x}_{1,} \quad (3)$$

where $\grave{X}$ is the subset of data matrix of d_1 compounds and complete samples corresponding to $x_{1,}$. With $\hat{\beta}_j$'s, the missing values can be predicted as

$$\hat{x}_{1,} := \hat{\beta}_1 + \hat{\beta}_2 \grave{x}_{2,} + \cdots + \hat{\beta}_{d_1} \grave{x}_{d_1}. \quad (4)$$

One major limitation with the previous general framework is that the sample size may be less than the number of predictors, i.e., $n - t_1 < d_1$, making it impossible to use Eq. (4) because $\grave{X}\grave{X}'$ is not invertible and commutable. In practice, singular value decomposition (SVD) [2] and Bayesian principal component analysis (BPCA) [7] are commonly used to overcome this issue. Specifically, in Eq. (1) SVD replaced d_1 predictors by k most significant right-singular vectors ($k < n - t_1 < d_1$). The right-singular vectors is matrix of V in the singular value decomposition of $\grave{X} := U\Sigma V^T$, and the k vectors are the first k columns of V. When k is less than the sample size, least square estimate can be consequently used (SVDImpute). BPCA shares a similar principal of SVD when choosing predictors, except the k predictors are chosen based on the right-singular vector of sample covariance matrix, $\frac{1}{n-t_1}\grave{X}'\grave{X}$ (instead of $\grave{X}$). The BPCA method relies on the normality assumption and uses Bayesian algorithm to fit posterior distribution to estimate β_j's.

More recently, Li [8] proposed a generalized two-step Lasso approach (TS-lasso) that can provide higher accuracy in imputation and exhibits stable performance across different missing patterns.

The first step of TS-Lasso replaced the least square objective (2) by solving lasso regression,

$$L_{\text{Lasso}}(\beta) = L(\beta) + \lambda \sum_{j=2}^{d_1} |\beta_j|, \tag{5}$$

where $\lambda > 0$ is a pre-specified value which controls the number of nonzero estimator of $\hat{\beta}_j$'s. Li et al proposed to use R package, glmet, to choose λ automatically by cross-validation hyper-parameter tuning. By introducing the term $\lambda \sum_{j=1}^{d_1} |\beta_j|$, some of the coefficients of x_j,'s are set to be zero, which reduces the number of predictors in the Eq. (1). With $\hat{\beta}_{\text{Lasso}}$, Eq. (4) can be similarly used to impute the missing values to obtain a dataset with no missing values. At the second step, LASSO is re-fitted to make the final imputation. The difference with the first step is that all compounds are used in the Eq. (5), i.e.,

$$L_{\text{Lasso}}(\beta) = \beta_1 + \sum_{j=2}^{p} \beta_j \hat{x}_j, + \lambda \sum_{j=2}^{p} |\beta_j|, \tag{6}$$

where $\hat{x}$ indicates the compound intensity with missing values imputed. With coefficients of β_j's estimated by (6), Eq. (4) is used to conduct the final imputation on the dataset. By utilizing the two-step procedure, higher level of imputation accuracy and more consistent performance were achieved across different missing patterns.

The techniques outlined above are parametric and rely on least squares (2), which are sensitive to outliers. A notable regression-based method specifically designed to be robust with outliers is proposed by Kumar et al. [9], called Kernel weighted least square approach. The basic idea is that each metabolite cell assumed to be a product of unknown row effect and column effect, i.e., $x_{j,i} = r_j c_i$. The weighted least square are used to estimate r_j's and c_i's, where weights are determined by non-parametric statistical quantities, e.g., median, to reduce the effect of outlier. With the fitted row and column effects, $\hat{x}_{j,i} = \hat{r}_j \hat{c}_i$ are used to impute missing values. It is worth to note that the method is supervised, namely utilizing sample group information. For example, the missing values imputation in a biological group is determined by non-missing values in the same biological group. While the degree of accuracy can possibly be improved, one should be careful about false positives introduced simply by using imputation techniques in the biomedical field.

3.4 Distribution-Based Sampling Methods

The above-mentioned methods are deterministic, i.e., a specific values are predicted and directly used to impute the missing values. Another method of compound-wise imputation is based on the sampling distribution, which fits the distributions of non-missing values first, and then imputes missing values through iterative and random samplings based on the estimated distribution. Most of the distribution-based methods use MCMC (Markov chain Monte Carlo) as the sampling technique to draw samples from distribution. In this session, we illustrate two most commonly used method, GSimp and BayesMetab.

GSimp is specifically designed for the MNAR cases, where missing values are usually blow the LOD. Gibbs sampler imputes the missing values by sampling from a truncated normal distribution, whose mean is based on the fitted values from a prediction model and standard deviation is based on root mean square deviation (RMSD) of the predicted values. We again use the first compound imputation to illustrate the procedure step by step.

1. Initialize missing values in $x_{1,i}$, for i in $1, \cdots, t_1$, by any imputation method, e.g., anyone mentioned above. Denote the imputed first compound intensity $\hat{x}_{1.}$
2. Build a predictive model, e.g., using Eq. (5), using $\hat{x}_{1,}$ as response and $x_{2,}\cdots, x_{d_1,}$ are predictors.
3. Use the built model to make prediction of missing values, denoted as $\hat{x}'_{1,i}$ for i in $1, \cdots, t_1$.
4. For the ith sample, draw a sample from a truncated normal distribution with mean $\mu_{1,i} := \hat{x}'_{1,i}$ and standard deviation $\sigma^2_{1,} := \frac{\sum_{i=1}^{t_1}(\hat{x}'_{1,i} - \hat{x}_{1,i})^2}{t_1}$ as imputed missing value for $x_{1,i}$, i.e.,

$$\hat{x}_{1,i} \sim N(\mu_{1,i}, \sigma^2_{1,}),$$

for i in $1, \cdots, t_1$.
5. Truncate $\hat{x}_{1,i}$ to be less than LOD, or an estimate of LOD (e.g., half-minimum of non-missing values): if $\hat{x}_{1,i}$ is larger than (estimated) LOD, then replace it with (estimated) LOD.
6. Update $\hat{x}'_{1,}$ by $\hat{x}_{1,}$.

The above step 2–6 are performed for each compound, i.e., $x_{j,}$, to finish one iteration for the entire dataset. The whole procedure is performed for a given number of iterations, e.g., 20 which was used in the paper. GSimp approaches half-minimum and kNN-TN using Normalized Root Mean Squared Error as criterion on both of the real-world and simulated metabolomics datasets and demonstrates the superiority of this method from different perspectives.

BayesMetab is another powerful distribution-based approach to impute the missing value. Different from the GSimp, which assumes MNAR, BayesMetab first discriminate between random and not random missing data, and then use Markov chain Mete Carlo to sample from estimated distribution. Shah, et al. assume that if true value of $x_{j,i}$ is less than the threshold of LOD, it will always be missing, and when $x_{j,i}$ is greater than the LOD, there is still a chance, $0 < \alpha < 1$ that it is missing. This can be summarized as

$$\Pr(R_{j,i} = 0 | x_{j,i}) = \begin{cases} \alpha, & x_{j,i} > \text{LOD} \\ 1, & x_{j,i} \leq \text{LOD} \end{cases},$$

where $R_{j,i}$ is missing indicator, and α is fitted from the data. Then for each missing value of $x_{j,i}$, an indicator variable $Z_{j,i}$ is introduced which determines whether $x_{j,i}$ will be below the threshold of LOD or above when being imputed. The probability of each $Z_{j,i}$ is estimated by a truncated conditional normal distribution. If $Z = 1$, i.e., the missing value is estimated to be below the LOD, then a left-truncated normal distribution will be used as the estimated distribution to draw imputed missing values, otherwise a right-truncated normal distribution will be used.

3.5 Imputation with Random Forest

Random forest imputation is a powerful machine learning algorithm that builds a predictive model by setting a particular target variable with non-missing values as the outcome and other variables as predictors, followed by a process of predicting the target variable with missing values. A random forest model is an ensemble of non-parametric tree models. While involving a random variable selection and tree split procedure, random forest is more robust with outliers and can tolerate with missing values. The RF imputation procedure is similar with Subheading 3.3 regression-based method but can provide more accuracy and more outlier-robust result. Equation (1) uses linear regression, which is not feasible when d_1 is greater than n. Random forest algorithm simply replaces Eq. (1) with random forest model, i.e.,

$$\grave{x}_{1,} = \mathrm{RF}(\grave{x}_{2,}, \cdots, \grave{x}_{d_1,}) + \varepsilon. \tag{7}$$

Once the random forest model is trained, $\widehat{\mathrm{RF}}(\grave{x}_{2,}, \cdots, \grave{x}_{d_1,})$ can be used to predict the missing value, i.e., $\hat{x}_1$. The Eq. (7) and above prediction are conducted for each of the compound to obtain a complete dataset. In training the random forest using Eq. (7), it is sometimes necessary to select hyper-parameter using cross-validation to avoid overfitting, because the same variables, $\grave{x}_{2,}, \cdots, \grave{x}_{d_1,}$ are used for both training and prediction. A comprehensive study of random forest-based imputation method showed that it performs best in most of the tested scenarios, including combinations of different types and rates of missingness.

4 Normalization Methods and Tools

To combat technical errors, several sample normalization strategies have been investigated. These can be categorized into three categories: (i) data-driven normalizations, (ii) internal standards (IS)-based normalizations, and (iii) quality control samples (QC)-based normalizations. Table 2 summarizes some of the most commonly used sample normalization approaches.

Table 2
A summary table of normalization approaches mentioned in the Subheading 4

Normalization method	Description	R package	Type
mTIC	Normalize compounds by the sum of all identified metabolites	metabox	data-driven
MSTUS	Normalize compounds by the sum of all compounds.	metabox	data-driven
Median	Normalize compounds by the median average intensity of all the compounds.	metabox	data-driven
Cubic Splines	Nonlinear baseline methods assuming the existence of nonlinear relationships between baseline and individual spectra	affy	data-driven
B-MIS	Normalizes peak areas by matching measured metabolites with isotope-labeled internal standards that behave similarly during the analysis.	GitHub-IngallsLabUW	IS-based
NOMIS	Linearly combine the intensity of multiple internal standards to optimize the normalization factor for each individual molecular species	metabolomics	IS-based
Batch-wise LOESS	LOESS curve for each batch using QC samples to calibrate the systematic variation	stat	QC-based
RFSC	Integration of random forest to learn the unwanted variations and predict the correction factor in the neighboring real samples responses	statTarget	QC-based
SVM	To reduce unwanted variations and integrate multiple batches in large-scale metabolomics studies	MetNormalizer	QC-based
SERRF	Utilize compound correlations to remove systematic error by random forest algorithm	GitHub-slfan2013	QC-based

4.1 Data-Driven Normalizations

Data-driven normalization methods are based on the assumption of *self-averaging property*, in which an increase in the concentration of one set of compounds is balanced by the decrease in the concentration of the other group of compounds, i.e.,

$$f(x_{,i}) = f(x_{,l}), \text{forall } i \text{ and } j,$$

where $f(\cdot)$ is a summarizing function of each sample. Examples include mass spectrum total useful signal (MSTUS), median, sum normalization with all annotated metabolites, and L_2 normalization, and they differ are based on the choice of function f. For example, median normalization used the 50th quantile as $f(\cdot)$, while MSTUS uses sum, i.e., $f(x_{,i}) := \sum_{j=1}^{p} x_{j,i}$ for the ith sample. A more common choice of $f(\cdot)$ is the sum of all annotated metabolites, i.e., or mTIC normalization. Untargeted metabolomics generated by MS can yield unknown compound which may not have as good as impurity as the identified ones, thus it is reasonable only use known compounds to normalize the whole dataset. A general framework of these methods can be summarized by the following procedure:

1. Calculate $f(x_i)$ for each sample.
2. Calculate the average of $f(x_i), \bar{f} := \sum_{i=1}^{n} f(x_i)$.
3. Normalize $x_{,i}$ by $\hat{x}_{,i} := x_{,i} \div \frac{f(x_i)}{\bar{f}}$ for each sample.

After normalization, $f(x_{,i})$'s are similar (or identical) to each. Note, the equation in the above step 3 can be changed to $\hat{x}_{,i} := x_{,i} - (f(x_i) - \bar{f})$, which uses difference instead of division. However, this may introduce negative values.

A major limitation of such methods is that the assumption of self-averaging may not always be true in metabolomics, since a specific systematic error may affect some compounds differently from others. Despite their limitations, they are simple and used as a complement to machine learning methods on small datasets where they may fail or overfit.

4.2 Internal Standard-Based Normalizations

The internal standard (IS)-based normalizations, including single IS [10], global IS [11], best-matched internal standard normalization (BMIS) [12], and normalization using optimal selection of multiple IS (NOMIS) [13], utilize internal and/or external standard compounds added to the subject samples to normalize the intensity of each metabolite. There should be a matching amount of internal standards added to each sample in terms of intensity. Any differences that occur in the IS intensities, therefore, will be due to systematic errors, and we can normalize each compound by making the IS intensities consistent. A common procedure of IS-based methods to normalize each compound is as follows:

1. For the jth compound, calculate $f(x_{,i}^{\mathrm{IS}})$, for each sample, where i in 1, $\cdots$, n. Here x^{IS} means internal standard compounds only.
2. Calculate the average of $f(x_{,i}^{\mathrm{IS}})$'s, for i in 1, $\cdots$, n, i.e., by $\bar{f} := \frac{1}{n}\sum_{i=1}^{n} f(x_{,i}^{\mathrm{IS}})$.
3. To normalize jth compound, calculate $\hat{x}_{j,i} = x_{j,i} \div \frac{f(x_{,i}^{\mathrm{IS}})}{\bar{f}}$.

Different IS-based methods use different functions of $f(\cdot)$. The single IS and BMIS use one internal standard for each compound. For single IS normalization, the internal standard is chosen based on domain knowledge, i.e., to normalize each compound based on an internal standard belonging to the same compound class. For BMIS method, the single internal standard is chosen based on variation matching. NOMIS, on the other hand, uses multiple internal standard for each compound. The concept of NOMIS is based on the idea of using the least square method in order to find coefficients of multiple internal standards that best predict the values of the compound being normalized. To be clearer, the $f(\cdot)$ function for NOMIS is linear regression, that is

$$\log x_{j,} = f(x^{\mathrm{IS}}) + \varepsilon_j = (\log x^{\mathrm{IS}})\beta_j + \varepsilon_j,$$

where log is used to transform dataset to be normally distributed, β_j is a vector of coefficient for the *j*th compound which can be fitted by least square technique, and ϵ_j is the residual following a normal distribution.

Although internal standard methods are common in normalizing untargeted metabolomics data, they suffer from the facts that (i) IS peak heights may not be indicative of all matrix effects, (ii) IS intensities are sensitive to their own obscuring variation, (iii) the IS can be influenced by the coelution of other compounds, and (iv) the structural properties of the IS may not cover all chemical species found in a metabolomics dataset.

4.3 Quality Control Samples-Based Normalizations

Quality control samples (QC)-based normalization approaches are becoming popular in recent years. A QC sample should have a matrix composition similar to the biological samples to be studied, which is normally achieved by pooling aliquots of the samples to be studied. A series of QC samples is then injected into the batch to evaluate the pretreatment performance, followed by QC-based normalization methods aimed at reducing variation in signal intensity. QC-based approaches to normalization aim to regress any unwanted systematic errors for each metabolite by using the intensity of the QCs to normalize the error accordingly. As a result, a key advantage of this approach is that undesirable technological variations can be accommodated without taking away the important biological variations.

We now introduce some notations that are necessary to illustrate the methods. Suppose we have n_{QC} QC's and B batches in the dataset. We denote QC samples as $x_{,i}^{\mathrm{QC}}$ for i in $1, \cdots, n_{\mathrm{QC}}$, and use $x_{,i}^{\mathrm{QC},b}$ and $x_{,i}^{,b}$ to indicate that the *i*th QC and biological sample belongs to batch *b*, respectively. Next, we introduce three most commonly used QC-based method in the literature.

4.3.1 Batch-Wise LOESS Normalization

Batch-wise LOESS (locally estimated scatterplot smoothing) normalization uses QCs in each batch to train a LOESS model and use the trained LOESS model to normalize the samples in each batch [14]. Specifically, for the *j*th compound, *B* LOESS models are trained, one for each batch, using QC samples in the corresponding batch. We denote the LOESS model for the *b*th batch as $f_{\mathrm{LOESS}}^{b}(t)$, where t is the sample acquisition time or injection order for the samples. The sample acquisition time or injection order is critical in the LOESS model as it assigns weights to the neighbor QCs to each study sample. The closer the QC to the sample, the higher the weight it has. The predicted systematic error of the sample at the t_i^{th} injection order in the *b*th batch can then be predicted as $\hat{f}^b_{\text{LOESS}}(t_i)$. Then calculate the average of the predicted systematic error of all biological samples, denoted as $\bar{f}$. The final normalization is conducted as

$$\hat{x}_{j,} = x_{j,} \div \frac{\hat{f}^b_{\text{LOESS}}(t_i)}{\bar{f}}.$$

Then one can apply similar procedures to all the compounds to achieve the final normalized dataset. Some other normalization methods, such as SVR and RFSC, also have similar procedures to the batch-wise LOESS normalization method. The only difference is that the SVR uses suppose vector regression, while RFSC uses random forest model, instead of LOESS model.

A drawback of LOESS normalization is that the model of LOESS is sensitive with outliers, which may introduce extra error in the biological samples. Furthermore, batch-wise LOESS, SVR, and RFSC are limited by the underlying assumption that the systematic errors in the variables are only related to the batch effect and injection order (or processing sequence), which are not adequate enough to summarize the systematic error trend.

4.3.2 SERRF Normalization

Systematic Error Removal Using Random Forest (SERRF) normalization was recently developed by Fan [15]. SERRF assumes that the systematic variation for each variable can be better predicted by the systematic variation of other compounds, in addition to batch effects and injection order numbers. Metabolomics datasets commonly exhibit this phenomenon, in which compounds are highly corrected with each other, and the technical drift behaves similarly. We use the ADNI dataset (Session 2) as an example to demonstrate the main advantages of SERRF. In Fig. 1, there is an obvious correlation between compounds 1–9 and compound 11, the raw intensity of Ceramide (d40:2). SERRF utilizes the correlation structure to conduct normalization as shown in Fig. 1 12), which has a better performance than LOESS normalization 10), which does not take advantage of the correlation. Before normalization (*see* Fig. 1 11)), the RSD was 33%. LOESS normalization reduced the QC-RSD to 19%, while SERRF further reduced down to 7%. Furthermore, random forest model are better than LOESS, because of the following advantages: (1) RF can be applied when there are more variables than samples ($p \ll n$), which fits the data structure of high-throughput untargeted metabolomics data, while other methods, e.g., LOESS, can only be applied to cases where $p \gg n$. Therefore, RF is a suitable model for using the correlation information of other metabolites when correcting for the metabolites. (2) Nonlinear trends observed in metabolomics can generally be fit using RF. (3) RF can handle the issue of multicollinearity (i.e.,

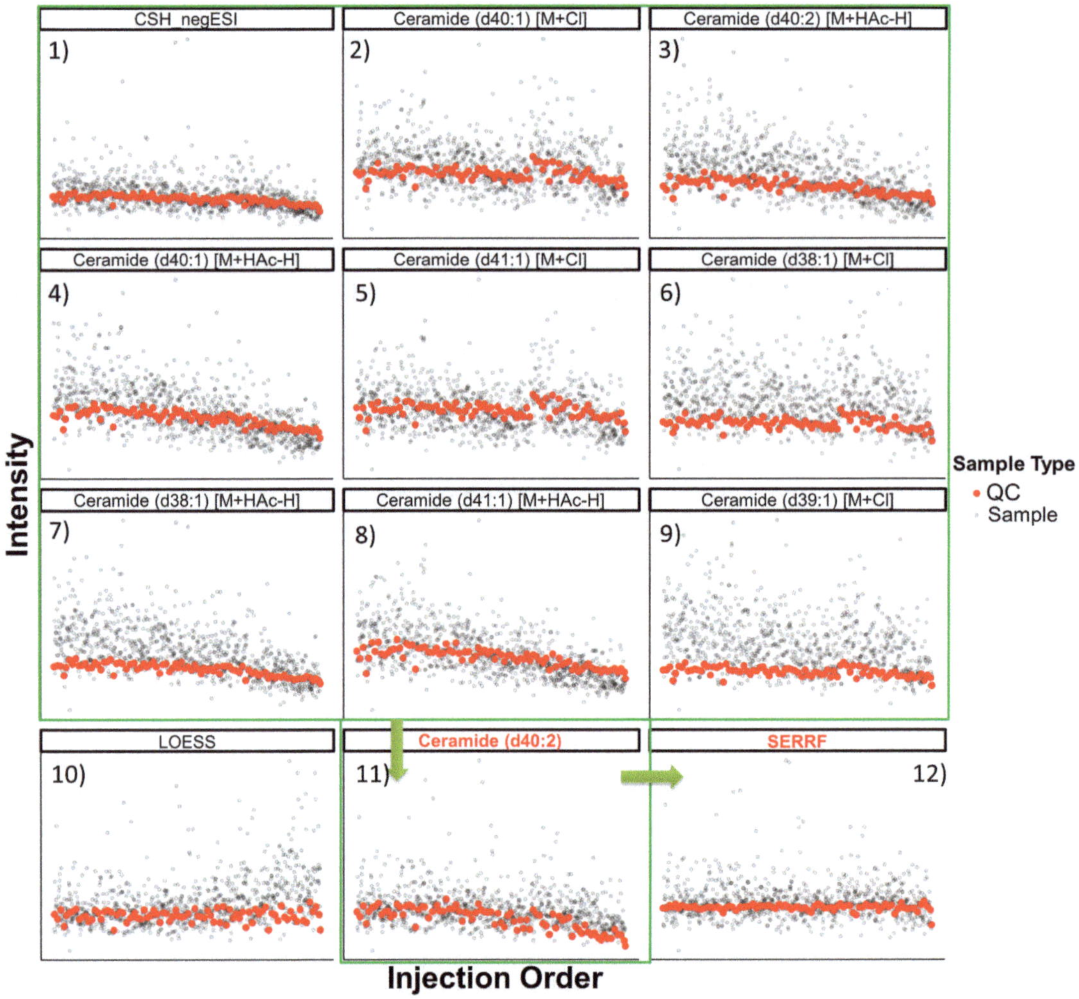

Fig. 1 Illustration of SERRF normalization. Ceramide (d40:2) in the ADNI dataset is used to demonstrate the performance of SERRF. There is an obvious correlation between 1–9 and 11. SERRF utilizes the correlation (12) and achieved a better normalization performance than LOESS (10)

high correlation among variables). (4) RF is robust with outliers and missing values. (5) RF is proven not to be overfitting when the number of trees increases. By utilizing SERRF, the systematic error quantified by relative standard deviation can be reduced below 5%.

In detail, SERRF assumes the systematic error trend of the *j*th compound, $x_{j,}$, is related to the batch effect, the sample acquisition time t (or injection order), and the intensity of the pooled QCs from the other metabolites $x_{-j,}^{QC}$. To construct SERRF algorithm, the author applied random forest (RF) model as follows:

1. Autoscale all the compounds of QCs and samples.
2. For each compound, train the RF model using the corresponding QC intensity as response and injection order,

batch effect b, and the intensity of the other metabolites as predictors to fit systematic variations. The model is trained as

$$x_{j,}^{\mathrm{QC}} \sim \widehat{\mathrm{RF}}\left(t, b, x_{-j,}^{\mathrm{QC}}\right),$$

where $\widehat{\mathrm{RF}}$ indicates a trained random forest model.

3. Normalize each compound by the predicted systematic error to the average variable intensity across all samples. The intensity of each compound was normalized by dividing the predicted systematic error

$$\hat{x}_{j,} := \frac{x_{j,}}{\widehat{\mathrm{RF}}(t, b, x_{-j,})} \times \bar{x}_{j,},$$

where $\bar{x}_{j,}$ is the average intensity of the jth compound before normalization.

4. Rescale back the normalized intensity to the original level.

One of the main advantages of SERRF over other commonly used normalization approaches is that it can effectively utilize information from all correlating compounds to normalize each metabolite. Based on the test data from three large-scale cohort studies, it has been demonstrated in the original paper that SERRF significantly improves the reproducibility of peak abundance of QC samples and increases the statistical power of detected compounds associated with phenotypes of interest.

5 Statistical Models for Metabolomics and Clinical Covariates

5.1 Metabolic Biomarkers

To detect differential abundant metabolites between phenotypes or exposure groups, one can use non-parametric approaches such as Kruskal-Wallis test and Mann–Whitney U test for multiple hypothesis testing. In a study that selects subjects or participants by pairs, a paired t-test and Wilcoxon signed-rank test can be used for differential abundance analysis. The association between individual metabolite and the risk of disease onset or prognosis can be modeled by the logistic regression or Cox proportional hazard model and tested based on the odds ratio or hazard ratio. For longitudinal metabolomics data, a trajectory analysis can be performed by linear mixed effect model implemented in lme4 or nonlinear mixed effect model implemented in edge [16].

5.2 Clustering of Metabolomes

Another type of analysis in metabolomics studies is the clustering of metabolome samples to discover latent subgroups of participants or subtypes of disease. Based on the multivariate Gaussian distribution of metabolites abundance, a common approach to cluster metabolomes is the Gaussian Mixture Model (GMM) that accounts for the correlation structure between metabolites or lipids and determines the number of subgroups by AIC or BIC. An example of applying GMM to metabolomics studies is [17], using R package mclust. In a longitudinal metabolomics study, clustering all the metabolome samples may only detect the age group instead of subgroup of participants because of the well-known age effect on metabolome and lipidome. Hence, an alternative approach is merging temporal metabolomes into one profile for each subject, and then used GMM to perform unsupervised clustering [18]. This approach requires every subject in the analysis to have metabolome profiled at each time point, which may reduce the sample size if there are missing samples. Hence, it is critical to develop a framework for the clustering at subject-level instead of sample-level in future research.

5.3 Modules of Metabolites

Clustering analysis can also be employed in the construction of metabolic networks. A popular method for network building is applying WGCNA to the metabolite abundance table. In this algorithm, a topographical overlap matrix (TOM), which measures the similarity between two elements and incorporates the similarity to all other elements [19], was computed based upon the rescaled Pearson correlation as an amalgamation of direct and indirect associations between two metabolites. The module of metabolites is identified by hierarchical clustering of TOM. We used the TCGA breast cancer metabolomics data in [20] as an example to illustrate the application of WGCNA in detecting metabolite modules in Fig. 2. The result from WGCNA showed that most of the amino acids related to the metabolism of branch chain amino acids and long chain fatty acids were grouped in the same clusters. However, some modules built by WGCNA were not consistent with the hierarchical distance between metabolites, as shown in the dendrogram (*see* Fig. 2). Hence, it is necessary to develop an algorithm to improve the consensus between module assignment and within-module distance.

5.4 Integrative Analysis of Metabolomics and Transcriptomics

The central dogma of biology is a principal that explains how genetic information influences the types of proteins produced and subsequently, in the case of enzymes, metabolite formation (5) (*see* Fig. 3a). To that end, separate parallel analysis can be conducted on each data type to identify elements that are related to a phenotype of interest, and then the relevant results can be linked across distinct data types by association analyses (*see* Fig. 3b).

Another approach to classify metabolites was recently proposed, which constructed the adjacency matrix by partial correlation coefficients in the sparse Gaussian Graphical Models (GGM) using graphical LASSO (GLASSO). To obtain the most direct associations between metabolites or genes, sparse partial correlations were estimated via the GLASSO method using the glassoFast R package. Once the sparse GGM was estimated, the topographical overlap matrix (TOM) [19] was computed, and modules were

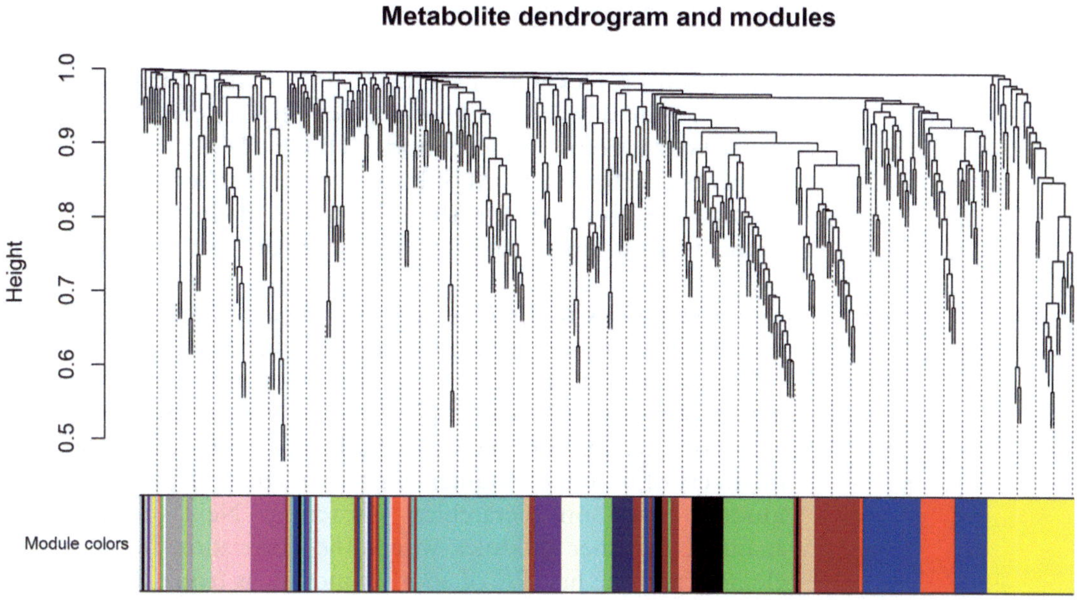

Fig. 2 Modules of metabolites generated by WGCNA from (n=23) TCGA breast cancer tumor samples

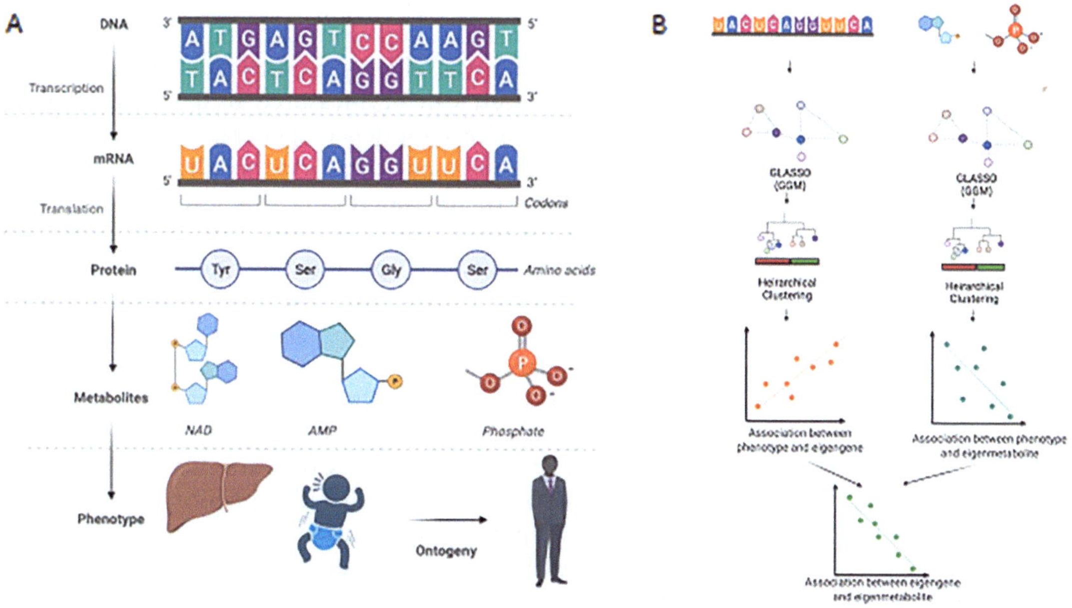

Fig. 3 Brief overview of the central dogma of biology (**a**), and summary of the proposed integrative data method which conducts analyses of transcriptomics and metabolomics to find phenotype-associated modules for each data type independently, and then evaluate association of phenotype-associated modules across data types (**b**)

formed with hierarchical clustering. The TOM distance measure ranges from 0 to 1, the larger value corresponding higher partial correlation profiles. On the other hand, smaller values of TOM signify a pair of elements that are not related/connected in the sparse GGM. For example, consider a binary TOM matrix; for two elements to achieve a TOM of 1, then their neighbor sets are proper subsets of each other. On the other hand, if the TOM between elements is 0, then the two elements do not have common neighbors.

Modules can be constructed by setting a static cut height across an entire dendrogram, or employing the dynamic tree cut to further refine assignments derived by the static cut height through analyzing the shape of each branch [21]. As many genes and metabolites are involved in multiple biological pathways, we allow features to be members of multiple modules. First, an element is assigned to the module based on hierarchical clustering. Next, elements are assigned to other modules where the correlation between that element and the first PC of another module is larger than correlation for the module to which it was assigned by hierarchical clustering. Finally, gene set or overrepresentation analysis is conducted to reveal biological themes for each module.

Once elements were assigned to modules, the modules are summarized by their first PC, i.e., eigen-gene or eigen-metabolite. The association between the eigen-gene or eigen-metabolite and the phenotype of interest can be measured using an appropriate statistical model (e.g., linear models or generalized linear model) with multiple testing adjustment completed to adjust for testing of multiple modules with the phenotype. In this study, we employed the Benjamini-Hochberg (BH) multiple testing procedure [22]. The association between age and the eigen-gene or eigen-metabolite was considered significant if the adjusted p-value <0.10. Association between eigen-metabolites and eigen-genes was completed using Spearman correlation.

Lastly, correlation tests were conducted between the modules that had significant association with the phenotype. This process is expected to result in sets of modules that are related across data types which are also associated with the phenotype. For these relationships to be valid, the correlation resulting from an eigen-gene and eigen-metabolite must have consistent direction of effect on the phenotype. That is, if both the eigen-gene and eigen-metabolite have effects with the same direction with the phenotype, then the correlation between the eigen-gene or eigen-metabolite must be positively correlated. Alternatively, if the eigen-gene and eigen-metabolite have an opposite direction of effect on the phenotype, then the correlation between the eigen-gene or eigen-metabolite must be negatively correlated.

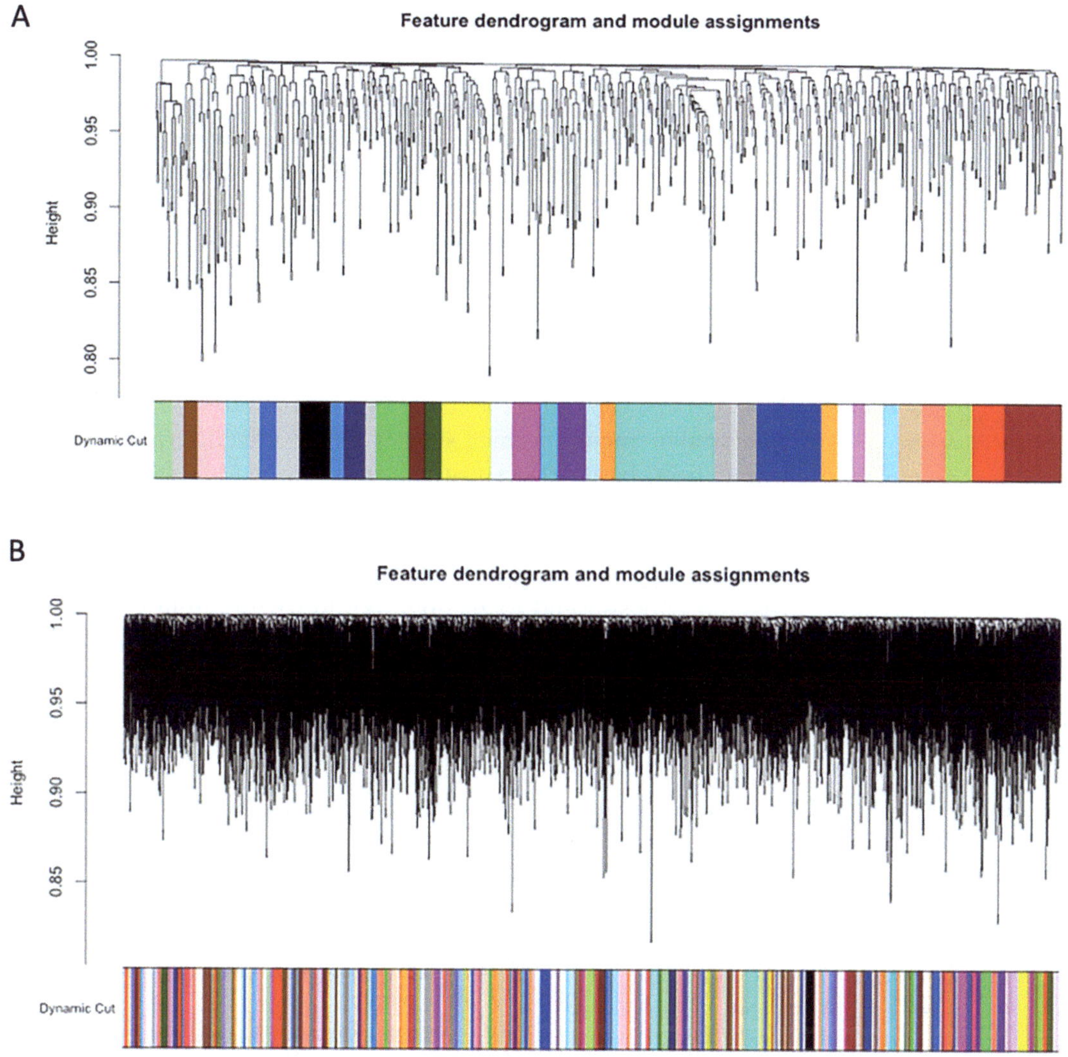

Fig. 4 Dendrogram of the constructed metabolomics (**a**) and transcriptomics (**b**) modules. Colored bars represent the different metabolite (N=32) and gene (N=260) modules

We applied this method to the multi-omic datasets profiled from postmortem pediatric livers sampled from four sources. Details of the overall study and metabolomics data are presented in [23], and the transcriptomic data are presented in [24]. This analysis only focused on the subjects that were less than 1 year old and were not stillborn or aborted resulting in 28 subjects (mean age = 0.087 years, range age = [0.003–0.9187] years). Modules were constructed for both the genes and metabolites resulting in 32 metabolite modules (average size 29.6 metabolites) (*see* Fig. 4a) and 260 gene expression modules (average size 38.26 genes) (*see* Fig. 4b). The association of eigen-features for each module with age found 102 eigen-genes (49.5%) and 12 eigen-metabolites (37.5%) significantly associated with age, with three significant and consistent with the central dogma of biology of these age-related eigen-genes and eigen-metabolites (p-value <0.10) (i.e., if the eigen-gene is positively associated with age and the eigen-metabolite is negatively associated with age, then the relationship between eigen-gene and eigen-metabolite needs to be negative to conform with the central dogma of biology). None of the age-associated gene modules was significantly enriched for a gene pathway after mapping onto KEGG and GO data.

6 Conclusion

Analyses in metabolomics studies usually include missing peaks imputation, batch effect normalization, individual biomarker discovery, subgroup identification, and multivariate association with phenotypes, which can be performed by various statistical and machine learning methods. Future methodological research in metabolomics should focus on longitudinal and multivariate modeling and integration with other omics data, such as microbiome, transcriptome, and proteome.

References

1. Barupal DK et al (2018) Generation and quality control of lipidomics data for the Alzheimer's disease neuroimaging initiative cohort. Scientific Data 5(1):1–13
2. Troyanskaya O et al (2001) Missing value estimation methods for DNA microarrays. Bioinformatics 17(6):520–525
3. Hu L-Y, Huang M-W, Ke S-W, Tsai C-F (2016) The distance function effect on k-nearest neighbor classification for medical datasets. SpringerPlus 5(1):1–9
4. Kim H, Golub GH, Park H (2005) Missing value estimation for DNA microarray gene expression data: local least squares imputation. Bioinformatics 21(2):187–198
5. Lee JY, Styczynski MP (2018) NS-kNN: a modified k-nearest neighbors approach for imputing metabolomics data. Metabolomics 14(12):1–12
6. Shah JS et al (2017) Distribution based nearest neighbor imputation for truncated high dimensional data with applications to pre-clinical and clinical metabolomics studies. BMC Bioinf 18(1):1–13
7. Nounou MN, Bakshi BR, Goel PK, Shen X (2002) Bayesian principal component analysis. Journal of Chemometrics: A Journal of the Chemometrics Society 16(11):576–595
8. Li Q et al (2020) GMSimpute: a generalized two-step lasso approach to impute missing values in label-free mass spectrum analysis. Bioinformatics 36(1):257–263
9. Kumar N, Hoque M, Sugimoto M et al (2021) Kernel weighted least square approach for imputing missing values of metabolomics data. Sci Rep 11(1):1–12
10. Bromke MA et al (2015) Metabolomic profiling of 13 diatom cultures and their adaptation to nitrate-limited growth conditions. PloS One 10(10):e0138965
11. Yang S, Sadilek M, Lidstrom ME (2010) Streamlined pentafluorophenylpropyl column liquid chromatography–tandem quadrupole mass spectrometry and global 13c-labeled internal standards improve performance for quantitative metabolomics in bacteria. J Chromatogr A 1217(47):7401–7410
12. Boysen AK, Heal KR, Carlson LT, Ingalls AE (2018) Best-matched internal standard normalization in liquid chromatography–mass spectrometry metabolomics applied to environmental samples. Anal Chem 90(2): 1363–1369
13. Sysi-Aho M, Katajamaa M, Yetukuri L, Orešič M (2007) Normalization method for

metabolomics data using optimal selection of multiple internal standards. BMC Bioinf 8(1): 1–17

14. Li B et al (2017) NOREVA: normalization and evaluation of MS-based metabolomics data. Nucleic Acids Res 45(W1):W162–W170
15. Fan S et al (2019) Systematic error removal using random forest for normalizing large-scale untargeted lipidomics data. Anal Chem 91(5):3590–3596
16. Storey JD, Xiao W, Leek JT, Tompkins RG, Davis RW (2005) Significance analysis of time course microarray experiments. Proc Natl Acad Sci 102(36):12837–12842
17. Orešič M et al (2012) Phospholipids and insulin resistance in psychosis: a lipidomics study of twin pairs discordant for schizophrenia. Genome Med 4(1):1–11
18. Li Q et al (2021) Plasma metabolome and circulating vitamins stratified onset age of an initial islet autoantibody and progression to type 1 diabetes: the teddy study Diabetes 70(1):282–292
19. Langfelder P, Horvath S (2008) WGCNA: an R package for weighted correlation network analysis. BMC Bioinf 9(1):1–13
20. Tang X et al (2014) A joint analysis of metabolomics and genetics of breast cancer. Breast Cancer Res 16(4):1–15
21. Langfelder P, Zhang B, Horvath S (2008) Defining clusters from a hierarchical cluster tree: the dynamic tree cut package for R. Bioinformatics 24(5):719–720
22. Benjamini Y, Hochberg Y (1995) Controlling the false discovery rate: a practical and powerful approach to multiple testing. J R Stat Soc Ser B Methodol 57(1):289–300
23. Wilson CM et al (2020) Ontogeny related changes in the pediatric liver metabolome. Front Pediatr 8:549
24. Meier R et al (2018) Ontogeny-related pharmacogene changes in the pediatric liver transcriptome. Pharmacogenet Genomics 28(3):86

Chapter 13

Statistical and Computational Methods for Proteogenomic Data Analysis

Xiaoyu Song

Abstract

Proteins are the functional molecules for almost all cellular and biological processes. They are also the targets of most drugs. Proteins employ complex, multilevel regulations, so their abundance levels do not well correlated with their mRNA expression levels. The structure, activity, and functional roles of proteins are affected by posttranslational modifications (PTM), which are even less correlated with mRNA expression levels than protein abundances. Comprehensive characterization of the proteomics data is critical for understanding the molecular and cellular mechanisms of biological systems and developing news therapeutics. Current large-scale proteomic profiling technologies, such as mass spectrometry, provide relative identification of peptides and proteins, with data vulnerable to outliers, batch effects, and nonrandom missingness. In order to perform high-quality proteomic data analysis, we will first introduce a data preprocessing and quality control pipeline that includes normalization, outlier detection and removal, batch effect identification and handling, and missing data imputation. Then, we will describe several statistical methods that leverage well-processed proteomic data to generate scientific discoveries, especially with an integration with genomics and transcriptomics. These methods cover topics like association analysis, network construction, clustering, and cell-type deconvolution. To demonstrate these methods, we will use the proteogenomic data from the lung squamous cell carcinoma study of the Clinical Proteomic Tumor Analysis Consortium and provide sample codes for data access and analyses.

Key words Proteomics, Mass spectrometry, Preprocessing and quality control, Integrative proteogenomic analysis

1 Introduction

1.1 Why Study Protein?

The protein-coding genes are transcribed from DNA to pre-mRNA, further processed to mature mRNA, and then translated into proteins. Proteins may be posttranslationally modified, which affects their activities, structures, and functional roles. Similar to transcriptomics, which is the study of transcriptome, proteomics is the study of proteome or, in other words, the study of complete set of proteins that are present in cells, tissues, and biofluids at a certain time [1]. Proteomic studies include both the

Brooke L. Fridley and Xuefeng Wang (eds.), *Statistical Genomics*,
Methods in Molecular Biology, vol. 2629, https://doi.org/10.1007/978-1-0716-2986-4_13,

identification and quantification of peptides, proteins, and their posttranslational modifications, and use the identified and quantified data for scientific discoveries [2].

In today's technologies [2], the mRNA expression levels of protein-coding genes are easier to quantify than their protein abundances, which causes most discoveries to be transcriptomics-based. However, proteins are the functional molecules for almost every single cellular and biological process. They are also the targets of most drugs [3]. As proteins employ complex, multilevel regulation that is not reflected at mRNA levels, they have substantial spatial, temporal, and quantitative differences from mRNA expression levels. Genome-wide mRNA-protein correlation is notoriously poor (e.g., ≤ 0.47) [4–6], even in isolated cells [7], indicating huge limitations for relying on mRNA expression levels in revealing the biological mechanisms of health and disease. Posttranslational modifications (PTM) add another layer of complexity after translation, so they are even less correlated with mRNA expression levels than the proteins [4–6]. Therefore, it is insufficient to rely solely on transcriptomics to understand the biological systems and to develop therapeutics.

The state-of-the-art technology for large-scale proteomic profiling is mass spectrometry (MS) [8]. This technology has a long history back to 1912 and is recently developed to use isobaric tags [9] to allow reliable protein quantifications from multiple samples. This incorporation of isobaric tags makes MS suitable for the proteomic profiling on samples in hundred and has revealed many novel disease mechanisms and candidate biomarkers [4–6, 10–12]. Unlike the technologies used for transcriptomic profiling, such as RNA sequencing methods, the MS technology is not quantitative in nature. One needs a good understanding of its design and data generation process to correctly analyze the data and interpret the results. In this chapter, we will first introduce the data generation procedures of MS technology, from a statistician's angle, in order to explain the data features. Then, we will use a publicly available proteomic dataset as an example to introduce several statistical and computational tools that are designed for preprocessing and quality control of proteomic data. Finally, we will describe several proteogenomic analysis methods that integrate proteomics with genomics, transcriptomics, and phenotype data for association, network, clustering, and deconvolution analysis. Note, in our chapter, we focus on the "discovery" proteomic studies that provide unbiased quantification of proteins for systematic characterization of the biological systems. Occasionally, targeted proteomic studies are implemented for the "verification" purposes, which use targeted assays to verify candidate proteins in biological and biomedical applications. Readers can rely on external articles to understand the analysis of targeted assays [13, 14].

1.2 Design of Mass Spectrometry-Based Proteomic Experiments

Understanding the design of medical and biological experiments is always the starting point of the analysis. Without a clear understanding of how the data are generated, all analyses are built upon no ground. Throughout this chapter, we will discuss the bottom-up proteomics, which first digests proteins into peptides and then analyzes peptides in MS. The top-down proteomics, which analyzes the intact proteins directly without the need for digestion, is an emerging but not mainstream technology, and we will rely on external reviews [15, 16] for interested readers to grasp the ideas. We will first describe how peptides are identified, quantified, and combined into proteins in MS experiments and then introduce labeling methods that enable high-quality proteomic quantifications for large samples. Lastly, we will extend the discussion to PTM.

Peptide Identification Figure 1a illustrates the peptide identification in the MS-based proteomic profiling experiments [17], in which sample labeling is not yet considered. In this type of experiments, proteins from each study sample are digested into peptides. Peptides are further fractionated into multiple samples based on their properties (e.g., charge, size, polarity, or hydrophobicity), and peptides in each sample go through the liquid chromatography (LC) for further separation. A group of closely located peptides on the LC are ionized via electrospray and injected into the mass spectrometer. Peptides are often distinguishable from each other in the mass spectrometer by evaluating their mass to charge ratio (m/z), and the mass spectrum of the intact peptides is called the MS1 spectrum. Sometimes, peptides cannot be well identified based on their m/z ratio alone. In that case, a peak in the MS1 spectrum, which corresponds to a peptide inside the mass spectrometer, is isolated and fragmented by collision with inert gas. The peptide fragments' m/z values are recorded in the MS2 spectrum, which, together with the precursor mass from the MS1-spectrum, are used to identify a peptide unambiguously.

Peptide and Protein Quantification The MS is an inherently nonquantitative technology. For example, the efficiency of digesting proteins into analyzable peptides and of ionizing peptides varies drastically from protein to protein [17]. Therefore, unlike the transcriptomic studies, the proteomic studies cannot directly obtain the absolute quantification of proteins. Studies can add spiked-in standards with known absolute concentration for the absolute quantifications [18]. As the quantification efficiency differs from protein to protein, these spiked-in studies provide absolute quantifications for limited proteins, instead of the entire proteome. Therefore, we need to rely on relative abundances of proteins for

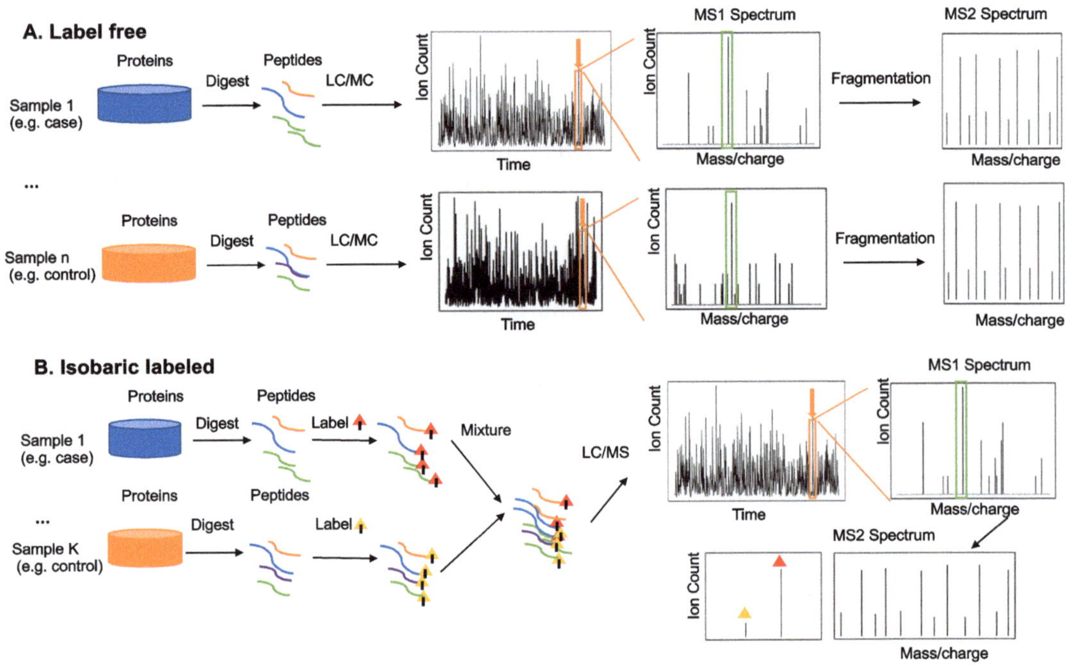

Fig. 1 Outline of the liquid chromatography–mass spectrometry (LC-MS) proteomic experiment. (**a**) The outline of the unlabeled proteomic experiment. Proteins from each sample are digested into peptides. Peptides are separated by liquid chromatography (LC), ionized via electrospray, and injected into the mass spectrometer (MS). At a given time in MS, multiple different peptides might coappear. The MS1 spectrum typically distinguishes these peptides by their mass-to-charge ratio (m/z). If unique peptide cannot be identified from the MS1 spectrum, a peak in MS1 spectrum, which corresponds to a peptide, will be isolated and fragmented by collision with inert gas. The MS2 spectrum records the peptide's fragment ions' m/z values, together with MS1 spectrum, identifies the peptide unambiguously. The peak m/z values in MS1 spectrum will be used as the quantification of the peptides. The resulting quantifications have limited measurement precision, and comparison across samples is difficult. (**b**) The outline of isobaric labeled proteomic experiment. Proteins from multiple samples are digested into peptides, labeled with isobaric tags, and mixed before entering liquid chromatography. Isobaric tags are comprised of two components, reporter group and mass balancer. The isobaric tags have the same total mass, so the same peptide of different samples in the mixture has the same peak in MS1 spectrum. The reporter groups of the isobaric tags, however, have different mass in different samples. The reporter group and mass balancer components are separated by fragmentation, so the same peptide of different samples has different peaks in the MS2 spectrum. Evaluating the peak m/z values in both MS1 and MS2 spectrum will quantify the relative peptide abundance from one sample to another in the mixture

most scientific investigations. In the MS experiment, each peak in the MS1-spectrum corresponds to a peptide, and the peak size proportional to the number of detected peptide ions is used as the quantification of peptide abundance. While MS2 spectra are used for peptide identification, their signal is often not used for quantification (unless in labeled approaches described later). Relative quantification of proteins is obtained by integrating signals from their peptides. This can be done in a variety of ways, such as by using the mean or median of (all or the top n) peptide ratios, taking a weighted average of peptide ratios based on signal intensity, calculating the ratio of total peptide ion counts, or using linear regression to fit a line through the peak intensities for each peptide [17].

It is critical to know, due to the design of the MS experiment, that it is not possible to directly compare the MS signals across peptides/proteins within a sample. It is possible, however, to compare peaks of the same peptide/protein across samples, which is what relative quantification is based on [19].

Labeling MS alone provides poor measurement precision and involves running different samples consecutively, making it especially difficult for comparisons across samples. The integration of MS with isobaric labeling methods [20], however, enables the global quantification of proteins across multiple samples in a single experiment. As illustrated in Fig. 1b, the isobaric tags are added to samples after the digestion and attached to the peptides. Different samples can be labeled with different tags and then mixed together for MS proteomic profiling. The isobaric tags have two components, reporter ion and mass balancer. Reporter ions from different isobaric tags have different masses, and the mass balancers take different masses to ensure the entire isobaric tags to have the same total mass across samples. As a result, the same peptide from different samples is labeled with tags of the same mass and has a single MS1 peak, which can be isolated for quantification of the total abundance across all samples. After a peptide is isolated on the MS1 spectrum, the two components of isobaric tags will be separated during fragmentation in MS2, and both show different masses in the MS2 spectrum and can be used for quantification of each sample relative to others. The isobaric tags enable comparable quantifications of the same peptides across samples that are profiled together.

On the other side, labeling samples also increase the complexity of the experiment. As the number of samples increases in a single experiment, the number of tag types with different reporter ions and mass balancer also need to increase. This limits the number of samples that can be compared in a single experiment. The most commonly used commercially available isobaric tags are absolute and relative quantification (iTRAQ) [21] and tandem mass tags (TMT) [22]. To our best knowledge, iTRAQ can encode up to 4 different samples, and TMT can encode up to 11 different samples at current stage. Therefore, when a large number of samples are investigated, multiple experiments need to perform. This is why we often call the proteomic experiments *multiplexed proteomics*. To enable comparisons of relative abundances across multiple multiplexes, the same samples should be put into each experiment for standardization. In other words, proteomic data from the same or different studies without common reference samples cannot be directly compared.

Protein PTM Quantification There are over 200 types of protein PTM that influence every aspect of normal cell biology and pathogenesis [23]. Several modifications, including phosphorylation, acetylation, glycosylation, and ubiquitylation, can be quantified by today's MS technologies [6]. The PTM quantifications need to overcome several additional technical challenges. Take phosphorylation, the most common type of PTM, as an example. Compared to their non-phosphorylated counterparts, phosphorylated peptides typically present in low abundance and have low ionization efficiency and thus are hard to detect [24]. PTM enrichment procedures, which aim at separating PTM proteins/peptides from non-PTM counterparts and inject PTM enriched proteins/peptides into the mass spectrometer, are essential to the success of PTM detection. In other words, PTM quantification includes another layer of relativity from enrichment procedures. Additionally, each protein/peptide might include multiple PTM of same and different types, and thus we may observe several PTM quantifications on different sites of the same protein/peptide.

2 Dataset

2.1 Large Data Repository

A wealth of proteomic data repositories is becoming increasingly available to the public. These databases provide excellent resources for secondary data analysis. By no means to be comprehensive, we introduce a few proteomic data repositories that host today's proteomic data for readers that might be interested to explore.

National Cancer Institute's (NCI) Proteomic Data Commons (PDC) The PDC represents the NCI's largest public repository of proteogenomic comprehensive tumor datasets (Fig. 2). It hosts validated datasets from NCI programs, such as Clinical Proteomic Tumor Analysis Consortium (CPTAC) and International Cancer Proteogenome Consortium (ICPC), in which data quality is assured. The proteomic data include not only the raw mass spectrometry-based files but also the mapped peptide sequences and assembled proteins from spectra with open standard computational pipelines. Data are organized by tumor type, study, and sub-proteome. Matched genetic, genomic, epigenomic, transcriptomic, and phenotypic data of the proteomic data can be accessed via NCI's Genomic Data Commons https://gdc.cancer.gov/.

Fig. 2 The main page of the National Cancer Institute's Proteomic Data Commons. This data repository hosts proteomic data of multiple cancer types from programs like Clinical Proteomic Tumor Analysis Consortium (CPTAC) and International Cancer Proteogenome Consortium (ICPC)

PDC is public accessible at https://proteomic.datacommons.cancer.gov/pdc/.

The PRIDE (PRoteomics IDEntifications) Database PRIDE is a user-populated proteomic data repository freely available at http://www.ebi.ac.uk/pride/. Users upload data generated by MS-based proteomics experiments that adhere to the PRIDE standards, including raw spectral data, peptides, and protein identifications/quantifications and associated statistics. Data may be designated as public or private. Public data are open to all PRIDE users without registering or logging in, while private data is restricted for access which can be granted by owners of the data. PRIDE is designed to be comprehensive that includes proteomic data from different species, tissue types, cell types, Gene Ontology terms, or diseases.

2.2 Dataset for Method Demonstration

To demonstrate the statistical and computational methods, we use a convenient subset of data from the CPTAC lung squamous cell carcinoma (LSCC) discovery study [6]. LSCC is a leading cause of cancer death with few therapeutic options [25]. In order to provide a deep exposition of LSCC biology for potential therapeutic implications, CPTAC characterized the proteogenomic landscape of LSCC using TMT11-based multiplexing linked by a common reference [6]. The data can be accessed through NCI's PDC. It can also be directly accessed through the publication. To avoid the hassles for users to register the account and submit the request (some data are protected), we show codes for loading the data directly from the supplementary tables of the manuscript. For the global proteomic data, we use gene-level data by loading data from the Supplementary Table 3C. For PTM data, we use phosphosite-level data by loading data from Supplementary Table 3E. To include experiment and clinical information, such TMT multiplex experiment number and tissue type (tumor vs adjacent normal), we also load the meta data from the Supplementary Table 1A. Lastly, as we will describe integrative proteogenomic analysis methods in later sections, we load somatic mutation data for top mutated genes, copy number variation (CNV), and RNA data from its Supplementary Table 1A, 2B and 2E. We narrow the analysis to a few hundred cancer-associated genes in Supplementary Table 1B for easy demonstration. The data include 207 samples (108 tumor and 99 adjacent normal) from LSCC patients.

```
library(readxl)
library(tidyverse)
# Load protein, phosphoprotein, clinical, RNA, CNV
and mutation data directly from the Supplementary
Tables of Satpathy et al 2021[6].
pro.dat =readxl::read_xlsx("data/1-s2.0-
S0092867421008576-mmc3.xlsx",
 sheet="Table S3C", skip=2) %>% column_to_rownames
(var = "id")
phos.dat =readxl::read_xlsx("data/1-s2.0-
S0092867421008576-mmc3.xlsx",
 sheet="Table S3E", skip=2)
clinical.dat=readxl::read_xlsx("data/1-s2.0-
S0092867421008576-mmc1.xlsx",
 sheet="Table S1A") %>% select(Sample.ID, Experi-
ment, Type, Age, Gender)
rna.dat=readxl::read_xlsx("data/1-s2.0-
S0092867421008576-mmc2.xlsx",
 sheet="Table S2E", skip=1)
cnv.dat=readxl::read_xlsx("data/1-s2.0-
S0092867421008576-mmc2.xlsx",
 sheet="Table S2B")
mut.dat=readxl::read_xlsx("data/1-s2.0-
S0092867421008576-mmc1.xlsx",
 sheet="Table S1A") %>% select(Sample.ID, 34:46)%>%
```

```
 rename_all(~str_replace(.,".mutation.sta-
tus",""))

# Narrow the analysis to these cancer associated
genes for a quick demonstration
cancer.gene=readxl::read_xlsx("data/1-s2.0-
S0092867421008576-mmc1.xlsx",
 sheet="Table S1B", skip=1) %>% select(geneSym-
bol='Gene Symbol')

# Clean protein data for demonstration in the next
section
pro.ex=cancer.gene %>% inner_join(pro.dat) %>% co-
lumn_to_rownames(var = "geneSymbol") %>%
 select(-(1:16)) %>% mutate_all(type.convert)
dim(pro.ex) # No. of genes * No. of samples
# [1] 551 207
mean(is.na(pro.ex)) # missing rate
# [1] 0.04995748

# Match clinical data with the proteomic data
ind.sample = match(colnames(pro.ex),clinical.dat
$Sample.ID);
clinical.match = data.frame(clinical.dat[ind.sam-
ple,]);
```

3 Preprocessing and Quality Control

Due to the nature of the discovery MS experiments, proteomics data are usually accompanied with quality issues including variational distributions, outliers, non-ignorable batch effects, and substantial missing values [26]. Here, we cover several statistical analysis topics that are often incorporated in the data preprocessing and data quality control pipelines in order to improve the quality of proteogenomic data. We provide sample codes using proteomic data as an example, while similar analyses can be performed for the PTM data.

3.1 Global Normalization

Peak intensity is used for quantification of peptide (posttranslational modified peptide) abundances, which is a relative ratio with large variation and non-Gaussian distribution (e.g., multi-modes, highly skewered). Logarithm transformation is usually performed during spectra matching to improve the quantification accuracy. The resulting peptide/protein/gene/PTM data are in log ratio scale and usually need to go through distribution check. For example, we can plot boxplot for each sample across all peptides/proteins/genes/PTM. If distributional differences are observed, global normalization is usually performed prior to other analyses. For example, given that data may be deviated from normal distribution, shifting each sample to the same median and scaling it to the same median absolute deviation are often considered for global normalization. Mean centering and standard deviation scaling might also be considered for downstream analyses that are based on z-scores.

```
# Global normalization function: median centering +
median absolute deviation (MAD) scaling
normalize.global = function(data,n.select = dim
(data)[1]) {
 data.select = apply(data,2,function(x){sort(x,de-
creasing = T)[1:n.select]})
 median.all = apply(data.select,2,median,na.rm=T);
 data.center = t(t(data.select)-median.all);
  mad.all = apply(data.center, 2, mad, na.rm=T,
center = 0, constant =1);
  data.n = t((t(data)-median.all)*(median(mad.all,
na.rm = T)/mad.all)+median(median.all,na.rm = T))
 return(data.n);
}
# Apply to the data
df.pro.n = normalize.global(ProEx);
```

3.2 Outlier Detection and Removal

Due to different peptide detections or incorrect peptide identifications in individual multiplex runs, the multiplexed proteomics data often contain multiplex level outlier data points. We recommend to use an "inter-plex t-test" to detect and remove outlier peptide/protein/gene/PTM (feature) abundances in samples of a given multiplex. In details, we performed t-test of the feature abundance of samples inside against outside of each multiplex. Features in multiplexes with a *p*-value lower than a predetermined cutoff (e.g., 10e-7) are flagged as outliers and removed from the dataset.

```
# Outlier detection function: get t.test p-value
for comparing in vs outside of batch mean values
for each gene;
outlier.detect = function(data,batch){
 apply(as.matrix(data),1,function(x){ # per gene
  p.all = sapply(unique(batch), function(i){ # per
batch
 if(length(table((batch==i)[(!is.na(x))]))==2 & &
  min(table((batch==i)[(!is.na(x))]))>1){ # True -
in batch & not NA; False -otherwise
  t.test(x~batch==i,var.equal = T)$p.value; # test
```

```
if batch i has the same mean with other batches
 }else if(length(table((batch==i)[(!is.na(x))]))
==2
 & & min(table((batch==i)[(!is.na(x))]))==1){1}
else NA;} # if only 1 T or F: mean comparison cannot
be performed and NA is returned
 )
 names(p.all) = unique(batch);
 return(p.all);
 })
}

# Outlier plot function:
outlier.plot = function(p.batch, fig.path=getwd())
{
# c-log-log transformation of p-value to obtain a
support on (-Inf, Inf); small p-value corresponds
to large llp values.
llp = c(log10(-log10(p.batch)));
# The density of llp is estimated using Kernel
density function.
# Kernel density estimation:
# output x: n coordinates of points where the den-
sity is estimated;
# output y: the estimated density values
d.llp = density(llp, na.rm = T);
 # mode + 3 * standard deviation as the outlier
cutoff
 mode=d.llp$x[which.max(d.llp$y)]
 cut.l = (mode+ 3*sd(c(llp[llp>mode], # llp greater
than mode
 2*mode-llp[llp>mode]), # use a symmetric version
of the right tail
 na.rm = T))
 pdf(paste0(fig.path, "/outlier.pdf"), width = 6,
height = 4)
 hist(llp,n = 200,main = 'inter-plex t-test',xlab =
'log10(log10(1/p_value))')
 abline(v = mode,col = 4, lty = 2,lwd = 2);
 abline(v = log10(ceiling(10^cut.l)),col = 2, lty =
2,lwd = 2);
 dev.off();
}

# Outlier removal function:
outlier.mv = function(data, p.batch, batch) {
 llp = c(log10(-log10(p.batch)));
```

```
 d.llp = density(llp, na.rm = T);
 mode=d.llp$x[which.max(d.llp$y)]
 cut.l = (mode+ 3*sd(c(llp[llp>mode], 2*mode-llp
[llp>mode]), na.rm = T))
 # remove data with small p-value
 batch.remove = p.batch[match(batch,unique
(batch)),]<10^(-ceiling(10^cut.l));
 batch.remove[is.na(batch.remove)] = FALSE;
 data.oulier.mv = data
 data.oulier.mv[t(batch.remove)] = NA
 return(data.oulier.mv)
}

# Apply the functions to the data
ttest.batch = outlier.detect(data = df.pro.n,batch
= clinical.match$Experiment)
df.pro.clean=outlier.mv(data= df.pro.n, p.batch=t-
test.batch,
 batch= clinical.match$Experiment, plot=T)

sum(is.na(df.pro.n))
#[1] 5698
sum(is.na(df.pro.clean))
#[1] 5775
sum(is.na(df.pro.clean))-sum(is.na(df.pro.n))
# [1] 77
outlier.plot(p.batch=ttest.batch, fig.path=getwd
()) # diagnosis plot
```

Outliers are observed from Fig. 3. Accordingly, 77 data points are identified as outliers and removed from the protein abundance data.

3.3 Batch Effect Identification and Correction

Next, we identify and remove the technical difference between different multiplexes. Plots from principal component analysis (PCA) are often used to visualize batch effects in the proteomic data. Here we provide sample codes for plotting the top 4 PCs.

```
# Function for plot PCs
plot.pc = function(data, col.batch, fig.path )
{
 library(ggplot2);
 library(grid);
 library(scales);

 pca_p = prcomp(t(data[rowMeans(is.na(data))
==0,]), scale. = F, center = F)
```

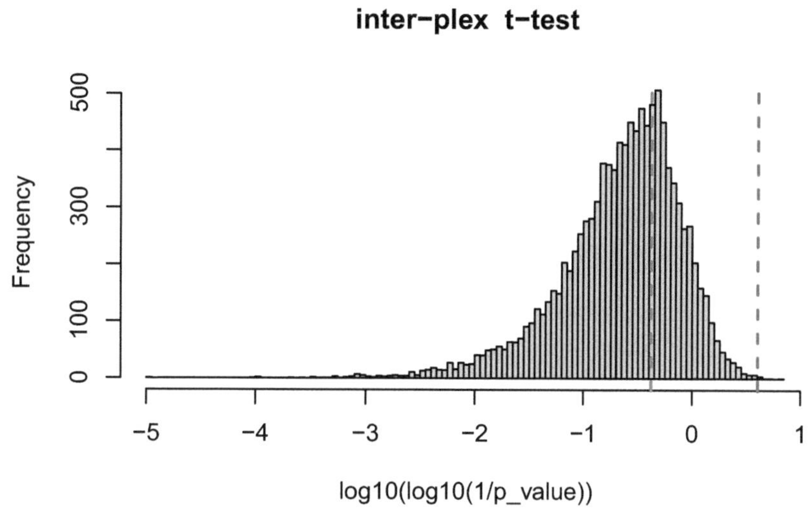

Fig. 3 The diagnostic plot for outliers. The histogram of c-log-log transformed p-values from inter-multiplex t-test is provided. The blue vertical line is the mode, and the red vertical line is the cutoff of outlier

```
gg23 = ggplot(as.data.frame(pca_p$x),aes(x = PC3,y
= PC2 ))+
 geom_text(aes(label=col.batch,col= rainbow(40)
[col.batch]))+
 theme(legend.position="none")

gg34 = ggplot(as.data.frame(pca_p$x),aes(x = PC3,y
= PC4 ))+
 geom_text(aes(label=col.batch,col= rainbow(40)
[col.batch]))+
 theme(legend.position="none")

gg24 = ggplot(as.data.frame(pca_p$x),aes(x = PC2,y
= PC4 ))+
 geom_text(aes(label=col.batch,col= rainbow(40)
[col.batch]))+
 theme(legend.position="none")

gg12 = ggplot(as.data.frame(pca_p$x),aes(x = PC1,y
= PC2 ))+
 geom_text(aes(label=col.batch,col= rainbow(40)
[col.batch]))+
 theme(legend.position="none")

gg13 = ggplot(as.data.frame(pca_p$x),aes(x = PC1,y
= PC3 ))+
 geom_text(aes(label=col.batch,col= rainbow(40)
[col.batch]))+
 theme(legend.position="none")
```

```
gg14 = ggplot(as.data.frame(pca_p$x),aes(x = PC1,y
= PC4 ))+
 geom_text(aes(label=col.batch,col= rainbow(40)
[col.batch]))+
 theme(legend.position="none")

 plotlist = list()
 plots <- c(list(gg12,gg13,gg14,gg23,gg24,gg34),
plotlist);
 # layout <- t(matrix(c(rep(1:2,each=3),rep(3:5,
each = 2)),ncol = 2));
 layout <- t(matrix(c(1,2,3,4,5,6),nrow = 3));
 numPlots = length(plots)

 pdf(fig.path,width = 12,height = 8)
 grid.newpage()
 pushViewport(viewport(layout = grid.layout(nrow
(layout), ncol(layout))))
 # Make each plot, in the correct location
 for (i in 1:numPlots) {
 # Get the i,j matrix positions of the regions that
contain this subplot
 matchidx <- as.data.frame(which(layout == i, arr.
ind = TRUE))

 print(plots[[i]], vp = viewport(layout.pos.row =
matchidx$row,
 layout.pos.col = matchidx$col))
 }
 dev.off();
}

# Apply the function to data to visualize potential
batch effect due to TMT multiplex
plot.pc(data = df.pro.clean,
 col.batch = clinical.match$Experiment, fig.path =
'plot/lscc_pro_batch.pdf')
```

Batch 22 is clustered together on the PC1 vs PC3 on Fig. 4, indicating potential batch effects in the proteomic data. We further quantify the batch effects using intraclass correlation coefficient (ICC). ICC estimates the fraction of total variances explained by the variations between groups. In this application, each batch is a group, and we evaluate for each gene, the proportion of variations due to variations between batches. If ICC is large (e.g., >0.5) for some genes, batch effect is generally believed to present in the data and needs to be corrected. Here are sample codes for calculating ICC:

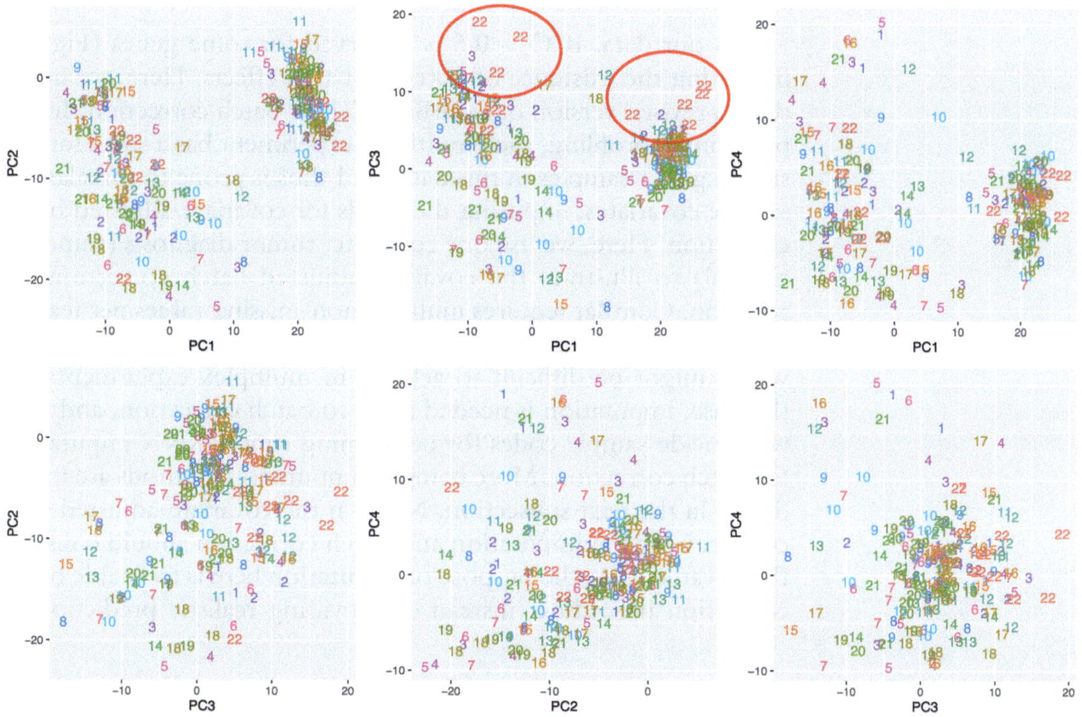

Fig. 4 The principal component (PC) plots for batches. Batch 22 is clustered together on PC1 vs PC3 plot, indicating there are probably batch effects in the data

```
# ICC function: Intraclass Correlation Coefficient
(the fraction of total variance due to variation
between groups)
# estimated from a one-way analysis of the variance
components
get.icc = function(data, batch){
 icc.p = c();
 for(i in 1:dim(data)[1]) # per gene
 {
 icc.p[i] = NA;
 if(length(table(batch[!is.na(data[i,])]))>1){
  icc.p[i] = ICCest(x = as.factor(batch),y = as.
numeric(as.character(unlist(data[i,]))),CI.type
'S')$ICC;
 }
 }
 return(icc.p)
}
# Apply function to data for batch check
icc.pro = get.icc(data = df.pro.n,batch = clinical.
match$Experiment) ### batch check
hist(icc.pro,n = 20,main = 'ICC',xlab = 'ICC
score')
```

In our data, ICC > 0.5 is observed for some genes (Fig. 5), indicating the existence of potential batch effects. Here, we introduce a revised version of ComBat [27] for batch correction. In the proteomics profiling, each multiplex experiment has a small sample size (e.g., 11 samples in this data) and thus is prone to unbalanced sample covariates, imposing the needs for covariate-adjusted batch correction. Here, we use one covariate, tumor diagnosis (tumor vs normal) to illustrate the covariate-adjusted batch correction. In addition, ComBat requires multiple non-missing values per feature (peptide/protein/gene/PTM) per batch for batch correction, which might be difficult to achieve in multiplex experiments. In this case, imputation is needed prior to batch correction, and here we provide sample codes for performing simple KNN imputation for batch correction. More complex imputation methods are introduced in the next subsection. Note, in the covariate-adjusted procedure, both the imputation and batch correction should consider the covariates. As the purpose of imputation here is to enable batch correction procedure, instead of providing realistic prediction of

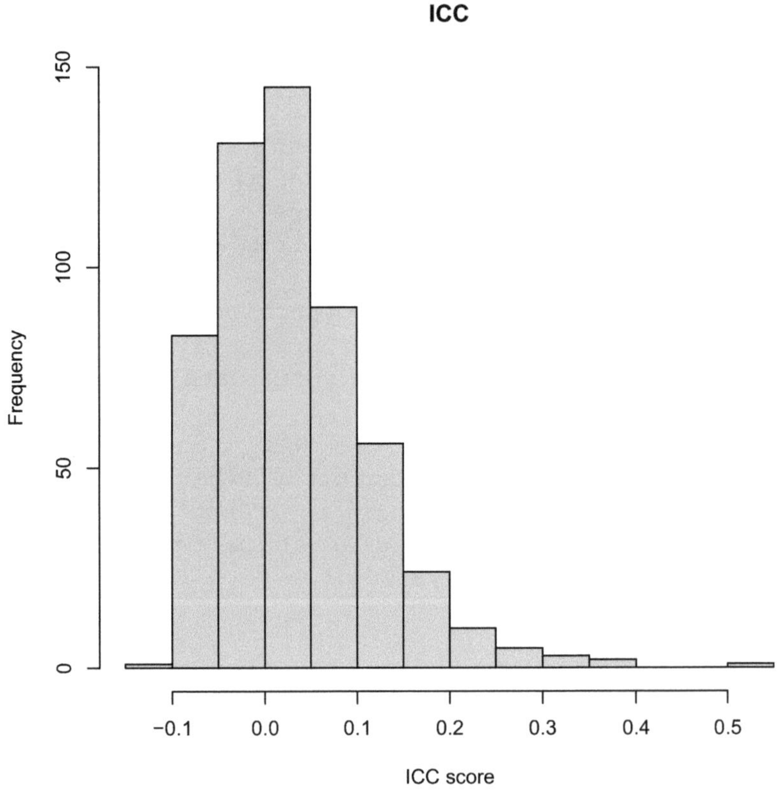

Fig. 5 The histogram of the Intraclass Correlation Coefficient across batches. ICC >0.5 for some genes is usually considered as the presentence of batch effects

missing values, we restore the missing data structure after batch correction.

```
library(sva)
library(impute)
median.sd.normlize = function(data)
{
 median.temp = apply(data,1,median,na.rm=T);
 sd.temp = apply(data,1,sd,na.rm=T);
 data.new = ((data-median.temp)/sd.temp);
 return(list(data = data.new, median = median.temp,
sd=sd.temp));
}

median.sd.normlize.rev = function(data, median.
old, sd.old) {(data*sd.old+median.old)}

scale.fixed.knn = function(data,k=10)
{
 norm.temp = median.sd.normlize((data));
 data.new = (norm.temp[[1]]);
 median.new = norm.temp[[2]];
 sd.new = norm.temp[[3]];
 impu.temp = impute.knn(data = data.new,k,rowmax =
0.99,colmax = 0.99);
 X1b.new = median.sd.normlize.rev((impu.temp
[[1]]), median.new, sd.new);
 return((X1b.new));
}

combat.knn.cov = function(data, batch, mod)
{
 data.n = data;
 data.n[rowMeans(is.na(data.n))>0.99,][is.na
(data.n[rowMeans(is.na(data.n))>0.99,])] = median
(data,na.rm = T);
 data.knn = scale.fixed.knn(data.n);

 data.b = ComBat(dat = data.knn, batch = batch, mod
= mod);

 data.b[is.na(data)] = NA;
 colnames(data.b) = colnames(data.knn);
 return(data.b);
}

# Apply the function for batch correction
data.b = combat.knn.cov(dat = as.matrix(df.pro.
clean),
```

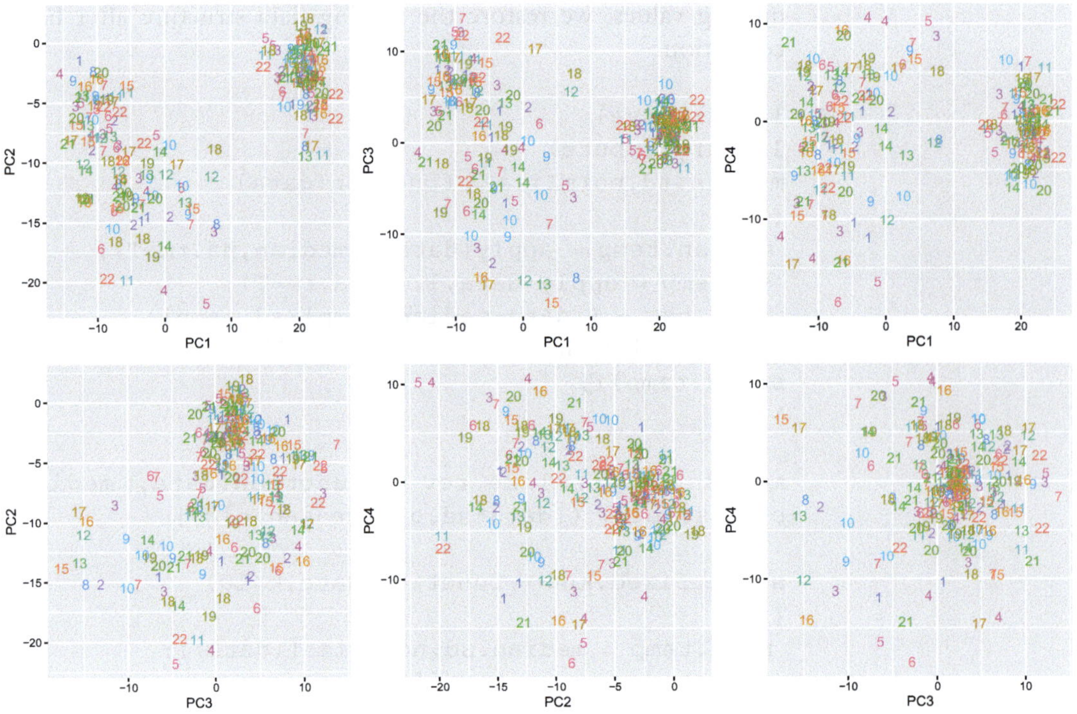

Fig. 6 The principal component (PC) plots of batches after batch correction with revised ComBat [27] method. Clusters by batches are less observed than before the correction

```
  batch = clinical.match$Experiment,
  mod=model.matrix(~Type, data=clinical.match))
plot.pc(data = df.pro.cb, col.batch = clinical.
match$Experiment, fig.path = 'plot/lscc_pro_cb.
pdf') ### after batch correction
```

It can be seen from Fig. 6 that no batches stand out after the batch correction procedure.

3.4 Missing Data Imputation

MS-based proteomic quantification is highly impacted by peptide-dependent ionization efficiency and peptide abundance and thus often has substantial missing rate. To make it worse, the missingness in MS-based proteomics data is not at random (NAT), wherein peptides/proteins with lower abundances tend to have higher probabilities of being missed, imposing the difficulties in proper imputing proteomics data. The NCI-CPTAC Dream Proteomics Imputation Challenge https://www.synapse.org/#!Synapse:syn8228304/wiki/413428used the crowdsource strategy to set state-of-the-art standard for proteomic data imputation. Its product DreamAI [28] (https://github.com/WangLab-MSSM/DreamAI) is an ensemble of six winning imputation methods (KNN, MissForest, ADMIN, Birnn, SpectroFM, RegImput) in the competition. The imputation accuracy of DreamAI is about

15–50% higher than previous tools for less abundant proteins with higher missing rates [28] and has been used in multiple CPTAC cancer studies [4, 5, 10, 11]. Note, to ensure imputation accuracy, the imputation is usually performed among features with <50% missing rate. Also, this procedure might be time-consuming, and the cloud computing if available is preferred for analysis. Here, we provide some sample codes for implementing the DreamAI imputation:

```
library(DreamAI)
df.pro.obs = df.pro.cb[!rowMeans(is.na(df.pro.cb))
>0.5,];
t0 = proc.time();
# For demonstration purpose, we only grow 20 trees.
In real data application, a large number of trees
(e.g. ntree=1000) should be used to ensure good
performance.
df.pro.imputed = DreamAI(data = df.pro.obs, k =
10, ntree = 20,
 maxnodes = 10, method = c("KNN","MissForest",
"ADMIN",
 "Birnn", "SpectroFM", "RegImpute"),
 out = c("Ensemble"))
t1 = proc.time();
```

4 Integrative Proteogenomic Analysis

After the preprocessing and quality control procedures, the proteomics data are often a G by N matrix for G features (peptides/proteins/genes/PTM) and N samples. This data structure has been widely observed in -omic data analysis field, such as from the normalized tissue-level transcriptomic data and from the M-value-based DNA methylation data. Abundant analytical tools have been developed for this data structure and can mostly be applied to the proteomic data for association, prediction, clustering, network, and other analyses. These methods have already been introduced in other chapters. Therefore, in this section, we will primarily focus on statistical and computational tools that are developed for integrative proteogenomic data analysis for a comprehensive characterization of multi-omic features of the tissue. For the cell-type deconvolution analysis, however, many transcriptomic-based methods cannot be transferred to proteomic data, and we will introduce methods that can be used for proteomic data analysis. This section has no intention to provide a comprehensive dictionary of all existing tools. Instead, we will highlight a few methods that are used in large consortiums, such as CPTAC and The Cancer Genome Atlas (TCGA), as examples to illustrate several popular ways for performing integrative proteogenomic analysis.

4.1 Multi-Omic Association Analysis

Association analysis using a single-omic data type (e.g., proteome) as outcome can be performed with many traditional methods, parametric or nonparametric, covariate adjusted or unadjusted [29]. One commonly implemented single-omic association analysis is differential analysis [26], which tests proteins with differential abundances between groups of samples and calls genome-wide significance using false discovery rate (FDR) or family-wide error rate (FWER). When multiple -omic data (e.g. mRNA, protein and PTM) are available in a dataset, applying these methods separately to each data type and combining the results in a "post hoc" manner will result in limited study power and compromised accuracy [30]. For example, proteomic researchers are often interested to differentiate the regulations unique at protein level from those shared with mRNA expression level. If an association is observed at protein level but not at its mRNA level using the "post hoc" approach, it's hard to distinguish whether this uniqueness is truly biological, or due to different study powers it in different data types. Therefore, it is important to have a systematic association analysis framework to jointly examine associations across -omic data types, to infer shared associations with enhanced power, and to identify unique associations with improved accuracy. In this section, we describe an example iProFun [30] to illustrate some efforts on this direction. iProFun starts with the regression on each -omic outcome (e.g., mRNA, protein and phosphoprotein) separately, and then uses the summary statistics as input to jointly infer the regulation patterns. It employs different analysis strategies for different predictors. For example, for CNV and DNA methylation that are variable across the genome, it leverages the parallel nature of the genes to jointly infer the association probabilities and boost study power for identifying associations on each data type. For somatic mutations that are present in few genes, it allows traditional analysis. It identifies associations not only based on *p*-values but also on biologically relevant association directions and association probabilities to allow highly confident calling of the signals. To demonstrate the method, we provide sample codes for gene-level -cis iProFun analysis that studies the impact of somatic mutation and CNV on their -cis mRNA, protein, and phosphoprotein levels in tumor, adjusting for age and gender as covariates.

```
library(iProFun)
# Prepare the data: narrow to cancer-associated
genes in tumor samples
rna= cancer.gene %>% inner_join(rna.dat) %>% co-
lumn_to_rownames(var = "geneSymbol") %>%
 select(-(1:4)) %>% mutate_all(type.convert) %>%
dplyr::select(!ends_with(".N"))%>%
 rownames_to_column(var="geneSymbol")
protein= df.pro.cb %>% as.data.frame() %>% dplyr::
```

```
select(!ends_with(".N"))%>%
 rownames_to_column(var="geneSymbol")
phospho= df.phos.cb %>% as.data.frame() %>% dplyr::
select(!ends_with(".N"))%>%
 rownames_to_column(var="id") %>% inner_join(phos.
info) %>% relocate(geneSymbol)
cnv= cancer.gene %>% inner_join(cnv.dat) %>%
 column_to_rownames(var = "geneSymbol") %>%
  select(-(1:3)) %>% mutate_all(type.convert) %>%
rownames_to_column(var="geneSymbol")
mut= mut.dat %>% column_to_rownames(var="Sample.
ID") %>%
t(.) %>% data.frame(.) %>% dplyr::select(!ends_-
with(".N")) %>% rownames_to_column(var="geneSym-
bol")
cov=clinical.dat %>% filter(Type=="Tumor") %>%
 mutate(Female=if_else(Gender=="male", 0,1)) %>%
 dplyr::select(Sample.ID, Age, Female) %>%
  column_to_rownames(var="Sample.ID") %>%t() %>%
as.data.frame()
# iProFun input:
yList = list(rna, protein, phospho) # three out-
comes
xList = list(mut, cnv) # two predictors
covariates = list(cov, cov, cov) # same covariates
are used for all three outcomes.
pi1 = 0.05 # Proportion of genes used to estimate
alternative density. Its choice has limited impacts
on the final results.

# iProFun is built upon linear regression on multi-
ple outcomes
reg.all=iProFun.reg(yList=yList, xList=xList, cov-
ariates=covariates, var.ID=c("geneSymbol"), var.
ID.additional=c("id"))
```

For predictor data types with many genes (e.g., CNV), we let iProFun identify associations based on (1) association probabilities >75%, (2) empirical FDR <0.1, and (3) the pass of association direction filtering. For data types with few genes (e.g., mutation), we let iProFun identify associations based on (1) FWER <0.1 and (2) the pass of association direction filtering. The association direction filtering is usually biologically driven. For example, scientists are usually interested in positive associations between CNV and their -omics outcomes, negative associations between DNA methylation quantified at the promoter and 5UTR regions and their -omics outcomes, and consistent associations (in both direction) between somatic mutation and their -omics outcomes.

```
# This function calculates FWER for somatic mutation.
FWER.all=iProFun.FWER(reg.all=reg.all, FWER.Index=c(1))
# This function calculates posterior association probability and eFDR rate for CNV.
eFDR.all=iProFun.eFDR(reg.all=reg.all, yList=yList, xList=xList, covariates=covariates, pi1=pi1,
 NoProbXIndex=c(1), permutate_number=2, var.ID=c
("geneSymbol"),
 var.ID.additional=c("id"), seed=123)
# This provides iProFun identifications with abovementioned criteria.
res=iProFun.detection(reg.all=reg.all, eFDR.all=eFDR.all, FWER.all=FWER.all, filter=c(0, 1),
 NoProbButFWERIndex=1,fdr.cutoff = 0.1, fwer.cutoff=0.1, PostPob.cutoff=0.75,
 xType=c("mutation", "cnv"), yType=c("rna", "protein", "phospho"))
# Plot a Venn diagram of the identified CNV events
subset = res %>% filter(iProFun.identification==1 &
xType=="cnv")
rna=subset %>% filter(yType=="rna") %>% select
(xName) %>% as.matrix() %>% as.character()
protein=subset %>% filter(yType=="protein") %>%
select(xName) %>% as.matrix() %>% as.character()
phospho=subset %>% filter(yType=="phospho")%>% select(xName) %>% as.matrix() %>% as.character()

VennDiagram::venn.diagram(
 list(rna=rna, protein=protein, phospho=phospho),
 filename = "plot/Venn.jpg",
 output=TRUE,
 disable.logging=T,
 # Output features
 height=1000, width=1000,resolution = 300,
 compression = "lzw")
```

Among the cancer-associated genes we investigate, 276 have -cis CNV-RNA/protein/phosphoprotein cascading effects that their CNV variations are associated with their mRNA, protein, and phosphoprotein abundance levels (Fig. 7). In cancer evolution,

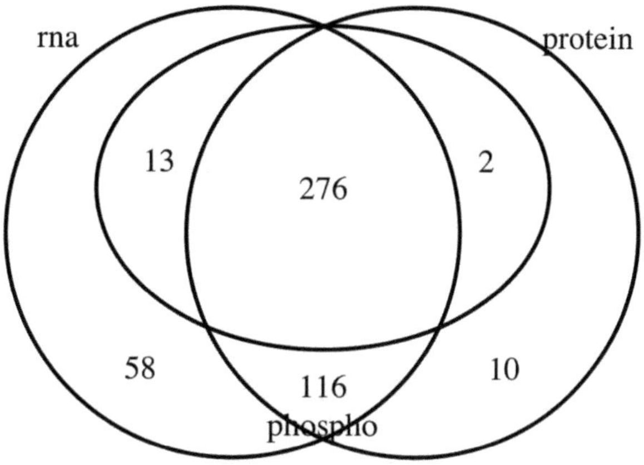

Fig. 7 The Venn diagram of genes with CNV-RNA/protein/phosphoprotein associations in iProFun [30] analysis

large DNA segments of the cells have similarly altered copy numbers, and thus it's difficult to distinguish CNV events that drive cancer progression from their neighboring events that are passengers of the evolution. The identification of the molecular functional consequences of CNV on multiple -omic outcomes at gene level helps to prioritize those likely to play a critical role in cancer mechanism [30].

4.2 Joint Network Construction

In the past decade, elucidating the topology of gene regulatory networks (GRNs) has become an energized research topic in computational systems biology, due to the large availability of transcriptomic data [31]. GRNs are composed of nodes, representing RNA, proteins, or other molecules, and edges, representing molecular relations such as gene–gene interactions. A collection of statistical and computational methods have emerged for constructing GRNs, and review of these methods is available elsewhere [32]. Briefly, most methods are designed to construct one network from one data type at a time. Presumably, if we have both RNA expression data and protein expression data for the same samples, we may construct two networks separately from the two data types and compare their similarity and differences. This strategy is suboptimal for two reasons: (1) it's difficult to tell whether the differences in topology structures are from true biological differences in the two data types or from the randomness of the data, and (2) RNA and protein may share some common structures in GRNs, which can be more accurately captured by joint modeling the two data types than by considering either of them. With accurate identifications of common edges, protein unique associations can also be captured with improved accuracy. To introduce some efforts on building multiple GRNs via joint model of different -omic data types, we describe the random forest-based method JRF [33] as an example. JRF outperformed the winning method of DREAM 5 network inference challenge GENIE3 [34] for network accuracy and estimate multiple related networks simultaneously. The key idea of JRF is to borrow information across different data by forcing the data-type-specific tree ensembles to use the same genes for the splitting rules [33]. To demonstrate this method, we apply *JRF* to build simultaneous RNA-based and protein-based GRNs for genes in the epithelial mesenchymal transition (EMT) pathway.

```
library(JRF)
library(GSA)
# load EMT pathway from Hallmark database (down-
loaded from http://www.gsea-msigdb.org/gsea/
msigdb/collections.jsp)
path.database=GSA::GSA.read.gmt("Hallmark.gmt")
gene.set=path.database$genesets[[which(path.data-
base$geneset.names=="HALLMARK_EPITHELIAL_MESENCH-
YMAL_TRANSITION")]]

# JRF does not handle missing data so we consider
genes with complete observations
rna.complete=rna[complete.cases(rna),]
protein.complete=protein[complete.cases(pro-
tein),]
# gene matching
genes.name=Reduce(intersect, list(gene.set, rna.
complete$geneSymbol, protein.complete$geneSym-
bol))
rna.overlap=rna[match(genes.name, rna$geneSym-
bol),-1]
protein.overlap=protein[match(genes.name, protein
$geneSymbol),-1]
# standardize variables to mean 0 and variance 1
rna.scale <- t(apply(rna.overlap, 1, function(x) {
(x - mean(x)) / sd(x) } ))
protein.scale <- t(apply(protein.overlap, 1, func-
tion(x) { (x - mean(x)) / sd(x) } ))

# Run JRF and obtain importance score of interac-
tions for RNA and protein
out<-JRF(list(rna.scale,protein.scale),mtry=round
(sqrt(length(genes.name)-1)),ntree=1000,genes.
name=genes.name)
# Run permutation procedure to identify
FDR-controlled network edges.
out.perm<-Run_permutation(list(rna.scale,protein.
scale),mtry=round(sqrt(length(genes.name)-1)),
ntree=1000,genes.name,M=10)
final.net<-JRF_network(out,out.perm,TH=0.1)
```

```
save(out, out.perm, final.net, file="JRF.net.rda")
# plot two networks on the same scale
library(igraph)
final.net[[1]][[1]]
rna.net = graph_from_data_frame(final.net[[1]],
directed = F)
protein.net = graph_from_data_frame(final.net
[[2]], directed = F)

# overwrite coordinates for shared nodes
lay1 <- layout.fruchterman.reingold(rna.net)
lay2 <- layout.fruchterman.reingold(protein.net)
lay2[which(V(protein.net)$name %in% V(rna.net)
$name), ] <-
  lay1[which(V(rna.net)$name %in% V(protein.net)
$name),]
xlim <- range(c(lay1[,1], lay2[,1]))
ylim <- range(c(lay1[,2], lay2[,2]))
# plot
par(mfrow=c(1,2))
plot(rna.net, vertex.color="white", vertex.frame.
color="white", layout=lay1, xlim=xlim, ylim=ylim,
rescale=FALSE)
title("RNA-based GRN",cex.main=2)
plot(protein.net , vertex.color="white", vertex.
frame.color="white", layout=lay2, xlim=xlim, yli-
m=ylim, rescale=FALSE)
title("Protein-based GRN",cex.main=2)
```

It can be noted from Fig. 8 that there is a common key hub comprised of PDGFRB, PRRX1, CDH11, COL3A1, COL1A1, and FBLN2 in RNA-based and protein-based GRNs. In the CPTAC LSCC study, PDGFRB has been identified to be highly correlated with the loss of epithelial cell-cell junction markers and upregulation of mesenchymal proteins involved in remodeling of extracellular matrix, induction of EMT, and promotion of cell migration, which mark the transition from epithelial to mesenchymal state in tumor [6]. The joint network modeling helps to identify network modules and hub genes critical to our understanding of the biological mechanisms of human health and disease.

4.3 Clustering

Single-omic-based clustering algorithms based on transcriptomics data have been introduced in Chap. 1, which can be largely applied for proteomics-based clustering analysis. Multi-omic clustering algorithms that model proteomics data jointly with transcriptomics and/or genomics data have been extensively developed in the past decade with review available elsewhere [35]. We introduce the

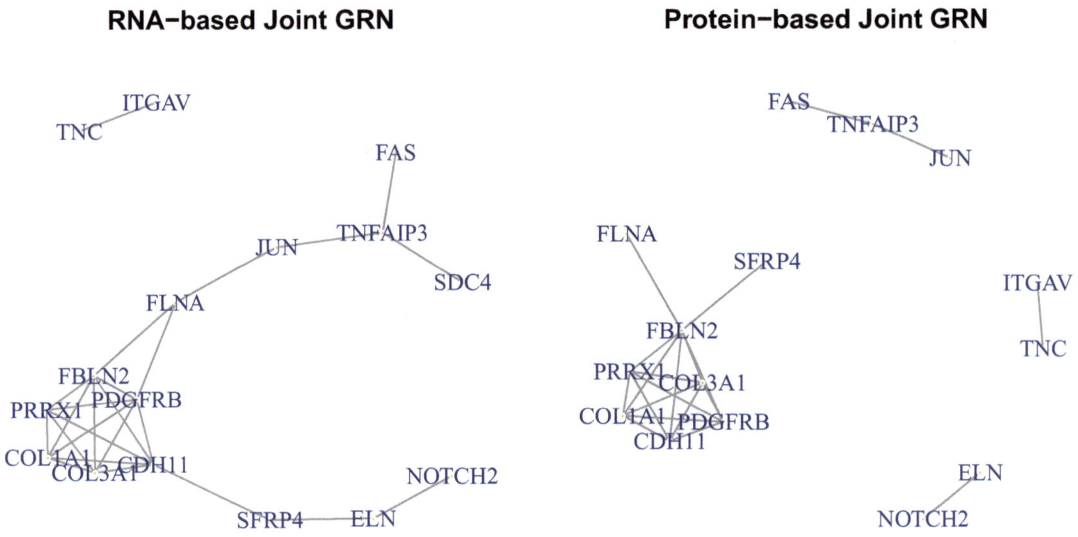

Fig. 8 The topology of the joint RNA-based and protein-based gene regulatory network obtained from the JRF method [33]

iClusterPlus [36] as an example for developing multi-omic clusters using proteomics data. An earlier version of iClusterPlus was applied by TCGA consortium to jointly cluster genomic, epigenomic, and transcriptomic data. Here are some sample codes for employing iClusterPlus method to jointly cluster proteomics and transcriptomics data:

```
library(iClusterPlus)
# iClusterPlus cannot handle missing data; we nar-
row the analysis to complete observations.
rna.complete=rna[complete.cases(rna),]
protein.complete=protein[complete.cases(pro-
tein),]
# iClusterPlus requires matching genes and samples
genes.name=Reduce(intersect, list(rna.complete
$geneSymbol, protein.complete$geneSymbol))
rna.overlap=as.matrix(rna[match(genes.name, rna
$geneSymbol),-1])
protein.overlap=protein[match(genes.name, protein
$geneSymbol),-1]
protein.sub.matching=as.matrix(protein.overlap[,
match(colnames(rna.overlap),
 colnames(protein.overlap))])
# iClusterPlus requires each row to be a sample and
each column to be a feature (gene).
rna.input=t(rna.overlap)
protein.input=t(protein.sub.matching)
# iClusterPlus is essentially a latent variable
```

```
regression with constraints, so tuning is needed
for good performance.
fit=iClusterPlus(dt1=rna.input,dt2=protein.input,
 type=c("gaussian", "gaussian"),
 lambda=c(1,1),K=1,maxiter=20)
# Tune lambda (penalty) and the number of clusters
from 2-6.
output=vector("list", length=5)
for(k in 1:5){
 cv.fit = tune.iClusterPlus(cpus=12,dt1=rna.input,
dt2=protein.input,
 type=c("gaussian","gaussian"),K=k, n.lambda=8,
 scale.lambda=c(1,1), maxiter=20)
 output[[k]]=cv.fit
}

nLambda = nrow(output[[1]]$lambda)
nK = length(output)
# The BIC value for each lambda and K
BIC = getBIC(output)
# The deviance ratio for each lambda and K
devR = getDevR(output)

# Get deviance ratio at minimal BIC for each K
minBICid = apply(BIC,2,which.min)
devRatMinBIC = rep(NA,nK)
for(i in 1:nK){
 devRatMinBIC[i] = devR[minBICid[i],i]
}
# Plot the deviance ratio at minimal BIC.
plot(1:(nK+1),c(0,devRatMinBIC),type="b",xlab="-
Number of clusters (K+1)",
 ylab="%Explained Variation")
```

The elbow in Fig. 9 usually serves as a good choice for number of clusters. Based on the figure, we pick three multi-omic clusters and plot the heat map for the two data types.

```
clusters=getClusters(output)
rownames(clusters)=rownames(gbm.exp)
colnames(clusters)=paste("K=",2:(length(output)
+1),sep="")
k=2
best.cluster=clusters[,k]
best.fit=output[[k]]$fit[[which.min(BIC[,k])]]
plotHeatmap(fit=best.fit,datasets=list(rna.input,
protein.input),
```

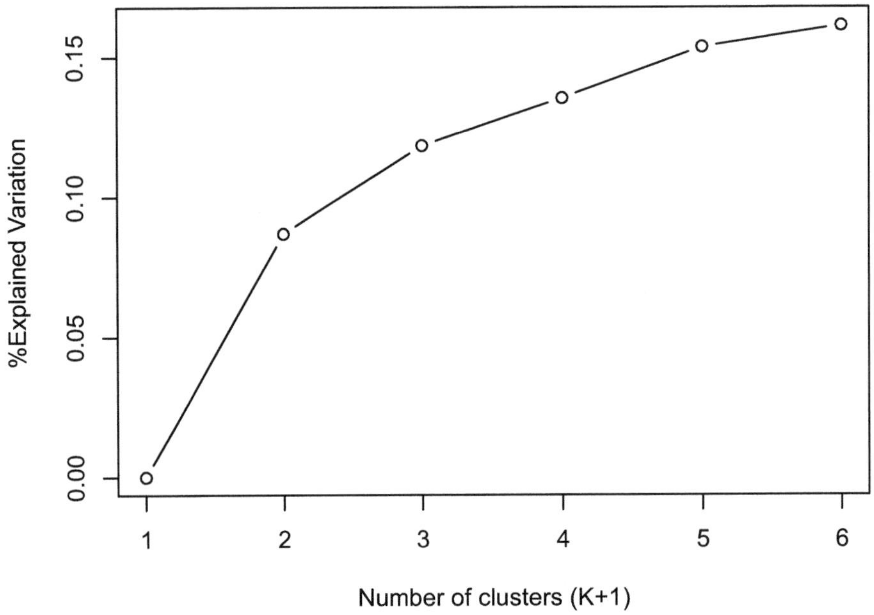

Fig. 9 The deviance ratio at minimal BIC

```
type=c("gaussian","gaussian"),
row.order=c(T,T),sparse=c(T,T))
```

From Fig. 10, we can see that the features vary across different clusters in both RNA and proteomics data. Block structure is easier observed in the proteomic-based heat map than in the RNA-based heat map, indicating that the clusters might better capture variations from proteomics data than from the transcriptomic data.

4.4 Cell-Type Deconvolution

One of the critical topics that worth mentioning in the proteogenomic data analysis is the cell-type deconvolution. While single-cell RNA sequencing (scRNAseq) technologies are becoming increasingly popular, the single-cell proteomic profiling is not yet mature for a wide application. Bulk profiling data have inherited vulnerability to the cell-type compositional heterogeneity of the tissues. For diseases that alter the cell-type composition, such as cancer and Alzheimer's disease, data analyses might be confounded by the cell-type heterogeneity. This issue has been widely acknowledged, and tremendous efforts have been put on estimating cell-type composition of tissues [37–45], mostly by deconvoluting the transcriptomic data. The resulting methods can be categorized into two major groups, reference-based and reference-free. Reference-free methods extract information purely from the data by employing advanced optimization algorithms [37, 44, 45], while reference-

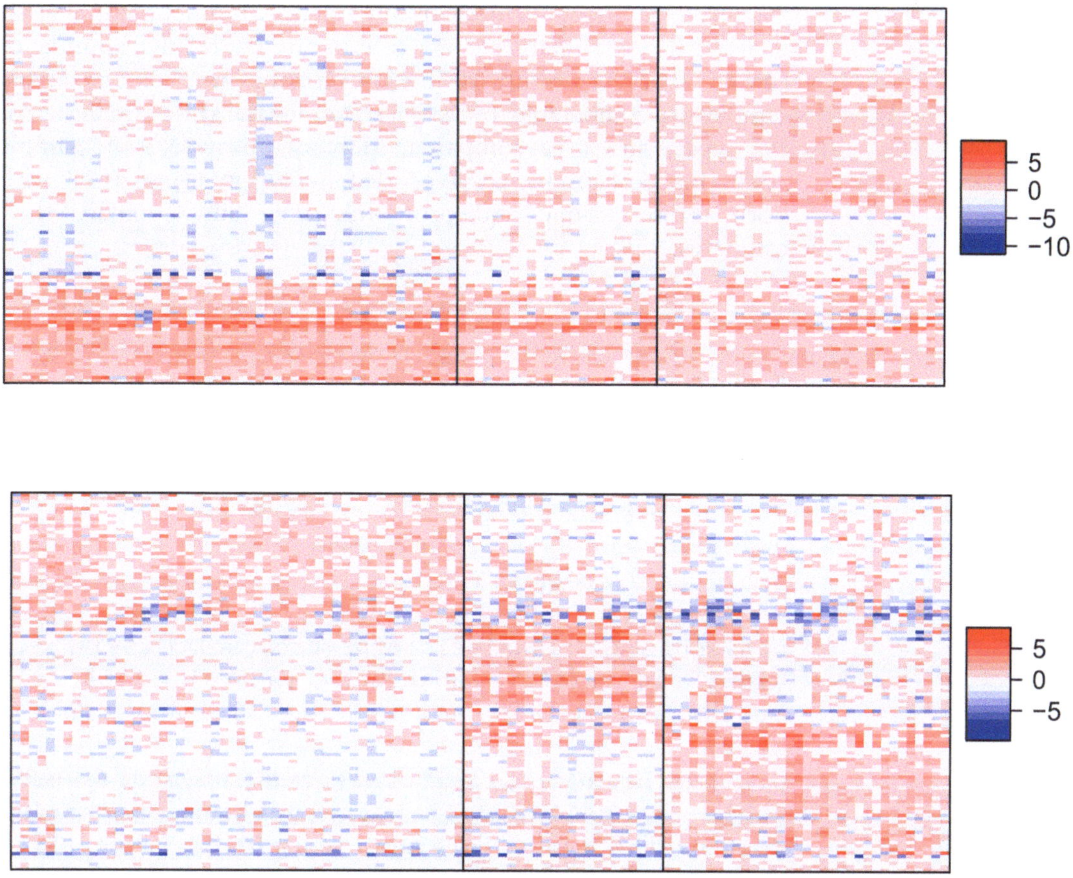

Fig. 10 The heat map of RNA and protein data based on the three clusters identified from iCluster method

based methods leverage prior knowledge/data on signature genes, cell-line expression, and scRNAseq [38–43]. Reference-based methods in general provide more accurate estimates; however, their application to proteomic data are inhibited limited proteomic-based references. Methods that use signature genes as reference might be transferrable to proteomic analysis, as signature gene list is not impacted by the scale and distribution of the data. Explorations that deconvolute proteomic data using transcriptomic-based gene signatures, by far, result in cell-type composition estimates with median correlations with estimates from transcriptomic-based deconvolution. Recently, a method emerged to jointly deconvolute transcriptomic and proteomic data [46], while its software is not yet publicly available for application. The joint cell-type composition estimation is an active direction that worth monitoring in the next few years.

In this section, we will illustrate the key idea of xCell [39] for estimating cell-type composition on proteomics data. xCell is a popular cell-type deconvolution method employed by the Genotype-Tissue Expression (GTEx) and CPTAC consortia [10, 47].

xCell [39] is essentially performing a gene-set enrichment analysis to use cell-type signature genes to estimate the relative abundance of a cell type. As a result, xCell scores for the same cell type across samples are comparable, but xCell scores for the same sample across cell types cannot be compared. Here, we provide some sample codes for performing cell-type deconvolution using xCell with both transcriptomic and proteomic data and compare their estimates:

```
library(xCell)

# xCell requires a large number of genes, so we use
all genes available for demonstration. In real data
analysis, most preprocessing procedures should be
applied prior to xCell analysis, but normalization
is not needed as xCell uses rank not the actual
values of the data for analysis.
pro.use= pro.dat %>% select(-1:-17)
# run xCell
pro.score=xCellAnalysis(expr=pro.use) # xCell pro-
vides the relative abundance (score) of 64 cell
types, and immune, stromal and microenvironment
score
pro.pvalue=xCellSignifcanceBetaDist(scores=pro.
score) # xCell provides the p-value of testing
whether the score is significantly different from
zero.

# We also run xCell again using RNA data for compar-
ison
rna.use= rna.dat %>% distinct(id, .keep_all = TRUE)
%>% column_to_rownames(var ="id") %>% select(-1:-
4)
rna.score=xCellAnalysis(expr=rna.use)

# Plot the correlation between RNA and protein
estimated CD4+ memory T-cells.

dat=pro.score %>% as.data.frame() %>% rownames_-
to_column(var="type") %>%
 pivot_longer(!type, names_to = "subject", values_-
to ="protein") %>%
 inner_join(
 rna.score %>% as.data.frame() %>% rownames_to_col-
umn(var="type") %>%
 pivot_longer(!type, names_to = "subject", values_-
to ="rna")
```

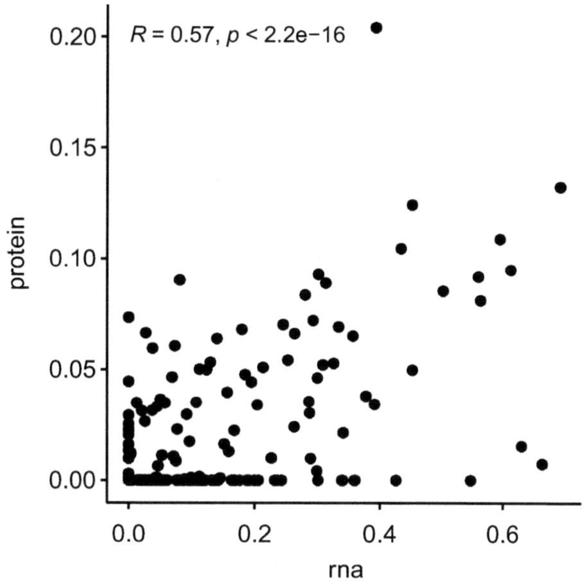

Fig. 11 The correlation between RNA and protein estimated cell-type relative abundance for CD4+ memory T-cells from the xCell method [39]

```
)
library(ggpubr)
ggscatter(dat[which(dat$type=="CD4+ memory
T-cells"),], x="rna", y="protein") +
 stat_cor(method = "pearson", label.x = 0.45, la-
bel.y = 0.2)
```

It can be seen from Fig. 11 that the RNA and protein-based correlation is not high ($R = 0.57$), which resonates the existing observations that RNA and protein data are not highly correlated to each other, and necessitates the development of protein-based analyses.

Immune Characteristics and Mechanisms Revealing the cell-type composition of a tissue will not only provide a foundation for eliminating its confounding effects in tissue-level analyses. It will also provide a great opportunity for understanding the cellular mechanisms of disease. For example, with the advent of cancer immunotherapies, it is critical to understand the relationship between the immune characteristics of the tumor and their responses to immunotherapies. Tumor response to the existing immunotherapies varies dramatically from patient to patient [48]. In most cancer types, some tumors respond completely to the immunotherapies, while others do not respond at all. The tumor microenvironment is playing a critical role on tumor responses, and the cell-type composition estimated from the deconvolution analyses contributes greatly to the understanding of the tumor microenvironment. For example, in the LSCC study [6], the RNA-based cell-type composition estimated from xCell has been used to characterize the tumor microenvironment. The resulting cell-type composition estimates of each tumor were used for identification of three different immune subtypes (hot, warm, and cold), in which immune hot tumors mostly likely benefit from immunotherapies.

References

1. Wilkins MR, Sanchez JC, Gooley AA et al (1996) Progress with proteome projects: why all proteins expressed by a genome should be identified and how to do it. Biotechnol Genet Eng Rev 13:19–50
2. Aslam B, Basit M, Nisar MA et al (2017) Proteomics: technologies and their applications. J Chromatogr Sci 55:182–196
3. Walgren JL, Thompson DC (2004) Application of proteomic technologies in the drug development process. Toxicol Lett 149:377–385
4. Clark DJ, Dhanasekaran SM, Petralia F et al (2019) Integrated proteogenomic characterization of clear cell renal cell carcinoma. Cell 179:964–983.e31
5. Gillette MA, Satpathy S, Cao S et al (2020) Proteogenomic characterization reveals therapeutic vulnerabilities in lung adenocarcinoma. Cell 182:200–225.e35
6. Satpathy S, Krug K, Jean Beltran PM et al (2021) A proteogenomic portrait of lung squamous cell carcinoma. Cell 184:4348–4371.e40
7. Reimegård J, Tarbier M, Danielsson M et al (2021) A combined approach for single-cell mRNA and intracellular protein expression analysis. Commun Biol 4:624
8. Domon B, Aebersold R (2006) Mass spectrometry and protein analysis. Science 312: 212–217
9. Rauniyar N, Yates JR (2014) Isobaric labeling-based relative quantification in shotgun proteomics. J Proteome Res 13:5293–5309
10. Petralia F, Tignor N, Reva B et al (2020) Integrated proteogenomic characterization across major histological types of pediatric brain cancer. Cell 183:1962–1985.e31
11. Wang L-B, Karpova A, Gritsenko MA et al (2021) Proteogenomic and metabolomic characterization of human glioblastoma. Cancer Cell 39:509–528.e20
12. Huang C, Chen L, Savage SR et al (2021) Proteogenomic insights into the biology and treatment of HPV-negative head and neck squamous cell carcinoma. Cancer Cell 39:361. https://doi.org/10.1016/j.ccell.2020.12.007
13. Song E, Gao Y, Wu C et al (2017) Targeted proteomic assays for quantitation of proteins identified by proteogenomic analysis of ovarian cancer. Sci Data 4:170091
14. Gao Y, Fillmore TL, Munoz N et al (2020) High-throughput large-scale targeted proteomics assays for quantifying pathway proteins in pseudomonas putida KT2440. Front Bioeng Biotechnol 8:603488
15. Toby TK, Fornelli L, Kelleher NL (2016) Progress in top-down proteomics and the analysis of proteoforms. Annu Rev Anal Chem Palo Alto Calif 9:499–519
16. Toby TK, Fornelli L, Srzentić K et al (2019) A comprehensive pipeline for translational top-down proteomics from a single blood draw. Nat Protoc 14:119–152
17. Pappireddi N, Martin L, Wühr M (2019) A review on quantitative multiplexed proteomics. Chembiochem Eur J Chem Biol 20:1210–1224
18. Gerber SA, Rush J, Stemman O et al (2003) Absolute quantification of proteins and phosphoproteins from cell lysates by tandem MS. Proc Natl Acad Sci U S A 100:6940–6945
19. Wang W, Zhou H, Lin H et al (2003) Quantification of proteins and metabolites by mass spectrometry without isotopic labeling or spiked standards. Anal Chem 75:4818–4826
20. Thompson A, Schäfer J, Kuhn K et al (2003) Tandem mass tags: a novel quantification strategy for comparative analysis of complex protein

mixtures by MS/MS. Anal Chem 75:1895–1904

21. Ross PL, Huang YN, Marchese JN et al (2004) Multiplexed protein quantitation in Saccharomyces cerevisiae using amine-reactive isobaric tagging reagents. Mol Cell Proteomics MCP 3: 1154–1169
22. Dayon L, Hainard A, Licker V et al (2008) Relative quantification of proteins in human cerebrospinal fluids by MS/MS using 6-plex isobaric tags. Anal Chem 80:2921–2931
23. Vellosillo P, Minguez P (2021) A global map of associations between types of protein post-translational modifications and human genetic diseases. iScience 24:102917
24. Hogrebe A, von Stechow L, Bekker-Jensen DB et al (2018) Benchmarking common quantification strategies for large-scale phosphoproteomics. Nat Commun 9:1045
25. Paik PK, Pillai RN, Lathan CS et al (2019) New treatment options in advanced squamous cell lung cancer. Am Soc Clin Oncol Educ Book Am Soc Clin Oncol Annu Meet 39:e198–e206
26. Urfer W, Grzegorczyk M, Jung K (2006) Statistics for proteomics: a review of tools for analyzing experimental data. Proteomics 6(Suppl 2):48–55
27. Johnson WE, Li C, Rabinovic A (2007) Adjusting batch effects in microarray expression data using empirical Bayes methods. Biostat Oxf Engl 8:118–127
28. Ma W, Kim S, Chowdhury S et al (2020) DreamAI: algorithm for the imputation of proteomics data. https://doi.org/10.1101/2020.07.21.214205
29. Nie L, Wu G, Culley DE et al (2007) Integrative analysis of transcriptomic and proteomic data: challenges, solutions and applications. Crit Rev Biotechnol 27:63–75
30. Song X, Ji J, Gleason KJ et al (2019) Insights into impact of DNA copy number alteration and methylation on the proteogenomic landscape of human ovarian cancer via a multi-omics integrative analysis. Mol Cell Proteomics MCP 18:S52–S65
31. Schadt EE (2009) Molecular networks as sensors and drivers of common human diseases. Nature 461:218–223
32. Delgado FM, Gómez-Vela F (2019) Computational methods for gene regulatory networks reconstruction and analysis: a review. Artif Intell Med 95:133–145
33. Petralia F, Song W-M, Tu Z et al (2016) New method for joint network analysis reveals common and different coexpression patterns among genes and proteins in breast cancer. J Proteome Res 15:743–754
34. Huynh-Thu VA, Irrthum A, Wehenkel L et al (2010) Inferring regulatory networks from expression data using tree-based methods. PLoS One 5:e12776
35. Rappoport N, Shamir R (2018) Multi-omic and multi-view clustering algorithms: review and cancer benchmark. Nucleic Acids Res 46: 10546–10562
36. Mo Q, Wang S, Seshan VE et al (2013) Pattern discovery and cancer gene identification in integrated cancer genomic data. Proc Natl Acad Sci U S A 110:4245–4250
37. Houseman EA, Molitor J, Marsit CJ (2014) Reference-free cell mixture adjustments in analysis of DNA methylation data. Bioinformatics 30:1431–1439
38. Newman AM, Liu CL, Green MR et al (2015) Robust enumeration of cell subsets from tissue expression profiles. Nat Methods 12:453–457
39. Aran D, Hu Z, Butte AJ (2017) xCell: digitally portraying the tissue cellular heterogeneity landscape. Genome Biol 18:220
40. Newman AM, Steen CB, Liu CL et al (2019) Determining cell type abundance and expression from bulk tissues with digital cytometry. Nat Biotechnol 37:773–782
41. Tsoucas D, Dong R, Chen H et al (2019) Accurate estimation of cell-type composition from gene expression data. Nat Commun 10: 2975
42. Wang X, Park J, Susztak K et al (2019) Bulk tissue cell type deconvolution with multi-subject single-cell expression reference. Nat Commun 10:380
43. Dong M, Thennavan A, Urrutia E et al (2021) SCDC: bulk gene expression deconvolution by multiple single-cell RNA sequencing references. Brief Bioinform 22:416–427
44. Li Z, Guo Z, Cheng Y et al (2020) Robust partial reference-free cell composition estimation from tissue expression. Bioinformatics 36: 3431–3438
45. Sun Q, Peng Y, Liu J (2021) A reference-free approach for cell type classification with scRNA-seq. iScience 24:102855
46. Petralia F, Calinawan AP, Feng S et al (2021) BayesDeBulk: a flexible Bayesian algorithm for the deconvolution of bulk tumor data. https://doi.org/10.1101/2021.06.25.449763
47. GTEx Consortium (2020) The GTEx Consortium atlas of genetic regulatory effects across human tissues. Science 369:1318–1330
48. Esfahani K, Roudaia L, Buhlaiga N et al (2020) A review of cancer immunotherapy: from the past, to the present, to the future. Curr Oncol Tor Ont 27:S87–S97

Chapter 14

Pharmacogenomic and Statistical Analysis

Haimeng Bai, Xueyi Zhang, and William S. Bush

Abstract

Genetic variants can alter response to drugs and other therapeutic interventions. The study of this phenomenon, called pharmacogenomics, is similar in many ways to other types of genetic studies but has distinct methodological and statistical considerations. Genetic variants involved in the processing of exogenous compounds exhibit great diversity and complexity, and the phenotypes studied in pharmacogenomics are also more complex than typical genetic studies. In this chapter, we review basic concepts in pharmacogenomic study designs, data generation techniques, statistical analysis approaches, and commonly used methods and briefly discuss the ultimate translation of findings to clinical care.

Key words Pharmacogenomics, Statistical genomics, Rare variants, High-throughput sequencing, Haplotypes, Quality control

1 Introduction

Among the human traits that are influenced by genetic variation, pharmacogenomic traits are perhaps the most useful to understand. Nearly two-thirds of all US adults take prescription drugs to treat clinical conditions [1], and genetic variants can significantly impact multiple aspects of drug response. For example, perhaps the most common are genetic variants that alter the metabolism of a drug, such as *CYP2C9* variants and warfarin [2]. Other examples are variants that change the generation of the active metabolite from a prodrug, such as in the case of *CYP2C19* variants and Clopidogrel [3]. Genetic variants can also alter risk of side effects or severe adverse events, such as *SLCO1B1* variants altering risk for myotoxicity [4] or rhabdomyolysis (in rare cases) [5, 6] in response to statin therapy. Due to the extensive need for these medications in the US population, predicting the correct dose and response while preventing adverse events using genetic information has great

Haimeng Bai, Xueyi Zhang and William S. Bush contributed equally with all other contributors.

Brooke L. Fridley and Xuefeng Wang (eds.), *Statistical Genomics*,
Methods in Molecular Biology, vol. 2629, https://doi.org/10.1007/978-1-0716-2986-4_14,

potential to increase the overall utility of a drug, by improving adherence and reducing severe reactions.

Pharmacogenomic traits are also useful to understand in the development and evaluation of new drugs. The rigorous standards of the Federal Drug Administration require that drugs have a proven effect within a clinically relevant population, but the therapeutic effect identified from most clinical trials is effectively an average over large numbers of individuals with varying genetic backgrounds. It is possible that a drug under evaluation has a limited effect across the entire population but may have a more potent effect in a genetic subset [7]. Since 2001, targeted therapies have become more common in cancer treatment, where the genetic profile of a patient tumor is used to identify drugs that specifically target cancer cells [8]. Despite these successes, there are far fewer examples of targeted systemic therapies for chronic conditions.

Unlike other areas of statistical genomics, the field of pharmacogenomics evolved from very targeted studies of xenobiotic metabolism genes (CYPs), which were explored clinically and functionally long before the advent of genetic sequencing [9–11]. Due to this history, the genetic polymorphisms for some genes have distinct designations referred to as "star-alleles" [12]. The molecular basis of these alleles was identified and characterized using gel electrophoresis many years later, and the star-alleles of CYP enzymes were correlated to splicing variants that influence electrophoretic mobility of the physical molecules [13]. The molecular effects of these alleles were experimentally determined in concert with investigations on the influence on drugs. After the large-scale adoption of genome-wide association studies [14], the statistical genomics of pharmacogenomic traits became very similar to other traits (see Ref. [15] for an example).

Due to their unique context, results from pharmacogenomic analyses are often immediately relevant and have direct clinical application. Knowing a priori that a patient will not respond (or will respond poorly) to a drug can steer clinicians to alternate therapies that will get a patient to the desired state faster. For several key pharmacogenomic effects, decision support systems have been implemented within electronic health record (EHR) systems to aid clinicians in prescribing drugs or ordering pharmacogenomic testing. The Pharmacogenomic Resource for Enhanced Decisions in Care & Treatment (PREDICT) initiative is an excellent example where preemptive genotyping was triggered by key clinical conditions, and results for actionable variants were included into patient records in anticipation of the need for anticoagulant therapies [16]. EHR alerts were triggered when a clinician attempted to prescribe a drug that would not be metabolized or therapeutically active for the patient. While this system was largely a demonstration of proof of concept, larger-scale implementation of pharmacogenomics into clinical care is a possibility and could have a significant effect on patient outcomes [17].

Recognizing the importance of pharmacogenomic information, and the fact that the field of genomics often moves too quickly for clinicians (or any other profession) to stay informed on current practices, the Clinical Pharmacogenomics Implementation Consortium (CPIC) was established to provide specific guidance on how to use genetic information for the prescription of drugs when relevant [18]. CPIC regularly reviews the pharmacogenomic literature and defines categories for the quality and strength of the evidence supporting a recommendation. A three-tier scheme is used where Level A or B provides the strong, consistent evidence from well-designed and conducted studies sufficient to recommend prescribing action, Level C indicates that evidence is sufficient to establish an effect, but the strength and consistency are insufficient to generalize to routine practice, and no prescribing action is recommended, and Level D denotes very limited evidence of the effect due to the quality and number of supporting studies with no actions recommended. As of 2019, 23 guidelines have been published covering 19 genes and 46 drugs within several therapeutic areas [19].

CPIC emerged as an extension from PharmGKB, the Pharmacogenomics Knowledge Base [20, 21], an online resource for identifying gene-drug relationships. PharmGKB has established tiers of "very important pharmacogenes" with detailed summaries of how the gene is involved in the metabolism or response to a drug. Summaries are typically published; some key examples are *CYP2C9* [22], *ACE* [23], and *TPMT* [24]. PharmGKB also assigns a variety of clinical annotations and biological pathways to genes, drugs, and genetic variants, with extensive literature citations.

2 Data Generation for Pharmacogenomic Discovery Analyses

Pharmacogenomic studies are often distinct from other types of genetic studies in a few different ways. First, because the phenotype is defined following exposure to a drug or therapy, pharmacogenomic studies usually have some component of a longitudinal rather than cross-sectional design. Second, exposure to a drug is generally only ethically acceptable when a condition indicates it; therefore, typical genetic study designs, such as segregation analysis, linkage analysis, or family-based association analyses, are not available for pharmacogenomic traits. Even for widely prescribed drugs, it can be difficult to identify related individuals with sufficiently similar clinical indications warranting drug prescription to power such analyses. Heritability estimates from twin or family studies are generally a key first step for establishing the genetic basis of a trait, but without the ability to enroll related individuals, heritability estimates for pharmacogenomic traits are complicated to conduct.

2.1 Population-Based Study Designs

Because of these limitations, population-based studies are the most common for pharmacogenomic traits, and as most drugs are prescribed in clinical settings, clinic-based study designs are typical. While sample sizes vary depending on the frequency of the drug prescription and the type of study design, most studies are smaller than a typical GWAS, with a mean sample size of ~900 samples and a median of ~500 [25]. However, it has also been noted that effect sizes for pharmacogenomic studies tend to be larger than typical disease-focused GWAS, with a median odds ratio nearly twice as large [26]. Many common diseases studied through genetics have clinically validated diagnostic criteria, but pharmacogenomic outcomes are often more complex, requiring more thought to the general phenotype definition [27]. Drug response, for example, may require pretreatment measurements, detailed diagnoses, and clear criteria for response versus nonresponse.

2.2 A Population-Based Study of Warfarin Sensitivity

Warfarin (branded Coumadin) is a widely used anticoagulant medication. Though efficacious, warfarin usage is often accompanied by severe adverse drug reactions, including over-anticoagulation and bleeding. Many pharmacogenomic studies have evaluated the warfarin dosing guidelines based on genotyping information. In a study conducted at pharmacist-run anticoagulation clinics [28], retrospective clinical data of patients who regularly attend the anticoagulant clinics were collected. The exon 3 and exon 7 of gene *CYP2C9* were genotyped. The primary outcome of this study was anticoagulation status, and the secondary outcome was the time until a severe adverse event. Two strata of patients were grouped by genotype, and Cox proportional hazards models were used to calculate the hazard ratio (HR) and 95% confidence interval (CI) comparing variant and wild-type genotype groups. This study identified the association between *CYP2C9* genotype and (1) warfarin maintenance dose, (2) temporal warfarin dosage changes, and (3) bleeding events. A more recent study used a register-based cohort study design (the PreMed study) linking data from multiple Finnish biobanks and national health registries to get a larger sample [29]. Based on *VKORC1* and *CYP2C9* genotypes, the participants were divided into three groups with different warfarin sensitivity. The primary outcome was bleeding events during warfarin exposure, and data on thromboembolic events and bleeding-related hospitalization were also extracted. The researchers investigated associations between the sensitivity group and the occurrence of outcome events using Cox proportional hazards models. They found no difference in the risk of bleeding complications among warfarin sensitivity groups, but sensitive and highly sensitive groups had a higher risk of thromboembolic events. The consistency of bleeding complications across groups can be interpreted by the lower daily doses in sensitive and highly sensitive groups, suggesting dose adjustment in clinical practices.

2.3 Cell-Line-Based Studies

Due to the difficulties in population-based study designs for pharmacogenomic traits, some researchers have utilized population-based cell lines to identify differences in cellular responses to drug exposures by genotype [30, 31]. Lymphoblastoid cell lines (LCLs) created for samples used in the International HapMap Project [32] are accessible for research and have extensive data collections publicly available, including genotype [33], whole-genome sequence [34], and gene expression data [35], among others [36, 37]. The pharmacogenomic Research Network also has an initiative to establish induced pluripotent stem cell libraries with accessible genotype data for similar experimental purposes [38]. Treating these cell lines with drugs (e.g., chemotherapeutic agents) then measuring outcomes like cell death provides an effective approach for establishing genetic determinants of response and toxicity [39]. Because cell lines are available for a variety of genetic backgrounds, this approach can provide pharmacogenomic results relevant to multiple populations without the issues of ascertainment bias [40, 41].

2.4 Cell-Line-Based Studies of Chemotherapeutic Toxicity and Response

Lymphoblastoid cell lines (LCLs) have been used to identify genetic variants associated with chemotherapeutic toxicity or response [31]. LCLs also allow for experimental manipulation to understand the underlying biology of genetic associations to drug treatment. Moreover, for drugs like chemotherapeutics which are too toxic to be given to unaffected individuals, LCLs provide an alternative to performing human studies. Paclitaxel is a chemotherapy medication used to treat several types of cancer, with chemotherapy-induced sensory peripheral neuropathy (CIPN) as an associated adverse effect. Researchers leveraged existing publicly available genetic data and the rich population diversity present in the HapMap and 1000 Genomes Project (1KG) populations to identify germline genetic variation associated with paclitaxel-induced peripheral neuropathy in East-Asian individuals [40] and European and African populations [42]. Cytotoxicity was measured in the paclitaxel-treated LCLs and was used as an outcome in the analyses. Genome-wide association analyses were stratified by population to identify variants associated with the cytotoxicity phenotype. For the East-Asian study, 183 cancer patients treated with monotherapy paclitaxel stored in Biobank Japan were also included in this study, and 31 variants were associated with both paclitaxel cytotoxicity and the grade of CIPN within the patients [40]. Another study expanded the scope of chemotherapy drugs evaluated using similar approaches [41].

2.5 Genotyping, Sequencing, and Quality Control (QC)

Array-based genotyping has matured into a reliable approach for generating highly accurate genotype calls on millions of genetic variants from nanogram quantities of DNA, and quality control of these data has been extensively covered elsewhere [43]. Array-based genotyping is limited to a targeted set of variants which may miss some clinically relevant DNA alterations. For studies implementing pharmacogenomic testing as part of clinical operations, genotype calling strategies will be different from population-based research studies due to the need to call genotypes for a single sample at a time. Because they are intensity-based, typical genotype calling algorithms that rely on batch-level processing to cluster genotype categories may not be applicable to calling for low numbers of samples typical of a clinical laboratory test. As a result, new calling algorithms were developed specifically for the targeted clinical application of specific pharmacogenomic variants. Both the Affymetrix DMET array and the VeraCode ADME core panel use proprietary algorithms to make genotype calls that do not rely as heavily on multi-sample clustering.

For some pharmacogenomic discovery analyses, whole-exome capture of targeted pharmacogenes [44] or whole-genome sequencing is attractive alternatives that provide a more comprehensive assessment of genetic variation. For pharmacogenomic implementation, sequencing provides broader information capture which may be useful for more rapidly expanding pharmacogenomic panels in a clinical setting; the underlying sequence data could be stored, while the processing pipelines are altered/updated as new actionable genes and variants are added.

An overview of the QC process for sequence-based data generation is shown in Fig. 1. Regardless of the technology, sequencing will produce collections of short-read sequences in FASTA/FASTQ file format. Basic sample quality control and template preparation are applied to sequence reads to ensure a successful sequencing run [45]. Once basic QC is complete, the data is ready for preparation. The first step is to trim off any adaptor or barcode sequences (a short piece of DNA sequence that is ligated to the end of a DNA fragment) that may have been introduced as part of the exome capture or library preparation steps. There are several widely used trimmers for this job, for example, Trimmomatic [46] and Trim-galore [47].

The next step is to align the sequence fragments to the reference; this requires choosing a human genome reference build, usually GRCh37 or GRCh38. There are several aligners that can perform this step including Bowtie [48] and Burrows Wheeler Aligner (BWA) [49]. Ultimately, the choice of aligner and genome reference is application-specific [50]. BWA-MEM (maximal exact match) [51] is recommended by the Genome Analysis Toolkit (GATK) software group [52].

The final step in sequencing processing is variant calling. Multiple genotype calling pipelines are available such as ATLAS [53], DRAGEN [54], and DEEPVARIANT [55]—see Ref. [56] for a comparison. In some scenarios, multiple pipelines may be applied to generate consensus calls [57]; however, the GATK pipeline [58] is the most widely used. The GATK website (https://gatk.

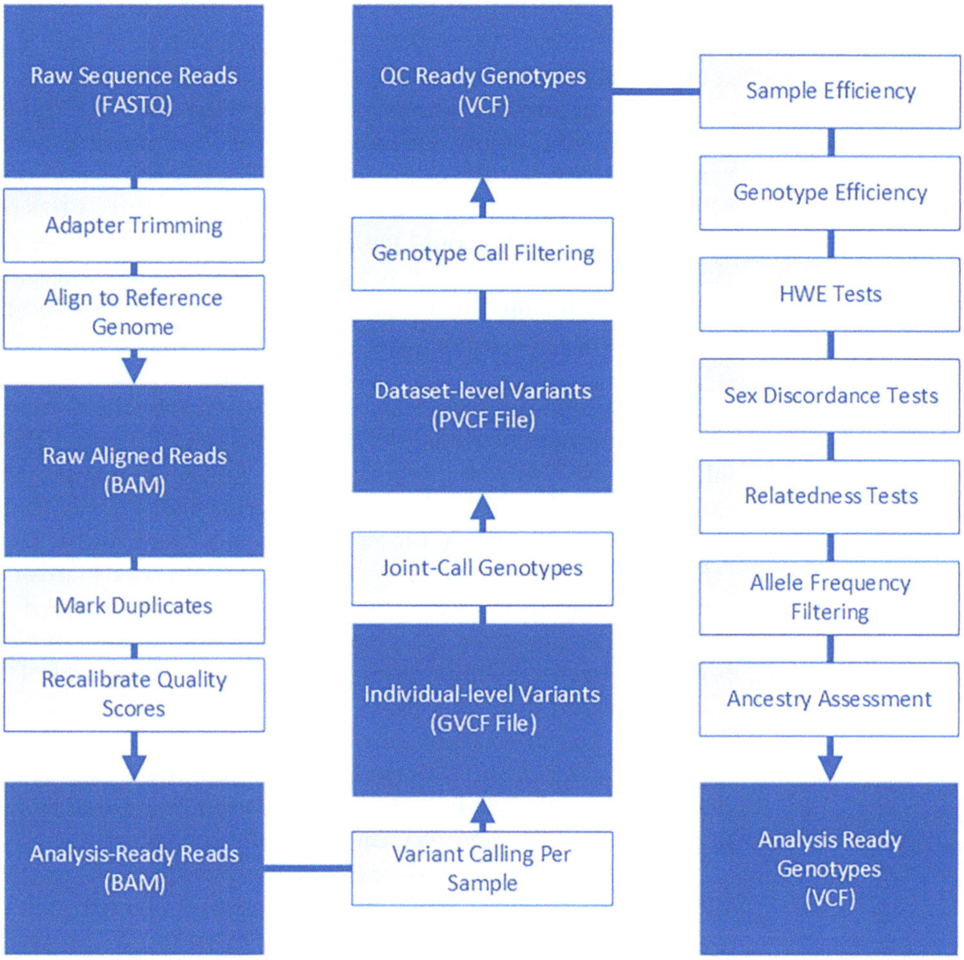

Fig. 1 Steps for data preparation from raw sequence reads to analysis-read genotypes. Derived data are shown in shaded squares and quality control steps are shown in open boxes. Data flow is indicated by arrows

broadinstitute.org/hc/en-us, https://github.com/gatk-workflows) provides various well-developed pipelines and tools for the genetic variant calling process, including a full pipeline for discovery of the commonly used germline single-nucleotide variants (SNVs) and indels. Pipelines are implemented as workflows (using workflow description language) and include detailed documentation about software installation and computing environment setup instructions for Docker and Google Cloud users.

Following variant calling, typical pre-analysis quality control of genotype calls includes several additional steps to ensure high-quality genotypes with limited bias. First, sample genotyping efficiency is evaluated; this refers to the proportion of overall variant calls for a sample. Low sample genotyping efficiency usually suggests poor quality of the sample DNA or issues in sample preparation. A threshold of 5% is commonly used for distinguishing good and bad DNA samples, and samples with an overall missing rate >5% should be excluded from further analysis [59].

Next, variant-level genotype call rates are checked. Similar to the sample genotyping efficiency, variant-level missing genotype rates are generated for each variant. A high missing rate of SNV implies poor genotyping quality and a high error rate at this particular variant site. This could indicate genomic complexity or repetitive sequences in the region that impacted alignment and sequence coverage, reducing the overall call rate. Thus, SNVs should be removed from the dataset at a genotype missingness threshold of 5%.

Depending on the study design, a minor allele frequency (MAF) filter may be desirable. The minor allele refers to the rarer allele of an SNV within a population. Low MAF dramatically limits the power of the statistical test at this marker site. Thus, if the study is not designed to study the rare variants, those SNVs with MAF <5% should be excluded from further analysis. The Hardy–Weinberg Equilibrium (HWE) principle describes the expected genotype probabilities of a marker assuming random mating in a largely homogenous population, among a few others. Violations of HWE can be checked by a goodness-of-fit χ^2-test [60] and may indicate biases toward heterozygotes or homozygotes in the genotype calling process. A threshold between $10e^{-5}$ and $10e^{-7}$ is usually used depending on the study and should be used as a guide to further investigate specific genotype calls for technical issues.

Sex discordance checks are another common QC step where the reported sex of a sample is compared to the genetically determined sex. Typically, males have one copy of the X chromosome, while females have two. Discordance checks are accomplished by calculating the X-chromosome homozygosity rate. The expected homozygosity rate is ~1 for male samples and <0.2 for female samples [61]. Deviation from these expectations implies sample contamination or incorrect sex information, and the sample data should be confirmed and removed from the analysis if discrepant. Sex discordance is also a common way to identify plate handling errors or other simple technical artifacts.

Except for family studies, GWAS samples are usually assumed to be independent. However, sometimes unexpected relatedness could occur, and this should be accounted for within or addressed prior to analysis to avoid violations of independence assumptions. This could be accomplished by estimating pairwise identity by descent (IBD) that measures the allele sharing level among two individuals' genomes from a common ancestor. An IBD equal to one indicates that the two samples are duplicated or monozygotic twins. Generally, higher IBD suggests closer relatedness. To eliminate relatedness, a conservative approach is to remove all individuals with IBD >0.1875 [61].

Table 1
Software and tools for data processing and quality control

Software/tool	Description
Trimmomatic	Sequencing barcode trimmer
BWA	Align sequence reads (FASTA, FASTQ) to references
Samtools	BAM/SAM files manipulation. Commonly used functions: view, sort, index, flagstat, etc.
Bcftools	VCF/BCF files manipulation and analysis
Vcftools	VCF file manipulation and analysis
Bedtools	BED files and its derived files manipulation
GATK	State-of-the-art sequencing data processing software. Commonly used functions: fastqc, MarkDuplicatesSpark, HaplotypeCaller, CombineGVCFs, GenotypeGVCFs, etc.
PLINK	Data manipulation of almost all genotype file formats including Plink (ped/map), Plink bfile (bed/bim/fam), Oxford (gen/sample), VCF/BCF files, etc. Also a powerful tool for genotype data QC with commonly used functions: --gen, --mind, --freq, --maf, --missing, --hardy, --genome, --pca, --check-sex, etc.
IMPUTE	Perform genetic imputation given references
SHAPEIT	Phase genotypes into haplotypes

Depending on the study design, QC may include generating ancestry-based principal components [62], defining ancestry categories for stratified analyses [63], and comparing genetic ancestry and self-reported or administratively assigned race/ethnicity. This is especially important when performing race or ancestry-stratified analyses, as the administrative assignment of race can have reporting biases [63, 64]. A full list of computational tools for QC and basic data analysis is shown in Table 1.

3 Statistical Analysis

3.1 Association Analysis for Common Variants

Due to the unique constraints of most pharmacogenomic studies, association analyses of common variants are conducted within the population or in clinical-based studies. However, sample sizes are often limited; therefore, additional methods for statistical modeling beyond what is used in typical GWAS are often used to optimize statistical power and to account for unique aspects of pharmacogenomic study designs.

Many analyses of pharmacogenomic traits apply classical approaches used in general genetic studies (like those outlined in other chapters). The genetic model used for analysis (additive, dominant, recessive) is generally assumed to be additive for pharmacogenomic traits as (1) prior simulation studies have shown this model is able to detect dominant and recessive effects [65] (though with reduced statistical power) and (2) the biological nature of the effect is assumed to follow a dose–response relationship per allele. Adjustment for appropriate covariates is another important consideration, especially when pharmacogenomic study designs may exhibit biases in ascertainment, follow-up, and measurement error due to the longitudinal and clinical nature of the studies. Given these considerations, linear or logistic regression models are generally used for quantitative or dichotomous outcomes, respectively. For these basic analyses, tools like PLINK [66] provide a fast, easy-to-use platform for statistical analysis. A list of commonly used tools for genomic data analysis is provided in Table 2.

3.2 A Common Missense Variant Associates to Leukopenia Adverse Events in Thiopurine Treatment

Thiopurines are immunosuppressive drugs that downregulate key components of T-cell responses that cause inflammation and are broadly used in the treatment of cancer, autoimmune diseases, and in organ transplantation. Immunosuppression for the treatment of inflammatory bowel disease is another example of the need to carefully balance dose response, as a common adverse event is leukopenia, a dangerously low white blood cell count [69]. Yang et al. used a case/control design with less than 1000 Korean Crohn's disease cases who used thiopurines (66 of whom developed leukopenia in the first 8 weeks of therapy, 280 who developed leukopenia after 8 weeks, and 632 controls) that were genotyped with an Illumina Immunochip array to identify pharmacogenomic variants associated with thiopurine toxicity [70]. From this limited sample size, a non-synonymous variant within the *NUDT15* gene was identified and was replicated in a secondary study of European-descent individuals with inflammatory bowel disease. All statistical analyses were conducted using PLINK software.

3.3 Phasing, Imputation, and Haplotype Methods

Many genes associated with pharmacogenomic traits exhibit genetic complexity, with insertions, deletions, or haplotype effects not explicitly captured by standard genotyping approaches. Haplotypes are combinations of groups of marker alleles on the same chromosome [71], and because genotyping typically measures a single site of both chromosomes simultaneously, they are not directly obtained from genotype data. For example, the star alleles of *CYP2C9* represent haplotypes of one missense variant in exon 3 and two missense variants in exon 7 [72], so in order to determine the effective function of this enzyme for an individual, the sequence of alleles on each of the two chromosomes for an individual must be understood. The process of mapping allele combinations (or genotypes) into haplotypes and inferring their chromosomal orientation is called *phasing*.

Methods used to infer the phasing information from unrelated individuals can be divided into two major categories: estimation maximization and Bayesian approaches. Methods based on the estimation-maximization (EM) algorithm estimate the population

Table 2
Packages and tools for statistical analysis

Software	Implementation	Usage
PLINK	Command line software (Linux/Windows/Mac)	Association analysis
		Haplotype analysis
lme4	R package	Mixed effects models for repeated measures or hierarchical models
survival	R package	Cox proportional hazards regression models
PheWAS [67]	R package https://www.vumc.org/cpm/center-precision-medicine-blog/phewas-r-package	Phenome-wide association analysis
METAL [68]	Command line tool https://genome.sph.umich.edu/wiki/METAL	Meta-analysis genome-wide association scans
SKAT	R package	Rare variants association test, including SKAT and SKAT-O
GCTA	Command line software (Linux/Windows/Mac)	Heritability estimation using mixed-effect models
nlme	R package	Linear and nonlinear mixed-effects models
PSCAN	R package	Protein-structure-guided rare variants test
POKEMON	Python package/Docker Container	Protein-structure-guided rare variants test
haplo.stats	R package	Haplotype association analysis
limma	R package	RNA-seq data analysis; Differential expression analysis; GSEA
DAVID	Website	GSEA
WEBGESTALT	Website	GSEA

haplotype frequencies by maximizing a likelihood function [71]. From genotype data, we observe the probability of genotypes given population haplotypes, and we want to model the probability of the haplotypes that our observations are conditioned on. To solve this, a likelihood function is calculated as the product of the probability of all haplotype pairs over the summation of the probabilities of all possible haplotypes [71, 73]. At each estimation step, the posterior probability for each possible haplotype is calculated, and at each maximization step, the calculated probabilities are aggregated to update the prior haplotype probabilities for the next iteration [71]. The EM approach is based on solid statistical theory and is robust to deviations from Hardy–Weinberg equilibrium. However, its performance is sensitive to initialization and, in some cases, may not converge. Also, because of the combinatorial nature of haplotypes, EM approaches cannot phase large sets of markers.

Methods that apply Bayesian approaches using Markov Chain Monte Carlo (MCMC) or Hidden Markov Models (HMM) treat haplotypes as states and infer the haplotype for an individual by modeling the probability of transition between states and the probability of emissions of insertion and deletions [74, 75]. Given a number of known phased haplotypes as a reference set, the transition probabilities at each variant position given the information of its previous can be calculated [75]. The phase of an unphased collection of genotypes can be inferred by applying these transition probabilities from beginning to end [75]. Emission probabilities are also estimated regarding SNP alleles that are not present in the references [75]. The Bayesian methods produce more stable estimates of the haplotype phase, but at the cost of increased computational complexity and burden. Efforts have been made to reduce the computational cost of these algorithms and apply them on a genome-wide scale; for example, SHAPEIT is a widely used algorithm that has successfully reduced the calculation complexity to the linear scale [76].

Once genotype data is phased, imputation approaches can be applied to estimate missing genotypes from adjacent marker data. Similar to Bayesian phasing, imputation also uses the HMM approach. The major difference is that in imputation, the position of the current state is unclear, and in such a way, the missing genotypes could be accounted for. There are several well-established whole-genome imputation references and pipelines [77]. The 1KG [78] and Haplotype Reference Consortium (HRC) [79] imputation pipelines are implemented on the Michigan Imputation Server [80]. The Trans-Omics for Precision Medicine (TOPMed) reference is based on whole-genome sequencing of over 100,000 individuals and is thus a richer reference panel [81]. Reference panels with a focus on African American populations also exist [82].

In addition to haplotypes, insertions, deletions, and other structural variants can be very difficult to characterize with typical genotype approaches. *CYP2D6*, for example, exhibits a variety of deletions and structural variants that influence enzyme activity [83, 84]. While some of these variants can be imputed from genotype data with reasonable accuracy, the error associated with imputation of these alleles likely requires additional sample sizes to account for inaccuracies [85]. Long read sequencing is an emerging technology that can be used to directly measure phased haplotypes. Well-known techniques are single-molecule real-time (SMRT) sequencing by Pacific Biosciences (PacBio) and nanopore sequencing by Oxford Nanopore Technologies Inc. (ONT). Several studies have shown the capability of generating more accurate phased haplotypes using long-read sequencing techniques relative to other approaches [86]. This technique is particularly useful to genotype some highly variable regions relevant to pharmacogenomic studies, such as *CYP2D6* and *HLA* [86]. However, important limitations exist, such as potential experimental bias [87] and higher error rates and costs [88].

3.4 Haplotype Effects in the VKORC1 Gene Associate to Warfarin Dose

The first study to identify common genetic variants within the vitamin K epoxide reductase complex 1 (*VKORC1*) gene that associate to warfarin dose is an excellent example of the potential utility of characterizing multiple functional variants within a pharmacogene. In this study, the authors performed targeted DNA sequencing across an 11 kilobase region surrounding the gene. Using approximately 500 clinic-based samples, 10 common variants in this *VKORC1* were identified and phased into 5 common haplotypes. Standard statistical approaches were applied to associate these haplotypes to warfarin dose, with two haplotypes requiring a lower warfarin dose and two haplotypes requiring an increased dose. This study further explored the haplotype diversity of this gene in multiple ancestral populations and demonstrated that *VKORC1* explains significantly more variation in warfarin dose than the previously established *CYP2C9* variants. Haplotype effects may explain some additional residual variation in dosing; however, later studies have demonstrated the effect of this gene is largely explained by a single variant [89].

3.5 Mixed Models in Pharmacogenomic Studies

Mixed linear models allow both typical fixed effects and random effects that are mainly used to account for nonindependence among observations in the data, such as scenarios where individuals have repeated measures, or where outcomes may be influenced by a hierarchical structure (e.g., accounting for patients that see the same doctor, etc.). In the context of pharmacogenomic research, mixed linear models often describe patient populations as a whole and patients as individuals [90]. Mixed linear models are also helpful in adjusting for clinical variability across a population-based sample. For example, we can define a random effect for billing codes or lab test values, similar to using principal components analysis to adjust for clinical conditions confounding blood pressure and hypertension associations as outlined in [91].

Nonlinear extensions of mixed effects models are also used to test the impact of gene variants on pharmacokinetic or pharmacodynamic responses in pharmacogenomic studies. A study on Atorvastatin dose response implemented the nonlinear mixed effects model (NLMEM) to estimate the dose–response curve parameters [92]. The Marshfield Clinic Personalized Medicine Research Project used EHR data linked to pharmacy databases to address compliance and adherence issues [93]. The authors constructed models

$$f(x_{ij}, \beta_i) = \beta_{Emax} + \frac{\beta_{E0} - \beta_{Emax}}{1 + \left(\frac{x_{ij}}{\beta_{EC50}}\right)^{\beta_{Slope}}}$$

$$y_{ij} = f(x_{ij}, \beta_i) + e_{ij}$$

$$\beta_i = A_i\beta_0 + B_i b_i$$

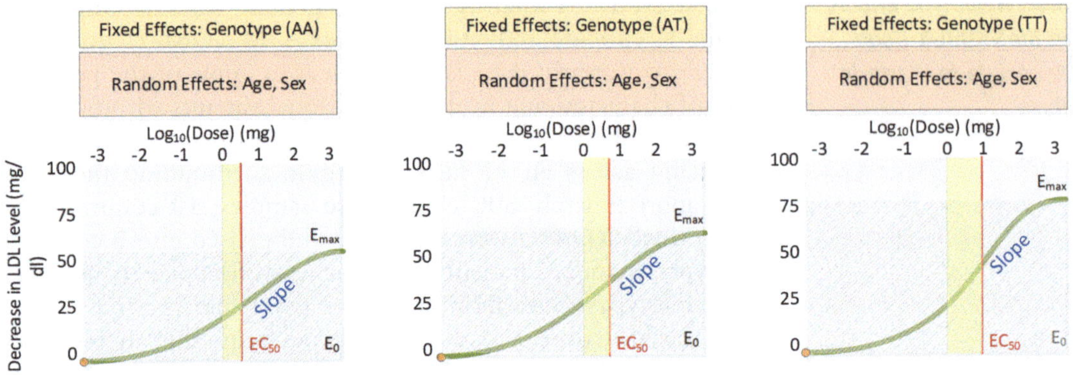

Fig. 2 Hypothetical example of a nonlinear mixed effects model of statin dose response. The dose of statins is often adjusted after initial prescription due to adverse events and other factors, therefore multiple drug doses and lipid levels within individuals could be used to estimate dose–response curves. In this example, y_{ij} represents the reduction in LDL in response to drug dose *j* for individual *i*. Four parameters of a logistic model, E_0 and E_{max} (colored gray), EC_{50} (colored red), and slope (colored blue) form a vector β_i that defines the logistic function (colored green) fit to each individual data point x_{ij}. The vector β_i is estimated from a mixed effect model containing fixed effects, in this case the genotype of an SNV (colored gold) and random effects accounting for individual-level variation including age, sex, and the within-individual repeated measures (colored orange). When fit, this model would estimate statin does-response curves by genotype, shown in the plots at the bottom of the figure

of full dose–response curves for atorvastatin as the phenotype with random patient effects. Raw data were fit with an NLMEM using a hierarchical modeling approach to build individual-level models [94]. A similar approach was used to estimate dose response for simvastatin within the BioVU biobank at Vanderbilt University [95].

A similar modeling strategy has been applied to cell lines using a four-parameter logistic curve [96], with parameters for response at zero concentration, response at infinite concentration, log of the response that is midway between these two points (also known as EC50), and the slope of the dose–response curve. This approach can be further parameterized in a hierarchical model to estimate the impact of a genotype as a fixed effect, accounting for variation in age, sex, and other covariates as random effects on the dose–response curve parameters (see Fig. 2). This approach was applied to model in vitro drug response for cancer cell lines [96].

Beyond dose–response modeling, mixed effects models form the basis for a software package called genome-wide complex trait analysis (GCTA) [97] which is designed to estimate the heritability of a complex trait within a population-based sample. GCTA fits the effects of all SNP genotypes as a random effect within a mixed linear model, essentially mimicking twin- and family-based heritability studies by empirically computing the genetic sharing among a set of population-based samples [98]. The GCTA approach estimates the narrow-sense heritability (i.e., the proportion of phenotype variability explained by additive genetic effects) by aggregating multiple associated SNPs with small effect sizes [99]. GCTA is vital in heritability estimation for pharmacogenomic phenotypes by offering an alternative testing approach considering polygenic genetic architecture. However, the small sample sizes of pharmacogenomic studies may limit the performance of GCTA. Mixed model approaches have also been used for time-to-event or survival analyses. A basic analysis of this type using pedigree-based kinship data was used to explore alcohol dependence [100], with more recent methods published to examine this for both genome-wide and gene-level effects [101]. Additional details on the methodological considerations in heritability estimation can be found in [98].

3.6 Examples of Heritability Estimation for Pharmacogenomic Traits

McGeachie et al. used GCTA to estimate the heritability of asthma pharmacogenomic traits from genome-wide SNPs [102]. Asthma and two asthma-related pharmacological phenotypes, bronchodilator response and airway hyperresponsiveness, were analyzed in this study. The most frequently used asthma drugs are bronchodilators, and response to bronchodilators are variable from patient to patient. Airway hyperresponsiveness is the characteristic airway obstruction produced in response to asthmatic stimuli, and this phenotype changes the most with steroid (anti-inflammatory) treatment. The study subjects were obtained from multiple asthma clinical trials and were matched with group controls from an external population-based study. From genome-wide genotype data, GCTA was used to estimate heritabilities of 28.6% for bronchodilator response, 51.1% for airway hyperresponsiveness, and 61.5% for asthma, which are comparable to prior estimates from twin and family studies.

3.7 Rare Variant Analysis for Pharmacogenomic Traits

Existing validated pharmacogenomic markers can explain only a small proportion of the observed clinical variability in the therapeutic outcome [103]. Similar to other traits being studied in statistical genomics, rare variants may account for a significant proportion of clinical and genetic variability in pharmacogenomic traits. With the advancement of sequencing technology, more and more rare variants are identified from genetic studies that may influence pharmacogenomic traits. For example, the Alzheimer's Disease Sequencing Project whole exome sequenced ~10 k individuals, and 97% of identified variants have a MAF of <1%, and 25% of those are singletons or doubletons [104]. A similar frequency pattern has also been observed in pharmacogenomic studies. In the February 2015 eMERGE-PGx data release based on 5639 clinical subjects, 82 pharmacogenes were sequenced from 9 eMERGE sites [105]. Among the 39,778 sequenced bi-allelic SNVs, there were 2102 common, 1230 low frequency, 9465 rare, 4606 doubleton, and 22,124 singleton variants. A searchable database of these variants is available through the eMERGE SPHINX resource [106].

A larger study involving 60,706 unrelated individuals who were part of the Exome Aggregation Consortium (ExAC) database analyzed the distribution of rare and common genetic variants in the 208 clinically relevant pharmacogenes that potentially modulate drug absorption, distribution, metabolism, and excretion (ADME) [107]. The importance of these variants for drug response was estimated using an ADME-optimized prediction framework [108]. This study found that rare pharmacogenomic variants were strongly enriched for mutations predicted to cause functional alterations in the gene and that the contribution of a rare variant to the prediction of drug response is highly gene- and drug-specific. This study also provides a proof of principle for scoring the relative importance of newly identified rare variants using computational models.

Despite the potential importance of rare genetic variants, the lack of statistical power can limit detection of associations between these variants and pharmacogenomic traits. Because the frequency of the predictor/exposure (in this case a rare variant) is so low across a sample, there is very little data to support a single-variant association test; therefore, several approaches have been developed to aggregate rare variants by gene (or other units) to perform association tests. The most widely used of these methods are the burden test [109], SKAT (sequence kernel association test) [110], and SKAT-O (SKAT optimal test) [111]. The burden test assumes all rare variants in the target region have effects on the phenotype in the same direction and of similar magnitude and is typically defined as a weighted sum of variants across the gene region within a single individual. SKAT is more powerful under the assumption that both protective and deleterious variants exist within the gene, along with null variants. SKAT-O provides an optimal test that combines both burden test and SKAT to optimize the power of detecting rare-variants-phenotype associations accounting for both scenarios. Application of these approaches to pharmacogenomic traits is generally similar to the analysis of other phenotypes.

More recently, approaches have been developed to leverage protein structure information in the analysis of rare variants. As drug mechanisms generally involve binding of a small molecule to a ligand-binding domain within a target protein to modulate a molecular response, approaches that incorporate protein structure information may be especially useful for pharmacogenomics. Protein-structure-guided approaches first integrate rare (and common) missense variants into the 3D space of a protein structure and then use Euclidean distances among all variants to compute a similarity matrix. This matrix can be used for clustering, density analysis, or as a component of a statistical test.

POINT is a single variant association test that borrows information from spatially neighboring rare variants to improve power [112]. POINT begins with a user-specified target variant, and the Euclidean distance within 3D structural space to all other variants is estimated and weighted by a parameter which scales the influence of neighboring variants relative to the target variant (this parameter is adaptively chosen based on the dataset). The matrix derived from these distances is then used in a SKAT-like test to determine the association of a variant collection to the outcome.

PSCAN (protein-structure-based scan) [113] is a hierarchical spatial scan test leveraging the spatial clustering of functional genetic variants in 3D protein space. The genetic variants are mapped to either experimentally determined protein structures or computationally predicted structures. PSCAN's scan windows are flexibly defined by the Euclidean distance in 3D protein space to accommodate varied topologies of variant positions in protein space. Successive variant sets are tested for disease association using a SKAT or burden test framework within each window, and spatial windows are then aggregated using a Cauchy distribution to define an optimal "signal region" within the 3D space of the protein. As PSCAN utilizes the SKAT and burden framework, it remains essentially a frequency-based test.

POKEMON is an alternative strategy that derives a test statistic more directly focused on characterizing differences between the spatial distribution of the variants by phenotype [114]. POKEMON is based on a matrix of variants that are weighted by their distance within a specific protein structure; spatially close variants are weighted higher, while spatially distant variants are weighted lower. The severity of each specific amino acid substitution is weighted via a BLOSUM62 matrix. A significant result from this test indicates that the spatial relationship among variants within the protein is related to the phenotype tested. While this test was explicitly designed for extremely low-frequency variants (singletons and doubletons), the kernel can be combined with the traditional beta-scaled variant frequency kernel employed in SKAT tests.

3.8 Protein-Based Analyses of Rare Variants Influencing Lipid Traits

Papers describing the POINT and PSCAN methods both demonstrate their utility by examining lipid traits, and the POINT study in particular identified variants within *PCSK9* that associate to low-density lipoprotein (LDL) levels. This is especially relevant for pharmacogenomic studies as PCSK9 inhibitors are a new class of lipid lowering medications [115]. POINT was applied to the

ACCORD clinical trial with over 10,000 type 2 diabetes patients from 77 centers across North America. The analysis identified five variants close to the protein-binding domain of PCSK9, several of which demonstrate altered LDLR binding stability through computational modeling. This study demonstrates the potential utility of protein structure information for identifying variants that influence ligand or drug biding.

3.9 Gene Set and Enrichment Analyses

Gene set analysis is used to detect potential pathways or other biological mechanisms affected by a set of genes or variants identified as significant from an association study. There are two major branches of gene set analysis: overrepresentation analysis and functional class scoring methods [116], both designed to serve as a downstream or post hoc analysis of differential gene expression analysis results in the first place. Within these two broad branches, there are many variants and adaptations of methods as well.

After obtaining a set of significant genes between case and controls, overrepresentation analysis detects highlighted pathways by determining if the significant genes belonging to the same pathway are overrepresented in the given set using a Fisher's exact test assuming a hypergeometric distribution of the set of genes [116]. The advantage of this approach is that it can be applied to any kind of data rather than expression data. However, the drawback is also obvious. The set of genes is determined by applying a p-value cutoff. All the information for the genes not passing that threshold was discarded, and those passing were treated equally [116].

A well-known algorithm for functional class scoring methods is the gene set enrichment analysis (GSEA) which was first designed for microarray expression data [117]. The functional class scoring methods calculate a weighted Kolmogorov–Smirnov statistic using an enrichment score computed from a correlation of the gene expression profile and associated phenotype profile [117]. Relative to overrepresentation approaches, these methods typically have a larger computational burden but produce more sensitive results.

Various databases have been used for gene set analysis. Among those, the most well-known are the Gene Ontology (GO), Kyoto Encyclopedia of Genes and Genomes (KEGG), and PharmGKB [118–122]. They are databases that provide computational representations of biological knowledge, from cellular components, molecular functions, and biological processes to metabolism pathways, interaction networks, diseases, drugs, organisms, and even ecosystems. Several convenient online tools are available for fast, small- to medium-scaled analysis, such as the Database for Annotation, Visualization, and Integrated Discovery (DAVID) [123] and the Web-based Gene Set AnaLysis Toolkit (WEBGESTALT) [124]. These tools are handy for researchers that have limited familiarity with programming.

4 Translation of Pharmacogenomic Findings and Future Approaches

Many human genetic studies end with the identification of new variants that are associated with disease risk, increasing the general understanding of disease etiology and perhaps improving disease prediction. In contrast, most pharmacogenomic studies have a more detailed and immediate end-goal—to translate actionable pharmacogenomic variants into clinical practice to improve patient care. The implementation of pharmacogenomic variants in clinical care requires several special considerations. First, unlike research-focused genotyping, sample handling and processing for genotype generation in a clinical setting requires compliance with Clinical Laboratory Improvement Amendments (CLIA) certifications for laboratory tests (see [125] for discussions of analytic validity of a genotyping assay). Total sample size and clinical throughput are considerations that can affect the accuracy of genotyping calling, so specialized platforms for generating genotype data may be needed [126]. Finally, the ultimate use of genotype information in the clinic should be considered; how is genotype information represented in electronic health record systems, and how should clinicians be informed about potential pharmacogenomic variants present within a patient? The primary route for incorporating often detailed pharmacogenomic information into clinical practice is through decision support systems embedded within the EHR [127]. Examples of such systems developed at Vanderbilt University Medical Center, University of Florida and Shands Hospital, Mayo Clinic, Mount Sinai Medical Center, and St. Jude Children's Research Hospital are outlined in [128]. Preemptive large-scale genotyping of all patients may ultimately prove useful, but in the interim, the PREDICT initiative at Vanderbilt University Medical Center [16] evaluated the utility of decision support tools to guide or recommend specific patients for pharmacogenomic genotyping with some success [129]. Despite these advances, pharmacogenomic testing remains difficult to support in the USA due to limited third-party payer support [130].

As with other genetic studies, the integration of additional "omics" measures with genomic data is becoming more common in the study of pharmacogenomic traits. Differential drug responses are often associated with protein sequence changes that can be studied using proteomic techniques [131]. Though study at the proteomic level may seem more accurate, it is harder to apply to implement in a clinical setting due to its increased costs and their low-throughput nature. Metabolomics/metabonomics has also been applied to better understand variability in drug metabolism through the direct measurement of metabolites (see Ref. [132] for a proof-of-concept study of acetaminophen toxicity). Integration of genomic and transcriptomic data has been used to personalize treatment for childhood acute lymphoblastic leukemia [133]. Each of these approaches may prove useful in the personalization of various therapies and may point to easier-to-measure genetic factors for clinical implementation.

Nearly 50 new drugs have gained FDA approval per year since 2017 [134], and each of those drugs will exhibit variability in individual response, risk for adverse events, and the optimal therapeutic dose. Genetic factors almost certainly play a role, and the need to understand and characterize their effects through statistical methods outlined here is growing ever more important to personalize clinical care.

References

1. Martin CB, Hales CM, Gu Q, Ogden CL (2019) Prescription drug use in the United States, 2015–2019
2. Jonas DE, McLeod HL (2009) Genetic and clinical factors relating to warfarin dosing. Trends Pharmacol Sci 30:375–386. https://doi.org/10.1016/J.TIPS.2009.05.001
3. Savi P, Combalbert J, Gaich C et al (1994) The antiaggregating activity of clopidogrel is due to a metabolic activation by the hepatic cytochrome P450-1A. Thromb Haemost 72: 313–317
4. Kee PS, Chin PKL, Kennedy MA, Maggo SDS (2020) Pharmacogenetics of statin-induced myotoxicity. Front Genet 11: 575678. https://doi.org/10.3389/fgene.2020.575678
5. Marciante KD, Durda JP, Heckbert SR et al (2011) Cerivastatin, genetic variants, and the risk of rhabdomyolysis. Pharmacogenet Genomics 21:280–288. https://doi.org/10.1097/FPC.0b013e328343dd7d
6. Mirošević Skvrce N, Macolić Šarinić V, Šimić I et al (2015) ABCG2 gene polymorphisms as risk factors for atorvastatin adverse reactions: a case-control study. Pharmacogenomics 16: 803–815. https://doi.org/10.2217/pgs.15.47
7. Joffe E, Iasonos A, Younes A (2017) Clinical trials in the genomic era. J Clin Oncol 35: 1011–1017. https://doi.org/10.1200/JCO.2016.70.8891
8. Waarts MR, Stonestrom AJ, Park YC, Levine RL (2022) Targeting mutations in cancer. J Clin Invest 132. https://doi.org/10.1172/JCI154943
9. Alexanderson B, Evans DAP, Sjöqvist F (1969) Steady-state plasma levels of nortriptyline in twins: influence of genetic factors and drug therapy. Br Med J 4:764–768. https://doi.org/10.1136/BMJ.4.5686.764
10. Mahgoub A, Dring LG, Idle JR et al (1977) Polymorphic hydroxylation of debrisoquine in man. Lancet 310:584–586. https://doi.org/10.1016/S0140-6736(77)91430-1
11. Eichelbaum M, Spannbrucker N, Dengler HJ (1975) Proceedings: N-oxidation of sparteine in man and its interindividual differences. Naunyn Schmiedeberg's Arch Pharmacol 287 Suppl:R94
12. Robarge JD, Li L, Desta Z et al (2007) The star-allele nomenclature: retooling for translational genomics. Clin Pharmacol Ther 82: 244–248. https://doi.org/10.1038/SJ.CLPT.6100284
13. Gonzalez FJ, Skodat RC, Kimura S et al (1988) Characterization of the common genetic defect in humans deficient in debrisoquine metabolism. Nature 331:442–446. https://doi.org/10.1038/331442A0
14. Bush WS, Moore JH (2012) Chapter 11: Genome-wide association studies. PLoS Comput Biol 8:e1002822. https://doi.org/10.1371/journal.pcbi.1002822
15. Takeuchi F, McGinnis R, Bourgeois S et al (2009) A genome-wide association study confirms VKORC1, CYP2C9, and CYP4F2 as principal genetic determinants of warfarin dose. PLoS Genet 5. https://doi.org/10.1371/journal.pgen.1000433
16. Pulley JM, Denny JC, Peterson JF et al (2012) Operational implementation of prospective genotyping for personalized medicine: the design of the Vanderbilt PREDICT project. Clin Pharmacol Ther 92:87–95. https://doi.org/10.1038/clpt.2011.371
17. Schildcrout JS, Denny JC, Bowton E et al (2012) Optimizing drug outcomes through

pharmacogenetics: a case for preemptive genotyping. Clin Pharmacol Ther 92:235–242. https://doi.org/10.1038/clpt.2012.66

18. Relling MV, Klein TE (2011) CPIC: Clinical pharmacogenetics implementation consortium of the pharmacogenomics research network. Clin Pharmacol Ther 89:464–467. https://doi.org/10.1038/clpt.2010.279
19. Relling MV, Klein TE, Gammal RS et al (2020) The clinical pharmacogenetics implementation consortium: 10 years later. Clin Pharmacol Ther 107:171–175. https://doi.org/10.1002/cpt.1651
20. Klein TE, Chang JT, Cho MK et al (2001) Integrating genotype and phenotype information: an overview of the PharmGKB project. Pharmacogenetics Research Network and Knowledge Base. Pharmacogenomics J 1: 167–170
21. Whirl-Carrillo M, McDonagh EM, Hebert JM et al (2012) Pharmacogenomics knowledge for personalized medicine. Clin Pharmacol Ther 92:414–417. https://doi.org/10.1038/clpt.2012.96
22. Van Booven D, Marsh S, McLeod H et al (2010) Cytochrome P450 2C9-CYP2C9. Pharmacogenet Genomics 20:277–281. https://doi.org/10.1097/FPC.0b013e3283349e84
23. Thorn CF, Klein TE, Altman RB (2010) PharmGKB summary: very important pharmacogene information for angiotensin-converting enzyme. Pharmacogenet Genomics 20:143–146. https://doi.org/10.1097/FPC.0b013e3283339bf3
24. Wang L, Pelleymounter L, Weinshilboum R et al (2010) Very important pharmacogene summary: thiopurine S-methyltransferase. Pharmacogenet Genomics 20:401–405. https://doi.org/10.1097/FPC.0b013e3283352860
25. Mapes B, El Charif O, Al-Sawwaf S, Dolan ME (2017) Genome-wide association studies of chemotherapeutic toxicities: genomics of inequality. Clin Cancer Res 23:4010–4019. https://doi.org/10.1158/1078-0432.CCR-17-0429
26. Maranville JC, Cox NJ (2016) Pharmacogenomic variants have larger effect sizes than genetic variants associated with other dichotomous complex traits. Pharmacogenomics J 16:388–392. https://doi.org/10.1038/tpj.2015.47
27. Zazuli Z, Otten LS, Drögemöller BI et al (2019) Outcome definition influences the relationship between genetic polymorphisms of ERCC1, ERCC2, SLC22A2 and cisplatin nephrotoxicity in adult testicular cancer patients. Genes (Basel) 10. https://doi.org/10.3390/genes10050364
28. Higashi MK, Veenstra DL, Midori Kondo L et al (2002) Association between CYP2C9 genetic variants and anticoagulation-related outcomes during warfarin therapy. JAMA 287:1690–1698. https://doi.org/10.1001/JAMA.287.13.1690
29. Vuorinen A-L, Lehto M, Niemi M et al (2021) Pharmacogenetics of anticoagulation and clinical events in warfarin-treated patients: a register-based cohort study with biobank data and national health registries in Finland. Clin Epidemiol 13:183–195. https://doi.org/10.2147/CLEP.S289031
30. Wheeler HE, Dolan ME (2012) Lymphoblastoid cell lines in pharmacogenomic discovery and clinical translation. Pharmacogenomics 13:55–70. https://doi.org/10.2217/pgs.11.121
31. Moen EL, Godley LA, Zhang W, Dolan ME (2012) Pharmacogenomics of chemotherapeutic susceptibility and toxicity. Genome Med 4. https://doi.org/10.1186/gm391
32. Belmont JW, Hardenbol P, Willis TD et al (2003) The international HapMap project. Nature 426:789–796. https://doi.org/10.1038/NATURE02168
33. International HapMap Consortium (2005) A haplotype map of the human genome. Nature 437:1299–1320. https://doi.org/10.1038/nature04226
34. Abecasis GR, Auton A, Brooks LD et al (2012) An integrated map of genetic variation from 1,092 human genomes. Nature 491:56–65. https://doi.org/10.1038/nature11632
35. Lappalainen T, Sammeth M, Friedländer MR et al (2013) Transcriptome and genome sequencing uncovers functional variation in humans. Nature 501:506–511. https://doi.org/10.1038/nature12531
36. Bell JT, Pai AA, Pickrell JK et al (2011) DNA methylation patterns associate with genetic and gene expression variation in HapMap cell lines. Genome Biol 12. https://doi.org/10.1186/gb-2011-12-1-r10
37. Gamazon ER, Im H-K, Duan S et al (2010) Exprtarget: an integrative approach to predicting human microRNA targets. PLoS One 5:e13534. https://doi.org/10.1371/journal.pone.0013534
38. Relling MV, Krauss RM, Roden DM et al (2017) New pharmacogenomics research network: an open community catalyzing research and translation in precision medicine. Clin

Pharmacol Ther 102:897–902. https://doi.org/10.1002/cpt.755
39. Huang RS, Duan S, Bleibel WK et al (2007) A genome-wide approach to identify genetic variants that contribute to etoposide-induced cytotoxicity. Proc Natl Acad Sci U S A 104: 9758–9763. https://doi.org/10.1073/PNAS.0703736104
40. Komatsu M, Wheeler HE, Chung S et al (2015) Pharmacoethnicity in paclitaxel-induced sensory peripheral neuropathy. Clin Cancer Res 21:4337–4346. https://doi.org/10.1158/1078-0432.CCR-15-0133
41. Mulford AJ, Wing C, Dolan ME, Wheeler HE (2021) Genetically regulated expression underlies cellular sensitivity to chemotherapy in diverse populations. Hum Mol Genet 30: 305–317. https://doi.org/10.1093/hmg/ddab029
42. Njiaju UO, Gamazon ER, Gorsic LK et al (2012) Whole-genome studies identify solute carrier transporters in cellular susceptibility to paclitaxel. Pharmacogenet Genomics 22:498–507. https://doi.org/10.1097/FPC.0b013e328352f436
43. Turner S, Armstrong LL, Bradford Y et al (2011) Quality control procedures for genome-wide association studies. Curr Protoc Hum Genet. https://doi.org/10.1002/0471142905.hg0119s68
44. Gordon AS, Fulton RS, Qin X et al (2016) PGRNseq: a targeted capture sequencing panel for pharmacogenetic research and implementation. Pharmacogenet Genomics 26:161–168. https://doi.org/10.1097/FPC.0000000000000202
45. Grada A, Weinbrecht K (2013) Next-generation sequencing: methodology and application. J Invest Dermatol 133. https://doi.org/10.1038/JID.2013.248
46. Bolger AM, Lohse M, Usadel B (2014) Trimmomatic: a flexible trimmer for Illumina sequence data. Bioinformatics 30:2114–2120. https://doi.org/10.1093/BIOINFORMATICS/BTU170
47. GitHub – FelixKrueger/TrimGalore: a wrapper around Cutadapt and FastQC to consistently apply adapter and quality trimming to FastQ files, with extra functionality for RRBS data. https://github.com/FelixKrueger/TrimGalore. Accessed 4 Jun 2020
48. Langmead B, Trapnell C, Pop M, Salzberg SL (2009) Ultrafast and memory-efficient alignment of short DNA sequences to the human genome. Genome Biol 10:R25. https://doi.org/10.1186/gb-2009-10-3-r25
49. Li H, Durbin R (2009) Fast and accurate short read alignment with Burrows-Wheeler transform. Bioinformatics 25:1754–1760. https://doi.org/10.1093/bioinformatics/btp324
50. Donato L, Scimone C, Rinaldi C et al (2021) New evaluation methods of read mapping by 17 aligners on simulated and empirical NGS data: an updated comparison of DNA- and RNA-Seq data from Illumina and Ion Torrent technologies. Neural Comput Appl 33:15669–15692. https://doi.org/10.1007/s00521-021-06188-z
51. Li H, Wren J (2014) Toward better understanding of artifacts in variant calling from high-coverage samples. Bioinformatics 30: 2843–2851. https://doi.org/10.1093/bioinformatics/btu356
52. McKenna A, Hanna M, Banks E et al (2010) The Genome Analysis Toolkit: a MapReduce framework for analyzing next-generation DNA sequencing data. Genome Res 20: 1297–1303. https://doi.org/10.1101/gr.107524.110
53. Challis D, Yu J, Evani US et al (2012) An integrative variant analysis suite for whole exome next-generation sequencing data. BMC Bioinform 13. https://doi.org/10.1186/1471-2105-13-8
54. Miller NA, Farrow EG, Gibson M et al (2015) A 26-hour system of highly sensitive whole genome sequencing for emergency management of genetic diseases. Genome Med 7. https://doi.org/10.1186/s13073-015-0221-8
55. Poplin R, Chang PC, Alexander D et al (2018) A universal snp and small-indel variant caller using deep neural networks. Nat Biotechnol 36:983. https://doi.org/10.1038/nbt.4235
56. Zhao S, Agafonov O, Azab A et al (2020) Accuracy and efficiency of germline variant calling pipelines for human genome data. Sci Rep 10. https://doi.org/10.1038/s41598-020-77218-4
57. Naj AC, Lin H, Vardarajan BN et al (2018) Quality control and integration of genotypes from two calling pipelines for whole genome sequence data in the Alzheimer's disease sequencing project. Genomics. https://doi.org/10.1016/j.ygeno.2018.05.004
58. Van der Auwera GA, Carneiro MO, Hartl C et al (2002) Current protocols in bioinformatics. Wiley, Hoboken, NJ
59. Anderson CA, Pettersson FH, Clarke GM et al (2010) Data quality control in genetic case-control association studies. Nat Protoc 5:

1564–1573. https://doi.org/10.1038/nprot.2010.116
60. Wittke-Thompson JK, Pluzhnikov A, Cox NJ (2005) Rational inferences about departures from Hardy-Weinberg equilibrium. Am J Hum Genet 76:967–986. https://doi.org/10.1086/430507
61. Wang MH, Cordell HJ, Van Steen K (2019) Statistical methods for genome-wide association studies. Semin Cancer Biol 55:53–60. https://doi.org/10.1016/j.semcancer.2018.04.008
62. Price AL, Patterson NJ, Plenge RM et al (2006) Principal components analysis corrects for stratification in genome-wide association studies. Nat Genet 38:904–909. https://doi.org/10.1038/ng1847
63. Hall JB, Dumitrescu L, Dilks HH et al (2014) Accuracy of administratively-assigned ancestry for diverse populations in an electronic medical record-linked biobank. PLoS One 9: e99161. https://doi.org/10.1371/journal.pone.0099161
64. Dumitrescu L, Ritchie MD, Brown-Gentry K et al (2010) Assessing the accuracy of observer-reported ancestry in a biorepository linked to electronic medical records. Genet Med 12:648–650. https://doi.org/10.1097/GIM.0b013e3181efe2df
65. Lettre G, Lange C, Hirschhorn JN (2007) Genetic model testing and statistical power in population-based association studies of quantitative traits. Genet Epidemiol 31:358–362. https://doi.org/10.1002/gepi.20217
66. Chang CC, Chow CC, Tellier LC et al (2015) Second-generation PLINK: rising to the challenge of larger and richer datasets. Gigascience 4:7. https://doi.org/10.1186/s13742-015-0047-8
67. Carroll RJ, Bastarache L, Denny JC (2014) R PheWAS: data analysis and plotting tools for phenome-wide association studies in the R environment. Bioinformatics 30(16):2375–2376. https://doi.org/10.1093/bioinformatics/btu197
68. Willer CJ, Li Y, Abecasis GR (2010) METAL: fast and efficient meta-analysis of genomewide association scans. Bioinformatics 26 (17):2190–2191. https://doi.org/10.1093/bioinformatics/btq340
69. Fraser AG, Orchard TR, Jewell DP (2002) The efficacy of azathioprine for the treatment of inflammatory bowel disease: a 30 year review. Gut 50:485–489. https://doi.org/10.1136/GUT.50.4.485
70. Yang SK, Hong M, Baek J et al (2014) A common missense variant in NUDT15 confers susceptibility to thiopurine-induced leukopenia. Nat Genet 46:1017–1020. https://doi.org/10.1038/NG.3060
71. Liu N, Zhang K, Zhao H (2008) Haplotype-association analysis. Adv Genet 60:335–405. https://doi.org/10.1016/S0065-2660(07)00414-2
72. Lee CR, Goldstein JA, Pieper JA (2002) Cytochrome P450 2C9 polymorphisms: a comprehensive review of the in-vitro and human data. Pharmacogenetics 12:251–263. https://doi.org/10.1097/00008571-200204000-00010
73. Excoffier L, Slatkin M (1995) Maximum-likelihood estimation of molecular haplotype frequencies in a diploid population. Mol Biol Evol 12:921–927. https://doi.org/10.1093/oxfordjournals.molbev.a040269
74. Stephens M, Smith NJ, Donnelly P (2001) A new statistical method for haplotype reconstruction from population data. Am J Hum Genet 68:978–989. https://doi.org/10.1086/319501
75. Li N, Stephens M (2003) Modeling linkage disequilibrium and identifying recombination hotspots using single-nucleotide polymorphism data. Genetics 165:2213–2233. https://doi.org/10.1093/genetics/165.4.2213
76. Delaneau O, Marchini J, Zagury JF (2012) A linear complexity phasing method for thousands of genomes. Nat Methods 9:179–181. https://doi.org/10.1038/nmeth.1785
77. Shi S, Yuan N, Yang M et al (2018) Comprehensive assessment of genotype imputation performance. Hum Hered 83:107–116. https://doi.org/10.1159/000489758
78. Clarke L, Fairley S, Zheng-Bradley X et al (2017) The international Genome sample resource (IGSR): a worldwide collection of genome variation incorporating the 1000 Genomes Project data. Nucleic Acids Res 45: D854–D859. https://doi.org/10.1093/nar/gkw829
79. McCarthy S, Das S, Kretzschmar W et al (2016) A reference panel of 64,976 haplotypes for genotype imputation. Nat Genet 48:1279–1283. https://doi.org/10.1038/ng.3643
80. Das S, Forer L, Schönherr S et al (2016) Next-generation genotype imputation service and methods. Nat Genet 48:1284–1287. https://doi.org/10.1038/ng.3656
81. Taliun D, Harris DN, Kessler MD et al (2021) Sequencing of 53,831 diverse genomes from the NHLBI TOPMed Program. Nature 590: 290–299. https://doi.org/10.1038/S41586-021-03205-Y

82. Vergara C, Parker MM, Franco L et al (2018) Genotype imputation performance of three reference panels using African ancestry individuals. Hum Genet 137:281–292. https://doi.org/10.1007/S00439-018-1881-4
83. Del Tredici AL, Malhotra A, Dedek M et al (2018) Frequency of CYP2D6 alleles including structural variants in the United States. Front Pharmacol 9:305. https://doi.org/10.3389/fphar.2018.00305
84. Molden E, Jukić MM (2021) CYP2D6 reduced function variants and genotype/phenotype translations of CYP2D6 intermediate metabolizers: implications for personalized drug dosing in psychiatry. Front Pharmacol 12:650750. https://doi.org/10.3389/fphar.2021.650750
85. Häkkinen K, Kiiski JI, Lähteenvuo M et al (2022) Implementation of CYP2D6 copy-number imputation panel and frequency of key pharmacogenetic variants in Finnish individuals with a psychotic disorder. Pharmacogenomics J. https://doi.org/10.1038/s41397-022-00270-y
86. Ammar R, Paton TA, Torti D et al (2015) Long read nanopore sequencing for detection of HLA and CYP2D6 variants and haplotypes. F1000Research 4:17. https://doi.org/10.12688/f1000research.6037.2
87. Laver TW, Caswell RC, Moore KA et al (2016) Pitfalls of haplotype phasing from amplicon-based long-read sequencing. Sci Rep 6. https://doi.org/10.1038/srep21746
88. Mantere T, Kersten S, Hoischen A (2019) Long-read sequencing emerging in medical genetics. Front Genet 10:426. https://doi.org/10.3389/fgene.2019.00426
89. Limdi NA, Wadelius M, Cavallari L et al (2010) Warfarin pharmacogenetics: a single VKORC1 polymorphism is predictive of dose across 3 racial groups. Blood 115:3827–3834. https://doi.org/10.1182/BLOOD-2009-12-255992
90. Diaz FJ, Yeh H-W, de Leon J (2012) Role of statistical random-effects linear models in personalized medicine. Curr Pharmacogenomics Person Med 10:22–32. https://doi.org/10.2174/1875692111201010022
91. Kaur H, Crawford DC, Liang J et al (2021) Replication of European hypertension associations in a case-control study of 9,534 African Americans. PLoS One 16. https://doi.org/10.1371/journal.pone.0259962
92. Wilke RA, Berg RL, Linneman JG et al (2008) Characterization of low-density lipoprotein cholesterol-lowering efficacy for atorvastatin in a population-based DNA biorepository. Basic Clin Pharmacol Toxicol 103:354–359. https://doi.org/10.1111/j.1742-7843.2008.00291.x
93. Peissig P, Sirohi E, Berg RL et al (2007) Construction of atorvastatin dose-response relationships using data from a large population-based DNA Biobank. Basic Clin Pharmacol Toxicol 100:286–288. https://doi.org/10.1111/j.1742-7843.2006.00035.x
94. Vis DJ, Bombardelli L, Lightfoot H et al (2016) Multilevel models improve precision and speed of IC50 estimates. Pharmacogenomics 17:691–700. https://doi.org/10.2217/pgs.16.15
95. Wei WQ, Feng Q, Jiang L et al (2014) Characterization of statin dose response in electronic medical records. Clin Pharmacol Ther 95:331–338. https://doi.org/10.1038/CLPT.2013.202
96. Abbas-Aghababazadeh F, Lu P, Fridley BL (2019) Nonlinear mixed-effects models for modeling in vitro drug response data to determine problematic cancer cell lines. Sci Rep 9. https://doi.org/10.1038/S41598-019-50936-0
97. Yang J, Lee SH, Goddard ME, Visscher PM (2011) GCTA: a tool for genome-wide complex trait analysis. Am J Hum Genet 88:76–82. https://doi.org/10.1016/j.ajhg.2010.11.011
98. Hall JB, Bush WS (2016) Analysis of heritability using genome-wide data. Curr Protoc Hum Genet 91:1.30.1–1.30.10. https://doi.org/10.1002/cphg.25
99. Yang J, Benyamin B, McEvoy BP et al (2010) Common SNPs explain a large proportion of the heritability for human height. Nat Genet 42:565–569. https://doi.org/10.1038/ng.608
100. Zhao JH (2005) Mixed-effects Cox models of alcohol dependence in extended families. BMC Genet 6. https://doi.org/10.1186/1471-2156-6-S1-S127
101. Yang CC, Zhang B, Wang S et al (2019) Gene-based association analysis of survival traits via functional regression-based mixed effect cox models for related samples. Genet Epidemiol 43:952–965. https://doi.org/10.1002/gepi.22254
102. Mcgeachie MJ, Stahl EA, Himes BE et al (2013) Polygenic heritability estimates in pharmacogenetics: focus on asthma and related phenotypes. Pharmacogenet Genomics 23:324–328. https://doi.org/10.1097/FPC.0b013e3283607acf
103. Cecchin E, Stocco G (2020) Pharmacogenomics and personalized medicine. Genes

(Basel) 11:1–5. https://doi.org/10.3390/genes11060679
104. Butkiewicz M, Blue EE et al (2018) Functional annotation of genomic variants in studies of late-onset Alzheimer's disease. Bioinformatics. https://doi.org/10.1093/bioinformatics/bty177
105. Bush WS, Crosslin DR, Obeng AO et al (2016) Genetic variation among 82 pharmacogenes: the PGRN-Seq data from the eMERGE network. Clin Pharmacol Ther. https://doi.org/10.1002/cpt.350
106. Rasmussen-Torvik LJ, Stallings SC, Gordon AS et al (2014) Design and anticipated outcomes of the eMERGE-PGx project: a multicenter pilot for preemptive pharmacogenomics in electronic health record systems. Clin Pharmacol Ther 96: 482–489. https://doi.org/10.1038/clpt.2014.137
107. Ingelman-Sundberg M, Mkrtchian S, Zhou Y, Lauschke VM (2018) Integrating rare genetic variants into pharmacogenetic drug response predictions. Hum Genomics 12. https://doi.org/10.1186/s40246-018-0157-3
108. Zhou Y, Mkrtchian S, Kumondai M et al (2019) An optimized prediction framework to assess the functional impact of pharmacogenetic variants. Pharmacogenomics J 19: 115–126. https://doi.org/10.1038/s41397-018-0044-2
109. Li B, Leal SM (2008) Methods for detecting associations with rare variants for common diseases: application to analysis of sequence data. Am J Hum Genet 83:311–321. https://doi.org/10.1016/j.ajhg.2008.06.024
110. Wu MC, Lee S, Cai T et al (2011) Rare-variant association testing for sequencing data with the sequence kernel association test. Am J Hum Genet 89:82–93. https://doi.org/10.1016/j.ajhg.2011.05.029
111. Lee S, Abecasis GR, Boehnke M, Lin X (2014) Rare-variant association analysis: study designs and statistical tests. Am J Hum Genet 95:5–23. https://doi.org/10.1016/j.ajhg.2014.06.009
112. West RM, Lu W, Rotroff DM et al (2019) Identifying individual risk rare variants using protein structure guided local tests (POINT). PLoS Comput Biol 15. https://doi.org/10.1371/JOURNAL.PCBI.1006722
113. Tang ZZ, Sliwoski GR, Chen G et al (2020) PSCAN: spatial scan tests guided by protein structures improve complex disease gene discovery and signal variant detection. Genome Biol 21. https://doi.org/10.1186/s13059-020-02121-0
114. Jin B, Capra JA, Benchek P et al (2022) An association test of the spatial distribution of rare missense variants within protein structures identifies Alzheimer's disease-related patterns. Genome Res 32:778–790. https://doi.org/10.1101/gr.276069.121
115. Pasta A, Cremonini AL, Pisciotta L et al (2020) PCSK9 inhibitors for treating hypercholesterolemia. Expert Opin Pharmacother 21:353–363. https://doi.org/10.1080/14656566.2019.1702970
116. Maleki F, Ovens K, Hogan DJ, Kusalik AJ (2020) Gene set analysis: challenges, opportunities, and future research. Front Genet 11: 654. https://doi.org/10.3389/fgene.2020.00654
117. Subramanian A, Tamayo P, Mootha VK et al (2005) Gene set enrichment analysis: a knowledge-based approach for interpreting genome-wide expression profiles. Proc Natl Acad Sci U S A 102:15545–15550. https://doi.org/10.1073/pnas.0506580102
118. Ashburner M, Ball CA, Blake JA et al (2000) Gene ontology: tool for the unification of biology. Gene Ontol Consortium Nat Genet 25:25–29. https://doi.org/10.1038/75556
119. Carbon S, Douglass E, Good BM et al (2021) The gene ontology resource: enriching a GOld mine. Nucleic Acids Res 49:D325–D334. https://doi.org/10.1093/nar/gkaa1113
120. Kanehisa M, Goto S (2000) KEGG: Kyoto encyclopedia of genes and genomes. Nucleic Acids Res 28:27–30. https://doi.org/10.1093/nar/28.1.27
121. Kanehisa M, Furumichi M, Sato Y et al (2021) KEGG: integrating viruses and cellular organisms. Nucleic Acids Res 49:D545–D551. https://doi.org/10.1093/nar/gkaa970
122. Whirl-Carrillo M, Huddart R, Gong L et al (2021) An evidence-based framework for evaluating pharmacogenomics knowledge for personalized medicine. Clin Pharmacol Ther 110:563–572. https://doi.org/10.1002/cpt.2350
123. Jiao X, Sherman BT, Huang DW et al (2012) DAVID-WS: a stateful web service to facilitate gene/protein list analysis. Bioinformatics 28: 1805–1806. https://doi.org/10.1093/BIOINFORMATICS/BTS251
124. Liao Y, Wang J, Jaehnig EJ et al (2019) WebGestalt 2019: gene set analysis toolkit with revamped UIs and APIs. Nucleic Acids Res

47:W199–W205. https://doi.org/10.1093/NAR/GKZ401

125. Collins KS, Pratt VM, Stansberry WM et al (2019) Analytical validity of a genotyping assay for use with personalized antihypertensive and chronic kidney disease therapy. Pharmacogenet Genomics 29:18–22. https://doi.org/10.1097/FPC.0000000000000361

126. Lin CH, Yeakley JM, McDaniel TK, Shen R (2009) Medium- to high-throughput SNP genotyping using VeraCode microbeads. Methods Mol Biol 496:129–142. https://doi.org/10.1007/978-1-59745-553-4_10

127. Goldspiel BR, Flegel WA, DiPatrizio G et al (2014) Integrating pharmacogenetic information and clinical decision support into the electronic health record. J Am Med Inform Assoc 21:522–528. https://doi.org/10.1136/AMIAJNL-2013-001873

128. Dunnenberger HM, Crews KR, Hoffman JM et al (2015) Preemptive clinical pharmacogenetics implementation: current programs in five US medical centers. Annu Rev Pharmacol Toxicol 55:89–106. https://doi.org/10.1146/ANNUREV-PHARMTOX-010814-124835

129. Schildcrout JS, Shi Y, Danciu I et al (2016) A prognostic model based on readily available clinical data enriched a pre-emptive pharmacogenetic testing program. J Clin Epidemiol 72:107–115. https://doi.org/10.1016/J.JCLINEPI.2015.08.028

130. Rogers SL, Keeling NJ, Giri J et al (2020) PARC report: a health-systems focus on reimbursement and patient access to pharmacogenomics testing. Pharmacogenomics 21:785–796. https://doi.org/10.2217/PGS-2019-0192

131. Jain KK (2016) Role of proteomics in the development of personalized medicine. Adv Protein Chem Struct Biol 102:41–52. https://doi.org/10.1016/BS.APCSB.2015.09.002

132. Clayton TA, Lindon JC, Cloarec O et al (2006) Pharmaco-metabonomic phenotyping and personalized drug treatment. Nature 440:1073–1077. https://doi.org/10.1038/NATURE04648

133. Pavlovic S, Kotur N, Stankovic B et al (2019) Pharmacogenomic and pharmacotranscriptomic profiling of childhood acute lymphoblastic Leukemia: paving the way to personalized treatment. Genes (Basel) 10. https://doi.org/10.3390/GENES10030191

134. Mullard A (2022) 2021 FDA approvals. Nat Rev Drug Discov 21:83–88. https://doi.org/10.1038/D41573-022-00001-9

Chapter 15

Statistical Methods for Disease Risk Prediction with Genotype Data

Xiaoxuan Xia, Yexian Zhang, Yingying Wei, and Maggie Haitian Wang

Abstract

Single-nucleotide polymorphism (SNP) is the basic unit to understand the heritability of complex traits. One attractive application of the susceptible SNPs is to construct prediction models for assessing disease risk. Here, we introduce prediction methods for human traits using SNPs data, including the polygenic risk score (PRS), linear mixed models (LMMs), penalized regressions, and methods for controlling population stratification.

Key words Single-nucleotide polymorphism, Complex disease prediction, Polygenic risk score, Linear mixed model, Penalized regression, Population stratification

1 Introduction

Most common human diseases have a genetic basis [1]. The basic unit to understand the heritable portion of human traits is the single-nucleotide polymorphism (SNP), which encodes genotype of subjects at a specific position of the DNA sequence and comprises at least one nucleotide substitution in a sampled population. Disease risk prediction was first explored by using SNPs with large effect, for example, the apolipoprotein E (*APOE*) in Alzheimer's disease [2], human leukocyte antigen (*HLA*) in type I diabetes [3], and *BRCA1* and *BRCA2* in breast and ovarian cancers [4]. However, only a small portion of heritability had been explained by genetic variables in large effect [5]. Advances in large-scale genome-wide association study (GWAS) discovered thousands of susceptible loci, most of which confer only modest association to disease. These findings facilitated the development of polygenic approaches to risk prediction. The disease prediction accuracy largely improved by including SNPs showing weak association. For example, only ~10% of the variance in adult height can be explained by near 200 loci identified from GWAS involving

Brooke L. Fridley and Xuefeng Wang (eds.), *Statistical Genomics*,
Methods in Molecular Biology, vol. 2629, https://doi.org/10.1007/978-1-0716-2986-4_15,

100,000 subjects [6], while up to 45% of the variance could be explained by adopting all common SNPs in the genome [7]. Similar results have also been observed in other complex diseases [8, 9]. The availability of biobank-level data also enabled researchers to improve disease prediction with inference drawn from sufficiently large samples [10]. Recently, risk scores derived from genotype data showed potential clinical utility for patient classification and complementing traditional diagnostic measures [11].

In this chapter, we focus on introducing the prediction methods for common human disease using genotype data. We first describe evaluation measures for prediction models and then introduce the common prediction methods including the polygenic risk score (PRS), LMMs, penalized regressions, and methods for controlling population stratification. Finally, we discuss the outlook for risk prediction in the post-GWAS era.

2 Quantifying Prediction Performance

2.1 Measurement of Prediction Performance for Continuous Phenotype

Performance of prediction models should be evaluated in independent samples from the model training data. This is often carried out in cross-validation (CV) groups, and the average prediction accuracy in testing CV groups is reported. For continuous traits, the prediction accuracy is commonly measured by the squared *sample correlation coefficient* (R^2) [12], in which the Pearson correlation coefficient R can be computed by:

$$R\left(\widehat{\Upsilon}, \Upsilon\right) = \frac{S_{\widehat{\Upsilon}\Upsilon}}{\sqrt{S_{\widehat{\Upsilon}\widehat{\Upsilon}} S_{\Upsilon\Upsilon}}} = \frac{\sum_{i=1}^{N}\left(\widehat{y}_i - \overline{\widehat{y}}\right)\left(y_i - \overline{y}\right)}{\sqrt{\left[\sum_{i=1}^{N}\left(\widehat{y}_i - \overline{\widehat{y}}\right)^2\right]\left[\sum_{i=1}^{N}\left(y_i - \overline{y}\right)^2\right]}} \in [0, 1], \quad (1)$$

where Υ is the trait, $\widehat{\Upsilon}$ is the predicted value; S is the sample variance or covariance; N is sample size; and $\overline{y}$ and $\overline{\widehat{y}}$ are the sample mean of observed and predicted value, respectively. In a linear regression with least-square estimates, R^2 is the proportion of variation in Υ explained by the predictor variables [12]. Another measure, the mean square error (MSE) [13], evaluates deviation of the predicted from observed Υ, that is:

$$\mathrm{MSE}\left(\widehat{\Upsilon}, \Upsilon\right) = \frac{1}{N}\sum_{i=1}^{N}\left(y_i - \widehat{y}_i\right)^2 \in R^+ \triangleq \{x : x \geq 0, x \in R\}. \quad (2)$$

The MSE of an estimator $\widehat{\theta}$ for prediction parameter θ can be written as $\text{MES}\left(\widehat{\theta}\right) = E_\theta\left[\left(\widehat{\theta} - \theta\right)^2\right]$. The MSE can be decomposed into the variance and bias of estimation as:

$$\text{MSE}\left(\widehat{\theta}\right) = E_\theta\left[\left(\widehat{\theta} - E_\theta\left[\widehat{\theta}\right]\right)^2\right] + \left(E_\theta\left[\widehat{\theta}\right] - \theta\right)^2 = \text{Var}_\theta\left(\widehat{\theta}\right) + \text{Bias}_\theta\left(\widehat{\theta}, \theta\right)^2.$$

2.2 Measurement of Prediction Performance for Binary Phenotype

For binary phenotype, prediction accuracy or error shall be measured from two aspects, the false-positive rate (FPR) and false-negative rate (FNR). The receiver operating characteristic (ROC) curve (Fig. 1) is a broadly applied method, as it considers both aspects. The ROC is drawn by connecting the points located by a pair of FPR and true positive rate (TPR), which is 1-FNR, measured at varying discrimination thresholds of a prediction model. While the ROC provides visualization of model prediction performances, the area under the curve (AUC) of the ROC can be used to summarize the prediction accuracy into a single value (Fig. 1) [14]. AUC = 1 indicates perfect prediction performance, and AUC = 0.5 indicates non-informative prediction with accuracy similar to the outcome of flipping a coin.

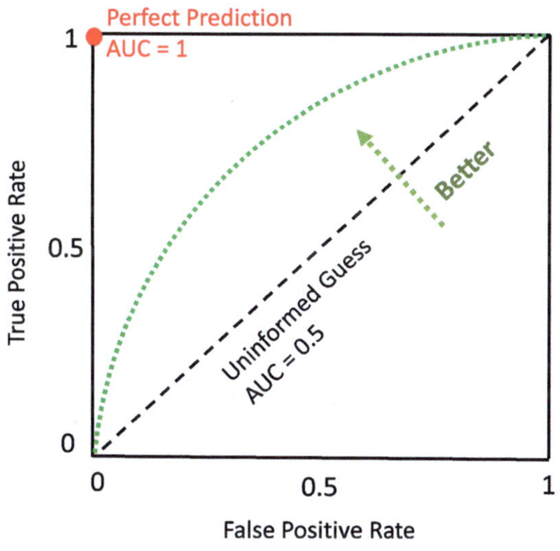

Fig. 1 Receiver operating characteristic (ROC) curve *x* and *y* axes are false-positive rate (FPR) and true positive rate (TPR), respectively. Each classification result (FPR, TPR) corresponds to one point in the ROC curve. The perfect prediction result approaches the (0, 1) point in the coordinates with AUC = 1. The diagonal line, or the nondiscrimination line, corresponds to uninformed guesses, of which the AUC = 0.5

3 Prediction by the Polygenic Risk Score

3.1 Basic Form of the PRS

The polygenic risk score (PRS) provides a summary statistic for genetic predisposition of an individual toward a phenotype [15]. The PRS can be calculated by the sum of genotype allele counts weighted by per-allele effect size [15], such that:

$$\mathrm{PRS} = \sum_p \widehat{\beta}_p X_p, p = 1, \cdots, P, \tag{3}$$

where X_p is standardized allele counts of the *pth* SNP, i.e., $X_p = \frac{G_p - 2f_p}{\sqrt{2f_p(1-f_p)}}$; G_p is minor allele count taking values 0, 1, or 2; f_p represents the corresponding allele frequency; $\widehat{\beta}_p$ is estimated effect size (or log odds ratio for binary traits); and P is total number of included SNPs. The PRS assumes additive genetic effect.

Disease prediction by the PRS can be carried out in the following procedure (Fig. 2). First, summary statistics of genetic effect are extracted from a *discovery data*, which is often comprised of large GWAS or biobank. Various effect size thresholding approaches can be implemented to improve the statistic, which are introduced in Subheading 3.2. The *target data* is where prediction is to be made and consisted of individual-level genotype and phenotype information. Quality control should be performed in the target data [16]. The PRS is often incorporated as the genetic predictor in a generalized linear model (GLM) to derive risk prediction. The GLM can be written as:

$$g[E(Y)] = \gamma \mathrm{PRS} + \boldsymbol{\alpha} \boldsymbol{W}, \tag{4}$$

where the left-hand side of the equation is: a transformation of the expected value of Y by link function $g()$ [17]. $\boldsymbol{W}$ is a vector of covariate variables; γ and $\boldsymbol{\alpha}$ are the coefficients. Thus, the PRS borrows information (SNP effect) from a large external dataset and is aggregated by considering the allele information of target subjects. The weighted sum form of the PRS alleviates the instability problem caused by high multicollinearity among the SNPs when estimating them simultaneously in a regression model, while the coefficient of the PRS calibrates the association relationship between genes and phenotype in the target data. The GLM also enables inclusion of individual-level covariates that are often highly informative to disease prediction, such as age and gender. The PRS can be performed by the "--score" function in PLINK [18].

3.2 Variations of the PRS

Since the SNP effects are usually estimated from single variable analysis, the potential LD structure in data and interactions may result in inflated estimates and render prediction robustness. To address this problem, one type of methods selects SNPs by LD and genetic association information, before including them in the PRS,

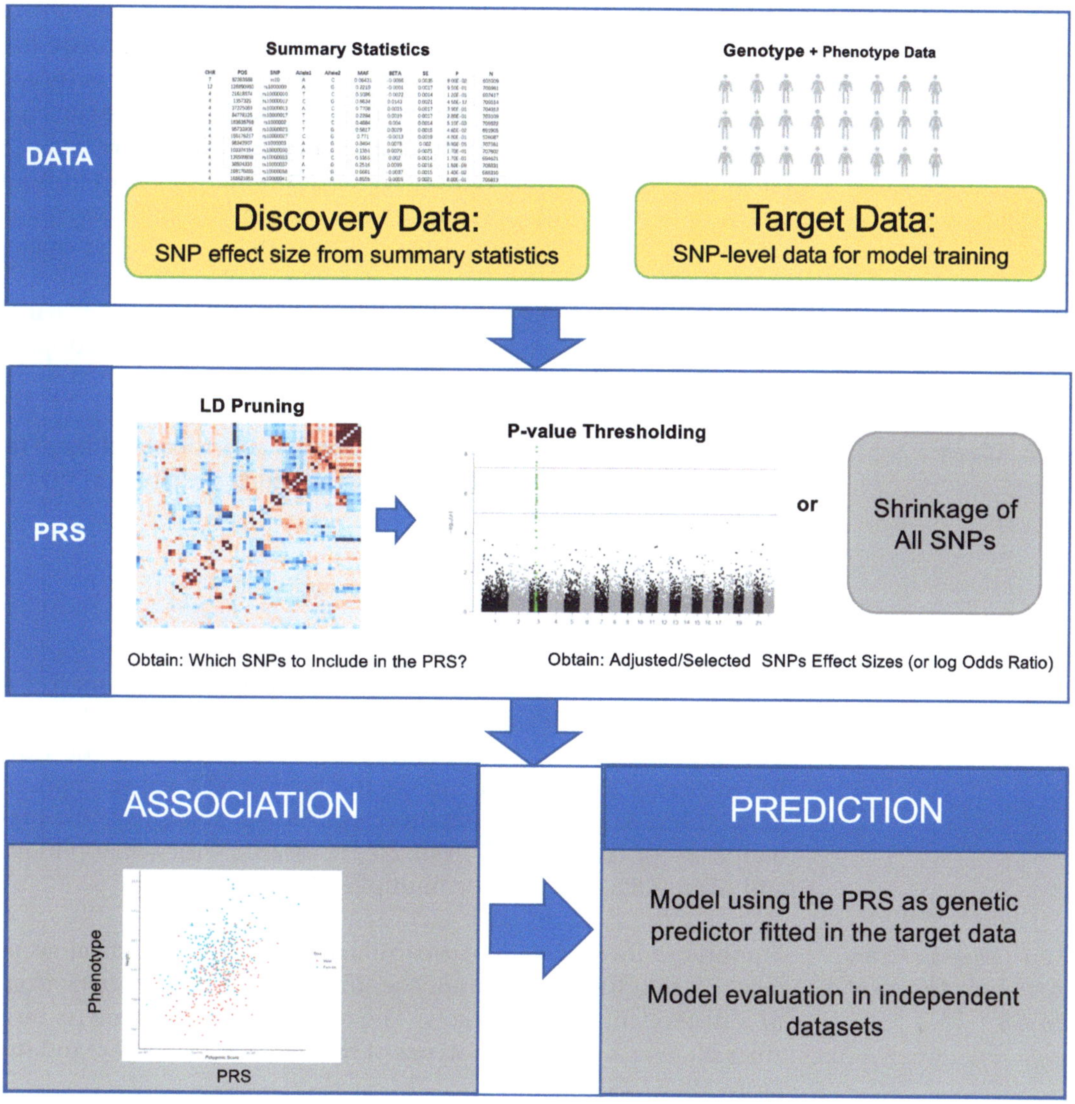

Fig. 2 The process of making prediction with the polygenic risk score (PRS)

which is introduced in Subheading 3.2.1. Another type of methods applies shrinkage thresholds to regulate SNP effects. This may be carried out by the Bayesian approach using shrinkage prior distributions (Subheading 3.2.2) or by the penalized regressions. The latter is described separately in Subheading 5 for its broad application. Next, we will briefly discuss factors influencing prediction accuracy of the PRS (Subheading 3.2.3), a real data example (Subheading 3.2.4), and finally its trans-ancestry application (Subheading 3.2.5).

3.2.1 LD-Pruning and P-Value Thresholding

The *LD-pruning and p-value thresholding* (P + T), or *clumping and thresholding* (C + T), is a widely adopted approach to improve the prediction accuracy of the PRS [15]. In the *pruning* step, independent (or weakly correlated) SNPs are extracted according to linkage disequilibrium (LD) information in the target dataset or a reference panel, such as the 1000 Genomes [19]. Among the SNPs in high LD, only the variable in strongest association to the phenotype is retained. In the *thresholding* step, SNPs with *p*-value lower than a certain threshold in the discovery data are used for PRS construction [20–22]. The "P + T" can be considered as a hard-thresholding approach because of the constant thresholds adopted in the model.

3.2.2 LDpred

Another type of shrinkage methods applies soft-thresholding. The LDpred [23] assumes the following probability distribution on the random effect of SNPs:

$$\beta_p \sim iid \begin{cases} N\left(0, \dfrac{h^2}{M\pi}\right) \text{with probability } \pi \\ 0 \text{ with probability } (1-\pi) \end{cases}, p = 1, \cdots, P, \quad (5)$$

where h^2 is disease heritability; π is the probability that a marker is drawn from a Gaussian distribution, i.e., the fraction of causal markers; and P is the total number of SNPs. LDpred models the LD structure by conditioning on the other SNPs via an approximate Gibbs Sampler. The method was shown to improve the predicted R^2 of the "P + T" from 20.1% to 25.3% for schizophrenia and from 9.8% to 12.0% for multiple sclerosis [23].

3.2.3 Influential Factors to Prediction Accuracy of the PRS

Dudbridge investigated factors influential to the prediction accuracy of the PRS [20]. Suppose the PRS is constructed from M independent causal markers. The relationships of genotype variables $X_1, \cdots, X_M$ and phenotypes of the discovery data (Y_1) and the target data (Y_2) are:

$$Y_1 = \sum_{p=1}^{M} \beta_{p,1} X_p + e_1, \quad \text{and } Y_2 = \sum_{p=1}^{M} \beta_{p,2} X_p + e_2, \quad (6)$$

respectively. X_p is the scaled allele count of the *pth* SNP; $\beta_{p,1}$ and $\beta_{p,2}$ are coefficients in the discovery and target data, respectively; and e_1 and e_2 are random errors. Suppose $\widehat{\beta}_p$ is the least square estimate of X_p on Y. The squared Pearson correlation R^2 of the PRS and the target phenotype Y_2 is:

$$R^2_{\text{PRS},Y_2} = \frac{M\tilde{n}\text{cov}\left(\widehat{\beta}_{p,1}, \beta_{p,2}\right)}{\text{var}\left(\widehat{\beta}_{p,1}\right)\tilde{n}\text{var}(Y_2)} = \sigma_1^2 \frac{\frac{N_1}{M}\sigma_1^2}{\left(\frac{N_1}{M}\sigma_1^2 + 1\right)\tilde{n}\text{var}(Y_2)}, \quad (7)$$

where σ_1^2 is the genetic variance in the discovery data, N_1 is the sample size, and $\text{var}(Y_2)$ is the variance of phenotype in target data. The above equation shows that predictive accuracy of the PRS is influenced by the genetic variance (σ_1^2), the ratio of sample size to the number of markers in the discovery data (N_1/M), and the phenotypic variance in the target data $\text{var}(Y_2)$. When P + T is implemented, effect of thresholding would contribute to the PRS through $\text{cov}\left(\widehat{\beta}_{p,1}, \beta_{p,2}\right)$ and $\text{var}\left(\widehat{\beta}_{p,1}\right)$ [20].

3.2.4 Real Data Example: Predicting Schizophrenia Risk by the PRS

The first successful application of the PRS was found on the GWAS of schizophrenia [15]. The discovery data consisted of 3322 European individuals with diagnosed schizophrenia and 3587 controls from 7 geographically distinct sample clusters [15]. After quality control, the SNPs were pruned using a sliding window of 200 markers at LD threshold 0.25 and p-value cutoffs of $p_T = 0.01$, 0.05, 0.1, 0.2, 0.3, 0.4, 0.5. In the target sample, prediction model was built by logistic regression containing the PRS predictor and covariates. The target data was consisted of the European-American (EA) and African-American (AA) [15] samples of the molecular genetics of schizophrenia (MGS) data and a UK cohort [24]. Analysis showed that the PRS was highly associated with schizophrenia in the European cohorts that it explained 3.2% of the phenotypic variation in MGS-EA ($p = 2 \times 10^{-28}$) and 2.3% of the variation in the UK sample ($p = 5 \times 10^{-11}$). In the MGS-AA population, only 0.4% ($p = 0.008$) of the phenotype variation was explained. The results showed that the genetic component underlying schizophrenia was shared between the two populations, yet its effect size was attenuated in the non-European population. This could be due to systematic difference of allele frequencies, LD pattern, and genetic heterogeneity across populations. The study also suggested that including a larger number of SNPs in the PRS would improve prediction. Similar results have also been observed in other complex traits including multiple sclerosis [25], height [6], body mass index [26], cardiovascular risk [27], and rheumatoid arthritis [28].

3.2.5 Trans-Ancestry Application of the PRS

Trans-ancestry application of the PRS usually exhibited reduced prediction accuracy, under the influence of heterogeneous genetic effect and architecture across populations [29–32]. Since most of the large GWASs were conducted in the European population, methods were developed to transfer summary statistics from a discovery data to a target population of different ancestry [33, 34]. In Cai et al. [33], the genetic relationship of two populations was modeled by the linear mixed model, in which the following variance structure of the random effect of SNPs was assumed,

$$\begin{pmatrix} \beta_{p,1} \\ \beta_{p,2} \end{pmatrix} \sim N(0, \Sigma_\beta) = N\left(\begin{matrix} 0 \\ 0 \end{matrix}, \begin{pmatrix} \sigma_1^2 & \rho\sigma_1\sigma_2 \\ \rho\sigma_1\sigma_2 & \sigma_2^2 \end{pmatrix}\right), p = 1, \cdots, P, \quad (8)$$

where σ_1^2 and σ_2^2 are the variances of genetic effects in two populations, respectively, and ρ is the correlation of trans-ancestry genetic effect for the same trait. When ρ is nonzero, the model was shown to improve the prediction by PRS [33].

3.3 Prediction for Related Phenotypes by the PRS

The PRS may also be applied in scenarios when phenotype of the target data is different from the discovery data. This application might be sought when summary statistics for the phenotype of interest are not available, but which for a related trait can be found. For example, PRS of the education year was used to predict cognitive performance [35, 36]. In some studies, a range of genetically related traits were analyzed jointly [37–41], and PRS constructed from these data was shown to facilitate the prediction for a trait falling within the disease spectrum [42–44].

3.4 Clinical Utility of the PRS

Clinical application of the PRS was explored for several common complex diseases including breast cancer, prostate cancer, coronary artery disease (CHD), obesity, type 1 diabetes, type 2 diabetes, and the Alzheimer's disease [11, 45], in the scenarios of screening high-risk patient group, estimating lifetime risk trajectories, and guiding treatment and clinical management. For example, for the CHD, PRS is often constructed from the summary statistics of the CardiogramPlusC4D study, and SNP effect modulation was carried out by LDpred [46, 47], lassosum [48], and meta-scoring approaches [49]. The CHD-PRS was shown to improve prediction for adverse cardiovascular events significantly. In addition, the PRS could be utilized together with the conventional risk predictors such as the Framingham risk score [50], pooled cohort equations [48, 49], and the QRISK. The exact benefit of the PRS for patients might be specific to health system [11].

4 Prediction by the Linear Mixed Model (LMM)

4.1 The Linear Mixed Model Basics and Variations

The general form of the linear mixed model can be written as:

$$y = W\alpha + X\beta + e, \tag{9}$$

where $y \in R^N$ is the phenotype vector, $X \in R^{N \times P}$ is the genotype matrix, $W \in R^{N \times Q}$ is the covariate matrix containing N individuals and Q fixed effect variables, $\alpha \in R^M$ is the vector of fixed-effect coefficients, and $\beta \in R^P$ is the vector of random effect variables. β follows a distribution with prespecified variance-covariance structure, and e is random error of individual, $e_i \sim N(0, \sigma_e^2)$. In disease prediction with the LMM, LD structure and genetic effect are controlled by random effect distribution [7, 51–54]. In the following, we introduce several representative random effect distributions adopted in various LMMs.

1. *Best linear unbiased predictor* (*BLUP*) [7, 55]: In this method, all SNPs are assumed to be causal. The random effect $\boldsymbol{\beta}$ is assumed to follow the distribution:

$$\beta_p\ i.i.d. \sim N\left(0, \sigma^2\right), p = 1, \cdots, P, \tag{10}$$

where σ^2 is the genetic variance explained by all SNPs.

2. *BayesC* [56]: This LMM incorporates a Bayesian framework with a spike and slab prior on the distribution of $\boldsymbol{\beta}$. The random coefficient is distributed as:

$$\beta_p\ i.i.d. = \begin{cases} 0 & \text{with probability } \pi \\ N\left(0, \sigma^2\right) & \text{with probability } (1-\pi), \end{cases} \tag{11}$$

where σ^2 is the genetic variance explained by causal SNPs.

3. *BayesR* [57]: In this LMM, the random effect of SNPs is assumed to follow a mixture of four zero-mean normal distributions, and the relative ratio of variance in each mixture component is constant:

$$\beta_p\ i.i.d. = \begin{cases} N\left(0, 0 \times \sigma^2\right) & \text{with probability } \pi_1 \\ N\left(0, 10^{-4} \times \sigma^2\right) & \text{with probability } \pi_2 \\ N\left(0, 10^{-3} \times \sigma^2\right) & \text{with probability } \pi_3 \\ N\left(0, 10^{-2} \times \sigma^2\right) & \text{with probability } \pi_4 \end{cases}. \tag{12}$$

Here, $\boldsymbol{\pi} = (\pi_1, \pi_2, \pi_3, \pi_4)$ are the mixture proportions that sum to unity. σ^2 is the common additive genetic variance shared by non-zero components.

4. *BayesS* [58]: In this LMM, the variance of the genetic effect distribution is modeled by individual SNP's MAF f_p through a parameter *S*, i.e., $\sigma_p^2 = \left(2f_p\left(1 - f_p\right)\right)^S \sigma^2$. The *S* accounts for the outcome of directional selection, and $S > 0$ indicates that the genetic effect size is positively related to the MAF. In implementation, the *S* is sampled from a posterior distribution by a gradient-based sampling approach.

5. *Dirichlet Process Regression* (*DPR*) [59]: This method adaptively estimates the random effect distribution through a nonparametric Dirichlet process prior and models the genetic effect with a range of distributions.

4.2 Prediction in Non-European Populations by the LMM

Direct application of the LMM in non-European populations showed reduced prediction performance compared to its application in the European cohorts, because of the relatively small non-European GWAS sample size [60–62]. Methods facilitating trans-ancestry prediction by LMM were also proposed. Coram et al. developed the two-component LMM, XP-BLUP [34], in which the first component ($\boldsymbol{g}_1$) consists of SNPs showing evidence of association in an auxiliary GWAS, and the second component ($\boldsymbol{g}_2$) includes all genotypes in the target data. The XP-BLUP model is written as:

$$Y = g_1 + g_2 + e, \quad (13)$$

where $g_k = X_k\beta_k$ is the genetic contribution of group k, $k = 1, 2$. X_k is the submatrix corresponding to SNPs in group k.

Xia et al. [63] proposed the Prism Vote (PV) method to infer the individualized risk by leveraging multiple subpopulation information. The framework evaluates the propensity of an individual to a spectrum of subpopulations by $\Pr(s \in k| x_s)$. For a binary Y, the predicted disease probability for subject s is:

$$\Pr(Y = 1|x_s) = \sum_{k=1}^{K} \Pr(Y = 1|s \in k, x_s)\Pr(s \in k|x_s), \quad (14)$$

where $\Pr(Y = 1| s \in k, x_s)$ is the stratum-specific disease risk of individual s and x_s is the genotype of subject s. The Prism Vote achieved improved accuracy (up to 12.1% improvement in R^2) and robustness (smaller standard deviation) of prediction outcome in independent testing data compared to the prediction without the framework.

4.3 Implementation Issues of the LMM and Solutions

Bayesian inference has been widely incorporated in various LMMs and effectively improved the prediction accuracy [56–58, 64, 65]. However, the estimation of Bayesian models using Markov Chain Monte Carlo (MCMC) largely increased the computing burden, especially in large-scale GWAS data (e.g., ~500,000 individuals and ~700,000 SNPs). Computation efficiency of LMM could be facilitated by paralleled calculation of partitioned data matrix. For example, Yang et al. [66] and Lloyd-Jones et al. [53] segmented the genotype data into approximately independent LD blocks that enabled multicore calculation. The Prism Vote [63] incorporated a subpopulation stratification design; thus, parallel-estimation could be carried out in population stratum, while the total genetic information extracted was maintained by integrating over multiple stratum in the final prediction step.

Most Bayesian LMMs require individual-level data that are often unavailable under restricted data sharing policy. Lloyd-Jones et al. [53] and Zhu et al. [67] developed the Bayesian LMMs based on summary statistics, which only required the main effect of SNPs, standard error, and LD information. As this type of approach could leverage large GWAS meta-analysis reports, e.g., $n \approx 700{,}000$ for height and BMI phenotype, it substantially improved the prediction performances [53].

5 Prediction by the Penalized Regression

In prediction with the genotype data, penalized regression was shown to effectively improve prediction accuracy, either in standalone application or in conjunction with other methods. In the following, we first introduce the basic penalized regression models and then their applications in genetic risk predictions.

For a continuous Y, the multiple linear regression model containing P predictor variables ($X_1, X_2, \cdots, X_P$) can be written as:

$$Y = \sum_{p=1}^{P} X_p \beta_p + \epsilon, \tag{15}$$

in which $\boldsymbol{\beta} = (\beta_1, \cdots, \beta_P)$ is the coefficients of SNPs and $\boldsymbol{\varepsilon} \sim N(0, \sigma^2)$. The coefficients can be estimated by minimizing the residual sum of squared (RSS):

$$L\left(\widehat{\boldsymbol{\beta}}\right)_{\mathrm{OLS}} = \sum_{i=1}^{N} \left(y_i - \boldsymbol{x}_i^T \widehat{\boldsymbol{\beta}}\right)^2 \tag{16}$$

High multicollinearity in the predictor variables would result in large standard deviations of coefficient estimates and hamper model stability. To counter this problem, Hoeral et al. [68] proposed the Ridge regression that adds a regularization term into the loss function to constrain the total squared coefficient size, that is:

$$L\left(\widehat{\boldsymbol{\beta}}\right)_{\mathrm{Ridge}} = \sum_{i=1}^{N} \left(y_i - \boldsymbol{x}_i^T \widehat{\boldsymbol{\beta}}\right)^2 + \lambda \left\|\widehat{\boldsymbol{\beta}}\right\|_2^2, \tag{17}$$

where $\left\|\widehat{\boldsymbol{\beta}}\right\|_2^2 = \sum_{p=1}^{P} \widehat{\beta}_p^2$ is the L2-norm penalty on $\widehat{\boldsymbol{\beta}}$ and $\lambda \geq 0$ is the parameter controlling for penalty strength. This loss function prefers a model with both a small RSS and coefficients in moderate effect size. Similar to the Ridge regression, Lasso regression incorporates the shrinkage by the sum of absolute estimated coefficients (L1 penalty) [69]. The loss function can be written as:

$$L\left(\widehat{\boldsymbol{\beta}}\right)_{\mathrm{LASSO}} = \sum_{i=1}^{N} \left(y^i - \boldsymbol{x}_i^T \widehat{\boldsymbol{\beta}}\right)^2 + \lambda \left\|\widehat{\boldsymbol{\beta}}\right\|_1, \tag{18}$$

where $\left\|\widehat{\boldsymbol{\beta}}\right\|_1 = \sum_{p=1}^{P} |\widehat{\beta}_p|$. When the penalty parameter λ is sufficiently large, the Lasso may shrink some coefficients to absolute zero, achieving a variable selection outcome. The elastic net [70] combines the L1 and L2 penalties in weighted form:

$$L\left(\widehat{\boldsymbol{\beta}}\right)_{\mathrm{ElasticNet}} = \sum_{i=1}^{N} \left(y^i - \boldsymbol{x}_i^T \widehat{\boldsymbol{\beta}}\right)^2 + \lambda_1 \left\|\widehat{\boldsymbol{\beta}}\right\|_2^2 + \lambda_2 \left\|\widehat{\boldsymbol{\beta}}\right\|_1 \tag{19}$$

R package glmnet provides a suite of functions to implement these regularizations in the GLM [71]. The biglasso [72], bigstatsr [73], and snpnet [74] were developed for GWAS-level implementation of the penalized models. In the UK Biobank data, the snpnet achieved competitive prediction accuracy compared to the PRS-based method, by using only a fraction of SNPs contained in the PRS [74].

The penalized regressions could also be jointly applied with the PRS. The lassosum adjusts the genetic effect of summary statistics considering LD structure and L1 penalty [75], using the objective function:

$$L\left(\widehat{\boldsymbol{\beta}}\right)_{\text{lassosum}} = \boldsymbol{y}^T\mathbf{y} + \widehat{\boldsymbol{\beta}}^T \boldsymbol{X}_r^T \boldsymbol{X}_r \widehat{\boldsymbol{\beta}} - 2\widehat{\boldsymbol{\beta}}^T \boldsymbol{X}^T \boldsymbol{y} + 2\lambda \left\| \widehat{\boldsymbol{\beta}} \right\|_1, \quad (20)$$

where $\boldsymbol{X}_r^T \boldsymbol{X}_r$ is the LD matrix and $\boldsymbol{X}^T\boldsymbol{y}$ is the correlation between SNPs and phenotype in the reference data. The adjusted genetic effects are then used to construct PRS for the follow-up GLM prediction.

In real data application of predicting ischemic stroke, Abraham et al. constructed the PRSs or genomic risk scores (GRSs) for 19 stroke-related traits in the European populations and then used the elastic net to combine these GRSs, forming the metaGRS predictor [76]. Similarly, Lu et al. constructed the metaPRS for stroke by incorporating GWASs of 14 stroke-related traits in the East Asian populations [77].

6 Population Stratification Control in Disease Prediction

6.1 Detecting Population Stratification in Genotype Data

In GWAS data, genetic relatedness among subjects is present even though individuals are not biological relatives. Systematic allele frequency differences between the cases and controls would cause spurious association due to ancestral variations and render power of prediction models. A common approach to evaluate the degree of population stratification is the genomic control (GC) factor, λ_{GC} [78]. When the trait is binary and the test statistic is chi-square distributed with one degree of freedom, a robust estimator of λ_{GC} [78] can be constructed using the median value of the statistics for all SNPs divided by 0.456 [79]. When no population stratification is present, $\lambda_{GC} = 1$ [78]. In a typical GWAS, λ_{GC} ranges between 1.01 and 1.11 [80, 81]. Sul et al. estimated that the 95% confidence interval of λ_{GC} [78] is 0.02 [80]; thus, a value greater than 1.03 would suggest evidence of population stratification.

6.2 Controlling Population Stratification by the Principal Component Analysis (PCA)

The PCA is a widely adopted approach to control for population stratification in genetic data analysis [82]. It can be carried out in the following procedure. Let $\boldsymbol{G}$ denote genotype matrix of N individuals and P SNPs. G_{ip} is the genotype of SNP p for individual i, coded by the number of minor alleles. The genotype matrix can be standardized by:

$$X_{ip} = \frac{G_{ip} - 2f_p}{\sqrt{2f_p\left(1 - f_p\right)}}, \quad (21)$$

where f_p is the minor allele frequency of SNP p. The genetic relationship matrix (GRM) $\boldsymbol{\Psi} \in \mathbb{R}^{N \times N}$ can be obtained by calculating [7]:

$$\Psi_{ij} = \frac{1}{P} \sum_{p=1}^{P} X_{ip} X_{jp}. \tag{22}$$

Ψ_{ij} is a measure of the average genetic similarity between individuals i and j. Common software, PLINK – make-rel [18] and GCTA – make-grm-alg 0 [83], calculate the GRM in this way. Next, orthogonal linear transformation is applied on the GRM in the direction of subjects. Eigen-decomposition gives:

$$\boldsymbol{\Psi} = \boldsymbol{V} \boldsymbol{D} \boldsymbol{V}^T, \tag{23}$$

where $\boldsymbol{V}$ is a $N \times N$ matrix with orthogonal column vectors. The columns of $\boldsymbol{V}$ are eigenvectors of the GRM $\boldsymbol{\Psi}$, and $\boldsymbol{D}$ is the diagonal matrix of eigenvalues. The k-th eigenvector of $\boldsymbol{\Psi}$ corresponds to the k-th largest eigenvalue. The i th component in the k-th eigenvector (V_{ik}) is individual i's variation along the k-th axis of ancestry [82]. The top PCs can be obtained by calculating the inner product of the top eigenvectors and the original data matrix, summarizing the major ancestral variations in the dataset. This procedure can be performed using the software EIGENSTRAT [82]. Figure 3 shows the PCA plot of 5 populations from 1000 Genomes Project Phase III data [84]: Africans (AFR), Admixed Americans (AMR), East Asians (EAS), Europeans (EUR), and South Asians (SAS). In disease risk prediction with the GLM, the top 10 or 20 PCs are often included as covariates together with the main genetic factors to control for extra variations of subject-relatedness.

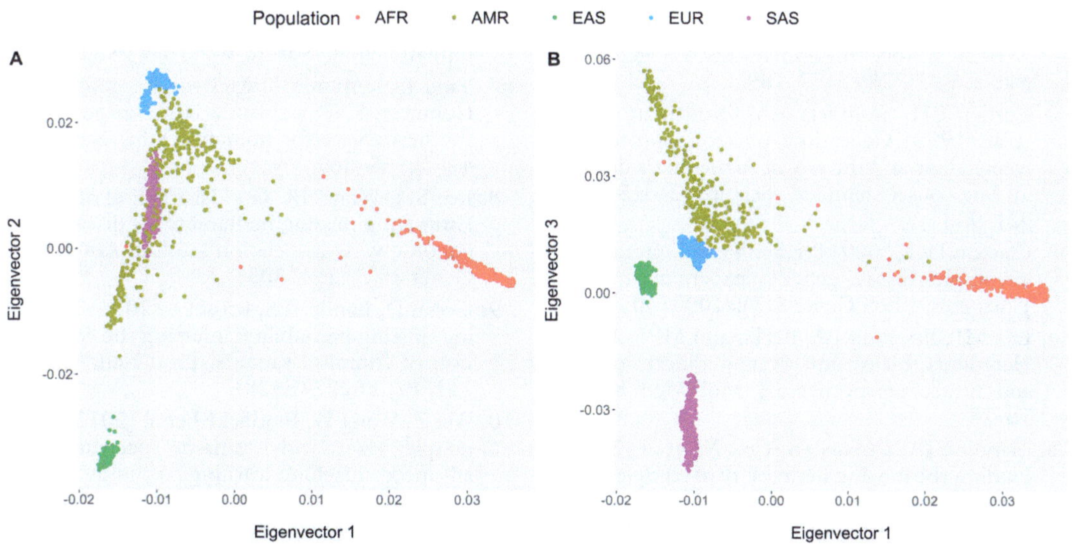

Fig. 3 PCA plot of 5 superpopulations from the 1000 Genomes Project

7 Summary and Outlook

Thus far, we have introduced several widely applied methods for disease risk prediction with genotype data. The PRS and LMM are featured with their marvelous ability to integrate over thousands of variables with moderate effects, which largely solved the heritability myth underlying human complex disorders, inferring a shared, polygenetic nature of genetic predispositions [1].

At present time, genetic risk predictions have drawn information from the genotype data and a few environmental variables. Recent development in genome-scale functional and expression databases such as the ENCODE [85], Roadmap Epigenomics [86], and GTEx [87] provides opportunities for pooling genomic data in predictive modeling. At clinical interface, electronic health records and image data offer valuable resources for connecting genetic predictors with patient clinical information. However, in the near future, due to the non-standardized patient records in different databases, it might be possible that only a few major clinical variables can be extracted for prediction application. Continuous improvement in database management, data standardization, security, and sharing would greatly facilitate the advancement of clinical-genetic data integration. Furthermore, the genomic and genetic samples of the non-European populations shall be expanded to improve the representativeness of genetic prediction. Finally, the future development of prediction methods might require even stronger interdisciplinary collaboration to solve the challenge of biobank-level computation, statistical innovations, and clinical interpretations for real-world model applications.

References

1. Claussnitzer M, Cho JH, Collins R et al (2020) A brief history of human disease genetics. Nature 577(7789):179–189
2. Corder EH, Saunders AM, Strittmatter WJ et al (1993) Gene dose of apolipoprotein E type 4 allele and the risk of Alzheimer's disease in late onset families. Science 261(5123): 921–923
3. Clayton DG (2009) Prediction and interaction in complex disease genetics: experience in type 1 diabetes. PLoS Genet 5(7):e1000540
4. Lux MP, Fasching PA, Beckmann MW (2006) Hereditary breast and ovarian cancer: review and future perspectives. J Mol Med 84(1): 16–28
5. Manolio TA, Collins FS, Cox NJ et al (2009) Finding the missing heritability of complex diseases. Nature 461(7265):747–753
6. Lango Allen H, Estrada K, Lettre G et al (2010) Hundreds of variants clustered in genomic loci and biological pathways affect human height. Nature 467(7317):832–838
7. Yang J, Benyamin B, McEvoy BP et al (2010) Common SNPs explain a large proportion of the heritability for human height. Nat Genet 42(7):565–569
8. Lee SH, Wray NR, Goddard ME et al (2011) Estimating missing heritability for disease from genome-wide association studies. Am J Hum Genet 88(3):294–305
9. Golan D, Lander ES, Rosset S (2014) Measuring missing heritability: inferring the contribution of common variants. Proc Natl Acad Sci 111(49):E5272–E5281
10. Wei Z, Wang W, Bradfield J et al (2013) Large sample size, wide variant spectrum, and advanced machine-learning technique boost risk prediction for inflammatory bowel disease. Am J Hum Genet 92(6):1008–1012

11. Lambert SA, Abraham G, Inouye M (2019) Towards clinical utility of polygenic risk scores. Hum Mol Genet 28(R2):R133–R142
12. Rencher AC, Schaalje GB (2008) Linear models in statistics. Wiley, Hoboken
13. Allen DM (1971) Mean square error of prediction as a criterion for selecting variables. Technometrics 13(3):469–475
14. Huang J, Ling CX (2005) Using AUC and accuracy in evaluating learning algorithms. IEEE Trans Knowl Data Eng 17(3):299–310
15. Visscher ISCMpPSMspmhebWNRSJL, Michael C. 6 Visscher Peter M. 5 PasWNRMSSPscmhedSPFOD, Gurling H et al (2009) Common polygenic variation contributes to risk of schizophrenia and bipolar disorder. Nature 460(7256):748–752
16. Anderson CA, Pettersson FH, Clarke GM et al (2010) Data quality control in genetic case-control association studies. Nat Protoc 5(9): 1564–1573
17. McCullagh P, Nelder JA (2019) Generalized linear models. Routledge, London
18. Chang CC, Chow CC, Tellier LC et al (2015) Second-generation PLINK: rising to the challenge of larger and richer datasets. Gigascience 4(1):s13742-13015-10047-13748
19. Clarke L, Fairley S, Zheng-Bradley X et al (2017) The international genome sample resource (IGSR): a worldwide collection of genome variation incorporating the 1000 genomes project data. Nucleic Acids Res 45(D1): D854–D859
20. Dudbridge F (2013) Power and predictive accuracy of polygenic risk scores. PLoS Genet 9(3):e1003348
21. Euesden J, Lewis CM, O'reilly PF (2015) PRSice: polygenic risk score software. Bioinformatics 31(9):1466–1468
22. Wray NR, Lee SH, Mehta D et al (2014) Research review: polygenic methods and their application to psychiatric traits. J Child Psychol Psychiatry 55(10):1068–1087
23. Vilhjálmsson BJ, Yang J, Finucane HK et al (2015) Modeling linkage disequilibrium increases accuracy of polygenic risk scores. Am J Hum Genet 97(4):576–592
24. O'donovan MC, Craddock N, Norton N et al (2008) Identification of loci associated with schizophrenia by genome-wide association and follow-up. Nat Genet 40(9):1053–1055
25. Consortium IMSG (2010) Evidence for polygenic susceptibility to multiple sclerosis—the shape of things to come. Am J Hum Genet 86(4):621–625
26. Speliotes EK, Willer CJ, Berndt SI et al (2010) Association analyses of 249,796 individuals reveal 18 new loci associated with body mass index. Nat Genet 42(11):937–948
27. Simonson MA, Wills AG, Keller MC et al (2011) Recent methods for polygenic analysis of genome-wide data implicate an important effect of common variants on cardiovascular disease risk. BMC Med Genet 12(1):1–9
28. Stahl EA, Wegmann D, Trynka G et al (2012) Bayesian inference analyses of the polygenic architecture of rheumatoid arthritis. Nat Genet 44(5):483–489
29. Duncan L, Shen H, Gelaye B et al (2019) Analysis of polygenic risk score usage and performance in diverse human populations. Nat Commun 10(1):1–9
30. Kim MS, Patel KP, Teng AK et al (2018) Genetic disease risks can be misestimated across global populations. Genome Biol 19(1):1–14
31. Martin AR, Gignoux CR, Walters RK et al (2017) Human demographic history impacts genetic risk prediction across diverse populations. Am J Hum Genet 100(4):635–649
32. Mostafavi H, Harpak A, Agarwal I et al (2020) Variable prediction accuracy of polygenic scores within an ancestry group. elife 9:e48376
33. Cai M, Xiao J, Zhang S et al (2021) A unified framework for cross-population trait prediction by leveraging the genetic correlation of polygenic traits. Am J Hum Genet 108(4):632–655
34. Coram MA, Fang H, Candille SI et al (2017) Leveraging multi-ethnic evidence for risk assessment of quantitative traits in minority populations. Am J Hum Genet 101(2): 218–226
35. Selzam S, Krapohl E, Von Stumm S et al (2017) Predicting educational achievement from DNA. Mol Psychiatry 22(2):267–272
36. Lee JJ, Wedow R, Okbay A et al (2018) Gene discovery and polygenic prediction from a genome-wide association study of educational attainment in 1.1 million individuals. Nat Genet 50(8):1112–1121
37. Zhang Y, Lu Q, Ye Y et al (2021) SUPERGNOVA: local genetic correlation analysis reveals heterogeneous etiologic sharing of complex traits. Genome Biol 22(1):1–30
38. Ruderfer DM, Fanous AH, Ripke S et al (2014) Polygenic dissection of diagnosis and clinical dimensions of bipolar disorder and schizophrenia. Mol Psychiatry 19(9): 1017–1024
39. Maier R, Moser G, Chen G-B et al (2015) Joint analysis of psychiatric disorders increases accuracy of risk prediction for schizophrenia, bipolar disorder, and major depressive disorder. Am J Hum Genet 96(2):283–294
40. Ruderfer DM, Ripke S, McQuillin A et al (2018) Genomic dissection of bipolar disorder

and schizophrenia, including 28 subphenotypes. Cell 173(7):1705–1715. e1716

41. Guo H, Li JJ, Lu Q et al (2021) Detecting local genetic correlations with scan statistics. Nat Commun 12(1):1–13
42. Krapohl E, Patel H, Newhouse S et al (2018) Multi-polygenic score approach to trait prediction. Mol Psychiatry 23(5):1368–1374
43. Maier RM, Zhu Z, Lee SH et al (2018) Improving genetic prediction by leveraging genetic correlations among human diseases and traits. Nat Commun 9(1):1–17
44. Grotzinger AD, Rhemtulla M, de Vlaming R et al (2019) Genomic structural equation modelling provides insights into the multivariate genetic architecture of complex traits. Nat Hum Behav 3(5):513–525
45. Wand H, Lambert SA, Tamburro C et al (2021) Improving reporting standards for polygenic scores in risk prediction studies. Nature 591(7849):211–219
46. Mars N, Koskela JT, Ripatti P et al (2020) Polygenic and clinical risk scores and their impact on age at onset and prediction of cardiometabolic diseases and common cancers. Nat Med 26(4):549–557
47. Khera AV, Chaffin M, Aragam KG et al (2018) Genome-wide polygenic scores for common diseases identify individuals with risk equivalent to monogenic mutations. Nat Genet 50(9): 1219–1224
48. Elliott J, Bodinier B, Bond TA et al (2020) Predictive accuracy of a polygenic risk score–enhanced prediction model vs a clinical risk score for coronary artery disease. JAMA 323(7):636–645
49. Inouye M, Abraham G, Nelson CP et al (2018) Genomic risk prediction of coronary artery disease in 480,000 adults: implications for primary prevention. J Am Coll Cardiol 72(16): 1883–1893
50. Abraham G, Havulinna AS, Bhalala OG et al (2016) Genomic prediction of coronary heart disease. Eur Heart J 37(43):3267–3278
51. Yang J, Zaitlen NA, Goddard ME et al (2014) Advantages and pitfalls in the application of mixed-model association methods. Nat Genet 46(2):100–106
52. Loh P-R, Tucker G, Bulik-Sullivan BK et al (2015) Efficient Bayesian mixed-model analysis increases association power in large cohorts. Nat Genet 47(3):284–290
53. Lloyd-Jones LR, Zeng J, Sidorenko J et al (2019) Improved polygenic prediction by Bayesian multiple regression on summary statistics. Nat Commun 10(1):1–11
54. Vilhjálmsson BJ, Nordborg M (2013) The nature of confounding in genome-wide association studies. Nat Rev Genet 14(1):1–2
55. Makowsky R, Pajewski NM, Klimentidis YC et al (2011) Beyond missing heritability: prediction of complex traits. PLoS Genet 7(4): e1002051
56. Habier D, Fernando RL, Kizilkaya K et al (2011) Extension of the Bayesian alphabet for genomic selection. BMC Bioinform 12(1): 1–12
57. Moser G, Lee SH, Hayes BJ et al (2015) Simultaneous discovery, estimation and prediction analysis of complex traits using a Bayesian mixture model. PLoS Genet 11(4):e1004969
58. Zeng J, De Vlaming R, Wu Y et al (2018) Signatures of negative selection in the genetic architecture of human complex traits. Nat Genet 50(5):746–753
59. Zeng P, Zhou X (2017) Non-parametric genetic prediction of complex traits with latent Dirichlet process regression models. Nat Commun 8(1):1–11
60. Durvasula A, Lohmueller KE (2021) Negative selection on complex traits limits phenotype prediction accuracy between populations. Am J Hum Genet 108(4):620–631
61. Shi H, Gazal S, Kanai M et al (2021) Population-specific causal disease effect sizes in functionally important regions impacted by selection. Nat Commun 12(1):1–15
62. Wang Y, Guo J, Ni G et al (2020) Theoretical and empirical quantification of the accuracy of polygenic scores in ancestry divergent populations. Nat Commun 11(1):1–9
63. Xia X, Sun R, Zhang Y et al (2022) A prism vote framework for individualized risk prediction of traits in genome-wide sequencing data of multiple populations. bioRxiv. https://doi.org/10.1101/2022.02.02.478767
64. Erbe M, Hayes B, Matukumalli L et al (2012) Improving accuracy of genomic predictions within and between dairy cattle breeds with imputed high-density single nucleotide polymorphism panels. J Dairy Sci 95(7): 4114–4129
65. Zhou X, Carbonetto P, Stephens M (2013) Polygenic modeling with Bayesian sparse linear mixed models. PLoS Genet 9(2):e1003264
66. Yang J, Fritsche LG, Zhou X et al (2017) A scalable Bayesian method for integrating functional information in genome-wide association studies. Am J Hum Genet 101(3):404–416
67. Zhu X, Stephens M (2017) Bayesian large-scale multiple regression with summary statistics from genome-wide association studies. Ann Appl Stat 11(3):1561

68. Hoerl AE, Kennard RW (1970) Ridge regression: biased estimation for nonorthogonal problems. Technometrics 12(1):55–67
69. Tibshirani R (1996) Regression shrinkage and selection via the lasso. J R Stat Soc Ser B Methodol 58(1):267–288
70. Zou H, Hastie T (2005) Regularization and variable selection via the elastic net. J R Stat Soc Ser B Stat Methodol 67(2):301–320
71. Friedman J, Hastie T, Tibshirani R (2010) Regularization paths for generalized linear models via coordinate descent. J Stat Softw 33(1):1
72. Zeng Y, Breheny P (2017) The biglasso package: a memory-and computation-efficient solver for lasso model fitting with big data in r. arXiv preprint arXiv:170105936
73. Privé F, Aschard H, Ziyatdinov A et al (2018) Efficient analysis of large-scale genome-wide data with two R packages: bigstatsr and bigsnpr. Bioinformatics 34(16):2781–2787
74. Qian J, Tanigawa Y, Du W et al (2020) A fast and scalable framework for large-scale and ultrahigh-dimensional sparse regression with application to the UK Biobank. PLoS Genet 16(10):e1009141
75. Mak TSH, Porsch RM, Choi SW et al (2017) Polygenic scores via penalized regression on summary statistics. Genet Epidemiol 41(6): 469–480
76. Abraham G, Malik R, Yonova-Doing E et al (2019) Genomic risk score offers predictive performance comparable to clinical risk factors for ischaemic stroke. Nat Commun 10(1):1–10
77. Lu X, Niu X, Shen C et al (2021) Development and validation of a polygenic risk score for stroke in the Chinese population. Neurology 97(6):e619–e628
78. Devlin B, Roeder K (1999) Genomic control for association studies. Biometrics 55(4): 997–1004
79. Devlin B, Roeder K, Wasserman L (2001) Genomic control, a new approach to genetic-based association studies. Theor Popul Biol 60(3):155–166
80. Sul JH, Martin LS, Eskin E (2018) Population structure in genetic studies: confounding factors and mixed models. PLoS Genet 14(12): e1007309
81. Clayton DG, Walker NM, Smyth DJ et al (2005) Population structure, differential bias and genomic control in a large-scale, case-control association study. Nat Genet 37(11): 1243–1246
82. Price AL, Patterson NJ, Plenge RM et al (2006) Principal components analysis corrects for stratification in genome-wide association studies. Nat Genet 38(8):904–909
83. Yang J, Lee SH, Goddard ME et al (2011) GCTA: a tool for genome-wide complex trait analysis. Am J Hum Genet 88(1): 76–82
84. Consortium GP (2015) A global reference for human genetic variation. Nature 526(7571): 68
85. Consortium EP (2012) An integrated encyclopedia of DNA elements in the human genome. Nature 489(7414):57
86. Bernstein BE, Stamatoyannopoulos JA, Costello JF et al (2010) The NIH roadmap epigenomics mapping consortium. Nat Biotechnol 28(10):1045–1048
87. Lonsdale J, Thomas J, Salvatore M et al (2013) The genotype-tissue expression (GTEx) project. Nat Genet 45(6):580–585

Chapter 16

Statistical Methods Inspired by Challenges in Pediatric Cancer Multi-omics

Xueyuan Cao, Abdelrahman H. Elsayed, and Stanley B. Pounds

Abstract

Pediatric cancer multi-omics is a uniquely rewarding and challenging domain of biomedical research. Public generosity bestows an abundance of resources for the study of extremely rare diseases; this unique dynamic creates a research environment in which problems with high-dimension and low sample size are commonplace. Here, we present a few statistical methods that we have developed for our research setting and believe will prove valuable in other biomedical research settings as well. The genomic random interval (GRIN) method evaluates the loci and frequency of genomic abnormalities in the DNA of tumors to identify genes that may drive the development of malignancies. The association of lesions with expression (ALEX) method evaluates the impact of genomic abnormalities on the RNA transcription of nearby genes to inform the formulation of biological hypotheses on molecular mechanisms. The projection onto the most interesting statistical evidence (PROMISE) method identifies omic features that consistently associate with better prognosis or consistently associate with worse prognosis across multiple measures of clinical outcome. We have shown that these methods are statistically robust and powerful in the statistical bioinformatic literature and successfully used these methods to make fundamental biological discoveries that have formed the scientific rationale for ongoing clinical trials. We describe these methods and illustrate their application on a publicly available T-cell acute lymphoblastic leukemia (T-ALL) data set. A companion github site (https://github.com/stjude/TALL-example) provides the R code and data necessary to recapitulate the example data analyses of this chapter.

Key words Gene expression, Genomic lesion, Genome-wide association, Random interval, Projection, T-ALL

1 Introduction

Statistical collaboration in pediatric cancer multi-omics research is both rewarding and challenging. It is very rewarding to find, evaluate, and establish cures for children with cancer that ensure their survival and improve their quality of life. The benefits of these cures are realized for decades as survivors mature into adults. At the same time, from a statistical perspective, it is very challenging to optimally utilize available information to advance cures for extremely

Brooke L. Fridley and Xuefeng Wang (eds.), *Statistical Genomics*,
Methods in Molecular Biology, vol. 2629, https://doi.org/10.1007/978-1-0716-2986-4_16,

rare diseases. One advantage of working in this domain is that pediatric oncology researchers strive to learn as much as possible from the experiences of each patient by collecting the most comprehensive clinical, pharmacology, molecular, imaging, psychometric, and quality-of-life data possible. This provides an inspirational motivation and rich environment to creatively address the classical statistical challenge of data with large dimension and small sample size.

In this unique research environment, we have developed innovative statistical methods for integrating multiple forms of omic data with multiple clinical outcome metrics. We believe these methods are also more broadly applicable as it becomes more common for other medical researchers to have multiple high-dimensional data sets on their patients as well. In this chapter, we describe three of these methods and illustrate how to implement them with a publicly available data set for T-cell acute lymphoblastic leukemia [1]. Each of these methods is briefly introduced immediately below and subsequently describe. We have also created a companion GitHub repository with a start-to-finish roadmap for readers to repeat the analyses of this book chapter (https://github.com/stjude/TALL-example).

Cancers are characterized and driven by *genomic lesions*, which are various abnormalities in tumor cell DNA. For example, a base pair may be inserted into or deleted from the DNA sequence of a gene (an indel) or replaced by another base pair (a single nucleotide variant). Also, long stretches of DNA may be duplicated, deleted, inverted, or rearranged. Genomic lesions naturally accumulate in cells over multiple generations of mitosis. Many of those changes are *passenger lesions* which have no meaningful biological impact, while others are *driver lesions* that accelerate the development and progression of a cancer. To help distinguish the passenger and driver lesions from one another, the *genomic random interval* (GRIN) method identifies genomic loci that are affected by genomic lesions in a statistically significant number of patients' tumors [2]. Lesions affecting those loci are then evaluated as candidate driver lesions in cell line and/or animal models of the disease.

Genomic lesions in tumor DNA can directly or indirectly alter the transcription of DNA to RNA. The abundance of RNA transcripts for a specific gene is commonly referred to as the RNA *expression* of that gene. The *associate lesions with expression* (ALEX) method identifies genes for which RNA expression levels are associated with genomic lesions. This provides a second line of evidence to discern between driver and passenger lesions. Genomic lesions that associate with RNA transcription are more likely to have downstream biological ramifications than are genomic lesions that show no association with RNA transcription.

Clearly, it is also important to better understand how genomic lesions or RNA transcription associate with patient outcomes. In most oncology clinical trials, multiple metrics of patient outcome are evaluated. It is common for an oncology trial to evaluate early response of tumor to therapy, a failure time outcome (such as event-free survival, the time from diagnosis to relapse, death, or other major clinical failures), and overall survival (time from diagnosis to death). The clinical outcome metrics have strong biological relationships and statistical associations with one another. In particular, a good early response (like complete disappearance of a tumor) usually associates with longer times until failure and longer times until death. The *projection onto the most interesting statistical evidence* (PROMISE) method leverages known relationships among patient outcome metrics to improve statistical power to identify individual prognostic omic features [3].

This suite of methods covers several important scientific questions that can be evaluated using multi-omic multi-outcome patient data. We use the following data set from a published study pediatric and young adult T-cell acute lymphoblastic leukemia [1] to illustrate the purpose, how to use, and how to interpret results of each of these methods. We also provide a companion GitHub repository (https://github.com/stjude/TALL-example) with code and data files for readers to reproduce the analysis results of this chapter.

2 T-ALL in Pediatric and Young Adults

Acute lymphoblastic leukemia (ALL) is a malignancy of lymphocytes that occurs in about 3000 children annually in the USA. The disease may occur in two different subtypes of lymphocytes, B-cells or T-cells; thus, ALL has two major histological subtypes: pre-B-cell ALL (B-ALL) or T-cell ALL (T-ALL). With modern therapy, over 90% of children and most young adults with ALL can be cured with 2 years of treatment with complex combinations and sequences of chemotherapy. A better understanding of the molecular basis of ALL development and prognosis could help develop effective cures for the minority of patients who relapse and/or develop less toxic treatments that maintain a good overall prognosis for these patients. Therefore, Liu et al. profiled the genomes and transcriptomes of 265 T-ALL patients treated on the Children's Oncology Group AALL0434 clinical trial between 2007 and 2011 [4] to better understand the molecular drivers of the development and prognosis of T-ALL. As participants in the clinical trial, the treatment protocol and ascertainment of clinical outcomes for these patients were consistent across institutions [1]. This data set is used for an example analysis for the statistical methods presented in this chapter. A GitHub repository (https://github.com/stjude/TALL-example/) provides code and data files for readers to replicate the analyses presented here.

3 Notation

Table 1 shows the mathematical notation to be used throughout this chapter. Let $i = 1, \ldots, n$ index the subjects in the study. These patients have tumors with $l = 1, \ldots, r$ genomic lesions of $w = 1, \ldots, d$ types (mutations, losses, gains, rearrangements, etc.) on one of $k = 1, \ldots, h$ chromosomes. Each lesion l is characterized by a five-tuple $(i_l, k_l, a_l, b_l, w_l)$ with the subject index i_l, chromosome k_l, base pair start position a_l, base pair end position b_l, and type w_l of the lesion. The positions of the lesions are compared to $v = 1, \ldots, m$ loci of scientific interest (typically including but not limited to genes). The loci of interest and lesions are located on one of $k = 1, \ldots, k$ chromosomes, and L_k is the length (number of base pairs) of chromosome k.

Additionally, let y_{ij} represent the RNA expression of transcript $j = 1, \ldots, g$ in the tumor of subject $i = 1, \ldots, n$. Also, let $c = 1, \ldots, d$ index metrics of clinical outcomes measured in the $i = 1, \ldots, n$ subjects. The indicator function $I(\cdot)$ equals 1 if the enclosed statement $\cdot$ is true and 0 otherwise.

4 Multiple-Testing Adjustments

Each analysis presented in this chapter involves performing a very large number of hypothesis tests. In each analysis, we address multiplicity by computing the q-value [5] with the estimate $\widehat{\pi}_0 = \min(2\overline{p}, 1)$ of the proportion π_0 of tests with a true null hypothesis [6], where $\overline{p}$ is the arithmetic mean of the p-values computed in that particular analysis. This conservative estimator is robust for discrete and continuous p-values and is motivated by the inequality $E(\overline{p}) \leq \frac{\pi_0}{2}$ that directly follows from the probability integral transform and the assumption that p-values of false nulls are stochastically less than uniform. Pounds and Cheng introduced $\widehat{\pi}_0 = \min(2\overline{p}, 1)$ as an intuitively simple and statistically robust estimate of π_0 that worked well for p-values with both discrete and continuous distributions [6]. The GitHub repository includes a file *qvalue-PC06.R* that defines an R function to compute this q-value. Some statistical tests had p-values that numerically evaluated to zero and were replaced with $p = 10^{-100}$ prior to calculating the q-values. In some cases, we choose to report $-\log_{10}q$ values for brevity and to better align the narrative with Manhattan plots of $-\log_{10}q$ values.

Table 1
Mathematical notation

Notation	Meaning
Subjects	
$i = 1, \ldots, n$	Indexes subjects (patients)
Genomic loci	
$k = 1, \ldots, h$	Indexes chromosomes
L_k	Length of chromosome k
$v = 1, \ldots, m$	Indexes genomic loci
(k_v, a_v, b_v)	Chromosome k_v, start a_v, and end b_v of locus v
Genomic lesions	
$l = 1, \ldots, r$	Indexes genomic lesions
$w = 1, \ldots, d$	Indexes types of lesions (mutation, loss, gain, etc.)
$(i_l, k_l, a_l, b_l, w_l)$	Subject i_l, chromosome k_l, start a_l, end b_l, and type w_l of lesion l
o_{lv}	Indicates that lesion l overlaps locus v with value 1 and 0 otherwise
λ_{ivw}	The null model probability that subject i has at least one lesion of type w that overlaps locus v
$N_{.vw}, n_{.vw}$	Random variable and observed value for the number $n_{.vw}$ of subjects with at least one lesion of type w overlapping locus v
RNA expression	
$j = 1, \ldots, g$	Indexes RNA transcripts
y_{ij}	Expression of RNA transcript j for subject i
Clinical outcomes	
$c = 1, \ldots, C$	Indexes clinical outcome metrics
s_{jc}	Measures the association of transcript j expression with outcome c
$\delta_c = \pm 1$	The sign of s_{jc} indicating that greater expression of transcript j associates with a better clinical outcome c
Statistical	
p	Unadjusted p-value
$\pi_0, \widehat{\pi}_0$	Proportion π_0 of hypothesis tests with a true null and its estimate $\widehat{\pi}_0$
q	q-value with $\widehat{\pi}_0 = \min(2\overline{p}, 1)$ estimating $\widehat{\pi}_0$ as in (Pounds and Cheng 2006)
$\beta, \widehat{\beta}$	Regression model coefficient β and its estimate $\widehat{\beta}$
Mathematical	
$I(\cdot)$	Equals 1 if the enclosed statement $\cdot$ is true and equals 0 otherwise
$\Pr_0(\cdot)$	Probability of the enclosed event $\cdot$ under a null model
Computing	
$b^* = 0, 1, \ldots, B$	$b^* = 0$ indexes observed data, $b^* = 1, \ldots, B$ indexes permuted data

5 Genomic Random Intervals

5.1 The Null Model of the GRIN Method

The genomic random interval (GRIN) method identifies genetic loci impacted by lesions in the tumor DNA of a statistically significant number of subjects. Statistical significance is evaluated against a null model in which lesions of the same size and type are uniformly located along the chromosome on which they occur. To do this, GRIN first compares the lesion data to the genetic loci to compute an indicator

$$o_{lvw} = \mathrm{I}(k_l = k_v)\mathrm{I}(b_l \geq a_v)\mathrm{I}(a_l \leq b_v)\mathrm{I}(l_w = w)$$

that lesion l is of type w and overlaps with locus v. In particular, $o_{lvw} = 1$ if lesion l overlaps locus v and is of type w and $o_{lvw} = 0$ otherwise. Sums of the o_{lvw} indicators are metrics of how frequently genomic lesions "target" genetic loci. These sums can be defined to determine the total number of lesions that overlap locus v, the number of lesions of type w that overlap locus v, the number of subjects with a lesion of type w overlapping locus v, and so forth. GRIN defines a null model for the location of the lesions, which then determines a null distribution for these sums. The observed value of the sums is then compared to the null distribution to determine statistical significance (i.e., compute a p-value).

This null model is defined as shown in Fig. 1 and described below. For each lesion l and locus v, let

$$\Pr_0(o_{lvw} = 1) = \min\left(\frac{b_l - a_l + b_v - a_v + 1}{L_{k_l}}, 1\right)\mathrm{I}(l_w = w).$$

be the null model probability that lesion l is of type w and overlaps locus v. For short loci and short lesions of type w, this equals the number of possible locations of the lesion on the chromosome such that the lesion overlaps the locus divided by the length of the chromosome. For a short lesion l and a short locus v, this is approximately the probability that an interval with a fixed size equal to that of lesion l and uniform random location along the same chromosome would overlap locus v. This null model probability linearly increases with the lesion length $b_l - a_l + 1$ and the locus length $b_v - a_v + 1$ until the sum of their lengths (minus one base-pair so that they overlap) $b_l - a_l + b_v - a_v + 1$ exceeds the length of the chromosome L_{k_l} on which they are located. If the sum of locus and lesion lengths exceeds the length of the chromosome, then the null probability of overlap is 1.

Under the null GRIN model, all lesions are stochastically independent. Therefore, the null probability distribution for the number of lesions of type w^* in the tumor genome of subject i^* that overlap locus v is given by the convolution of $\Pr_0(o_{lv})$ over all l such that $i_l = i^*$, $k_l = k_v$, and $w_l = w^*$. There is a (possibly degenrate)

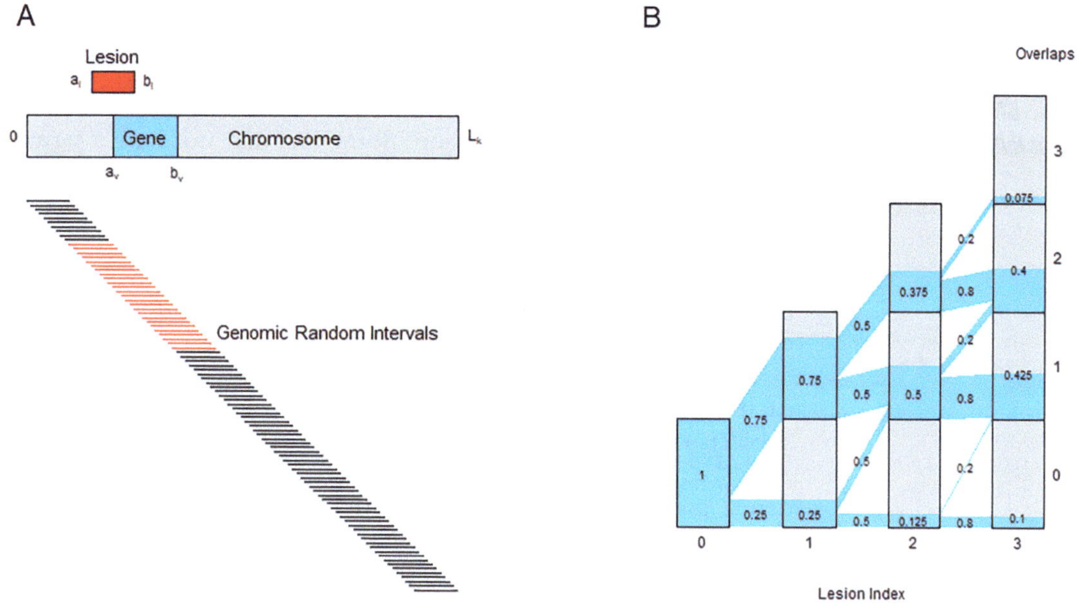

Fig. 1 The null model of the genomic random interval method. (**a**) In the null model, lesions are represented as intervals of fixed length but uniform random location along the chromosome on which they were observed. The probability that this kind of random interval overlaps the locus of a gene is a simple function of the size of the gene, the lesion, and the chromosome. (**b**) Lesions are assumed to be stochastically independent, so the distribution for the number of lesions overlapping a gene is computed by a convolution of nonidentical Bernoulli random variables. This panel shows the convolution of three Bernoulli random variables with probabilities of 0.75, 0.5, and 0.2

convolution (with $\text{Pr}_0(0) = 1$) like this for each locus v and each subject i. Each of these convolutions defines the null probability

$$\lambda_{ivw} = \text{Pr}_0\left(\sum_{l_i = i} o_{ivw} > 0\right)$$

that at least one lesion of type w for subject i overlaps locus v. Thus, for each subject i, there is a Bernoulli(λ_{ivw}) distribution for the occurrence of at least one lesion of type w in locus v. The null distribution $\text{Pr}_0(N_{.vw})$ for the total number $N_{.vw}$ of subjects with a lesion of type w in locus v is the convolution of the independent but nonidentical Bernoulli(λ_{ivw}) distributions over all subjects i. Therefore, given data to compute the observed number $n_{.vw}$ of subjects with at least one lesion of type w overlapping locus v,

$$p_{vw} = \text{Pr}_0(N_{.vw} \geq n_{.vw})$$

is a p-value testing the null hypothesis that the probability that a subject has a lesion of type w overlapping locus v is less than or equal to what is expected by the GRIN null model. For each type of lesion w, this gives a set of p-values p_{vw}, one for each locus $v = 1, \ldots, m$. For each type of lesion w, the p-values $p_{1w}, \ldots, p_{mw}$ are used to compute q-values $q_{1w}, \ldots, q_{1w}$ as described above to address multiplicity.

5.2 GRIN Constellation Tests

Some important loci in pediatric (or adult) cancer may be affected by more than one type of lesion. Some loci may not have a significant overabundance of any one type of lesion but may instead be affected by different types of lesions in a cumulatively significant number of subjects. These different lesions may have a similar biological impact in the development of the disease. Thus, it is helpful to combine information across types of lesions to more effectively identify these loci statistically.

After computing the GRIN p-values as described above, for each locus v, there is a set of p-values $p_{v1}, \ldots, p_{vd}$ for the number of subjects of each type of lesion $w = 1, \ldots, d$. For each locus v, let $p_{v(1)} \leq p_{v(2)} \cdots \leq p_{v(d)}$ be the ordered p-values. Under the independence assumption of the GRIN null model, these p-values are independent uniform random variables. Therefore, for each locus v and each ordering index (w), the ordered p-value $p_{v(w)}$ follows a beta $((w), d-(w)+1)$ distribution (Example 5.4.5 of Casella and Berger [7]). A $(w)^{\text{th}}$ *order constellation p-value* $\tilde{p}_{v(w)}$ can be computed as $\mathcal{B}\left(p_{v(w)}; (w), d-(w)+1\right)$, the cumulative distribution function of the beta$((w), d-(w)+1)$ distribution evaluated at $p_{v(w)}$. For each $(w) = 1, \ldots, d$, the (w)-order constellation p-value $\tilde{p}_{v(w)}$ represents the evidence that at least (w) type(s) of lesions overlap with locus v in a significant number of subjects' tumors. The first-order constellation test can identify loci that are affected by at least one type of lesion, thereby providing a metric to order loci that are affected by different types of lesions. The second-, third-, and higher-order constellation tests can identify genes that are affected by at least two, three, or more types of lesions, respectively.

5.3 GRIN Analysis of T-ALL Data Set

We applied GRIN to the example T-ALL data set [1]. This data set included the loci of $r = 6,887$ lesions of $d = 4$ types (gain, loss, mutation, or fusion) in the tumors of $n = 265$ patients. The center panel of Fig. 2 shows all lesions for all subjects across the entire genome. The right-hand side shows a transposed Manhattan plot of the number of subjects with each type of lesion at each of $m = 6887$ loci across the entire genome. The left-hand side shows a similar figure with $-\log_{10} q$ for each type of lesion affecting each locus.

The figure shows that gains (shown in red) and losses (shown in blue) of entire arms or chromosomes are common in this disease. However, these very large abnormalities have essentially no impact on statistical significance in the GRIN analysis. Under the GRIN null model of chance, very large lesions have a very great probability of affecting loci on that chromosome. This definition of statistical significance also aligns well with a biological interpretation of the

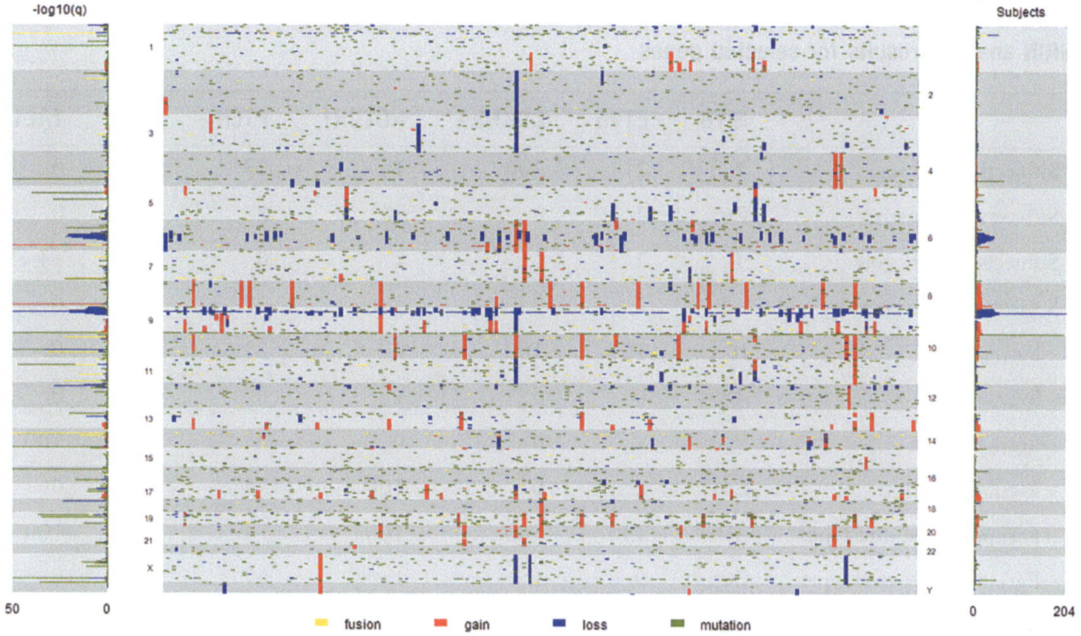

Fig. 2 Genome-wide GRIN result. The large center panel shows the genomic lesion data for the cohort. Genetic loci are depicted as rows and subjects as columns. The chromosomes are shown as regions of alternating dark and light gray background. Red, blue, green, and yellow rectangles, respectively, depict regions of gain, regions of loss, mutation loci, and fusion breakpoint loci. The panel on the right shows the number of subjects (out of 265) with each type of lesion at each locus with the same colors. The panel on the left is a transposed Manhattan plot of $-\log_{10}q$ values for the statistical significance of each type of lesion at each locus using the same color legend

data. Large lesions are not very informative as to which individual genes are most relevant biologically.

The $-\log_{10}q$ plot shows that some less common but very focal lesions (mutations and fusion break points) are very statistically significant. This also aligns with the interpretation of biological specificity. If recurrent lesions are very concentrated in one locus, this very clearly indicates that abnormalities affecting that locus may be important for development of the disease. Table 2 gives results for some selected genes with very significant results in the GRIN analysis. Losses overlapped the locus of CDKN2A in the tumors of $n = 200$ subjects; this finding was incredibly significant ($-\log_{10}q > 100$). In the tumors of $n = 29$ subjects, gains overlapped AHI1 ($-\log_{10}q = 51.4$). TAL1 had the most significant frequency of fusion ($n = 54$ subjects; $-\log_{10}q = 100$), and NOTCH1 had the most significant frequency of mutation ($n = 196$ subjects; $-\log_{10}q > 100$). The biological and clinical significance of these four genes in leukemia and other cancers is extensively documented in the literature. Table 3 gives some comments and supporting

Table 2
GRIN analysis results for selected genes

	MYB	ETV6	CDKN2A	NOTCH1	PTEN	AHI1	TAL1
Subjects with loss	2	21	200	1	23	2	0
Subjects with gain	26	2	0	13	8	29	0
Subjects with fusion	4	2	1	1	0	7	54
Subjects with mutation	13	7	2	196	37	1	0
$-\log_{10}q$ loss	0	15	100	0	33	0	0
$-\log_{10}q$ gain	48	0	0	1	0	51	0
$-\log_{10}q$ fusion	7	1	0	0	0	11	100
$-\log_{10}q$ mutation	24	4	0	100	74	0	0
$-\log_{10}q$ at least one type	49	14	100	100	74	52	100
$-\log_{10}q$ at least two types	50	9	1	1	67	24	0
$-\log_{10}q$ at least 3 types	26	7	2	1	0	0	0

Pubmed references (as PubMed identifiers: PMID) for these and other genes described in this chapter.

The constellation tests identify additional genes of biological and clinical relevance. Among the genes shown in Table 2, PTEN has the most significant evidence of being impacted by at least two types of lesions ($-\log_{10}q = 66.8$). PTEN was affected by loss in 23 subjects ($-\log_{10}q = 33.2$), gain in 8 subjects ($-\log_{10}q = 0$), and mutation in 37 subjects ($-\log_{10}q = 73.9$). MYB has the most significant evidence of being impacted by at least three types of lesions ($-\log_{10}q = 26.2$). MYB was affected by loss in 2 subjects ($-\log_{10}q = 0$), gain in 26 subjects ($-\log_{10}q = 48.2$), fusion in 4 subjects ($-\log_{10}q = 7.1$), and mutation in 13 subjects ($-\log_{10}q = 24.1$). Both of these genes also have well-documented relevance for leukemia and other cancers (Table 3). The constellation tests help bring attention to these genes that don't stand out as much when considering only one type of lesion.

The constellation tests provide a quantitative approach to ordering genes that may not be impacted by the same types of lesions or that may be impacted by more than one type of lesion. The first-order constellation test is extremely significant for all the genes listed in Table 2. PTEN is affected by copy number loss and mutation in 23 and 37 patients ($q \approx 10^{-23}$ and 10^{-37}), respectively. The second-order constellation test finds very compelling evidence that PTEN is impacted by at least two types of lesions ($q = 10^{-135}$). PTEN stands out much more significant in the second-order constellation test than those genes that are significantly impacted by

Table 3
Some publications describing biological or clinical knowledge of selected genes identified by statistical analysis

Gene	Comment (abbreviated title)	Reference
AHI1	Activated by insertion mutations in mouse models of leukemia	[15]
	A AHI-1-BCR-ABL-DNM2 complex regulates CML properties	[16]
CDKN2A	CDKN2 alteration simplifies with tumor growth in adult T-ALL	[17]
	Inactivation of Cdkn2a associated with adult T-ALL progression	[31]
ETV6	ETV6-NCOA2: a novel fusion gene in acute leukemia	[18]
	Alternative ETV6-JAK2 fusions associated with T-ALL in zebrafish	[19]
LYL1	Multiple mechanisms induce LYL1 expression in T-ALL cell lines	[20]
	TAL1, TAL2, and LYL1: a family of proteins implicated in T-ALL	[32]
MYB	RUNX1 is required for oncogenic Myb and Myc activity in T-ALL	[21]
	Duplication of the MYB oncogene in T-ALL	[22]
	C-MYB translocations and mutations define a pediatric T-ALL subtype	[14]
NOTCH1	NOTCH1 mutations in T-ALL prognostics and leukemogenesis	[23]
	Prognostic NOTCH1 and FBXW7 mutations in pediatric T-ALL	pubmed.gov/20683149
PHF6	Phf6 loss drives tumor initiation in T-ALL	[24]
		[33]
PTEN	High frequency of PTEN, PI3K, and AKT abnormalities in T-ALL	[25]
	Negative prognostic impact of PTEN mutation in pediatric T-ALL	[26]
	NOTCH1/FBXW7/RAS/PTEN risk classification of adult T-ALL	[27]
TAL1	Oncogenic transcriptional program driven by TAL1 in T-ALL	[28]
	Origins of STIL-TAL1 fusion genes in children who developed T-ALL	[29]
	An interchromosomal interaction mediates TAL1 in T-ALL	[30]

only one type of lesion. The third-order constellation test finds that MYB is significantly affected by at least three types of lesions ($q \approx 10^{-26}$), with losses observed in two subjects ($q = 1$), gains observed in 26 subjects ($q \approx 10^{-48}$), fusions observed in four subjects ($q \approx 10^{-7}$), and mutations observed in 13 subjects ($q \approx 10^{-24}$). MYB is not among the top genes for being impacted any one type of lesion, but it is the top gene in the third-order constellation test. Still, MYB is an important gene in many cancers, including T-ALL, with one article arguing that genomic alterations

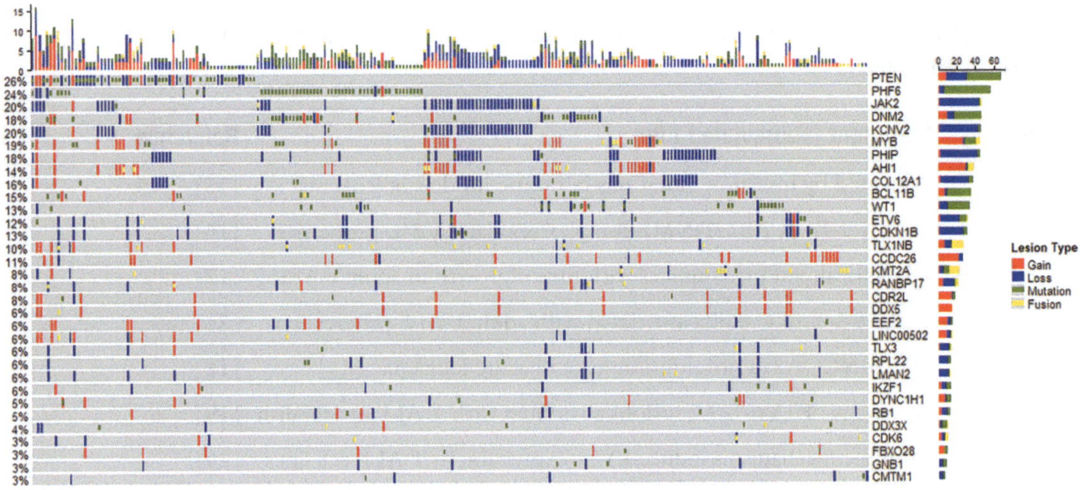

Fig. 3 OncoPrint of top genes in second-order constellation GRIN test. The figure shows genes as rows and subjects as columns. Gain, loss, mutation, or fusion break point within the locus of the gene is indicated by the colors as shown in the legend. The stacked colored barplots on the right indicate the number of subjects with each type of lesion in the gene; the stacked colored barplots on the top indicate the number of each type of lesion observed in these genes in the tumors of each subject

involving MYB define a unique subtype in pediatric T-ALL [14]. The constellation test highlights these important genes that aren't at the very top of lists ordered by significance of being affected by one specific type of lesion.

Figure 3 provides an OncoPrint visualization of the data for the genes with the most significant evidence of being affected by at least two types of lesions. Many of these genes have known relevance in leukemia and other cancers.

6 Association of Lesions with Expression

6.1 Association of Lesions with Expression (ALEX)

GRIN identifies loci with a statistically significant overabundance of genomic lesions. Additional insights can be gained by evaluating whether lesions associate with altered RNA expression of that locus. The association of lesions with expression (ALEX) method is a simple and effective way to statistically evaluate this biologically important question given genomic lesion data and RNA transcript expression data for tumors on the same set of subjects.

For each transcript $g = 1, \ldots, m$, let $v_g = 1, \ldots, m_g$ index the loci that are annotated to transcript g by a bioinformatic knowledge base and/or by proximity in terms of genomic location. For each locus v_g annotated to transcript g, assign each subject i into one distinct subgroup s_{iv_g} as follows:

- Let $s_{iv_g} = 0$ for each subject i with no lesions overlapping locus v_g.

- For each type of lesion $w = 1, \ldots, d$, let $s_{iv_g} = w$ if the subject has lesions *only of type* w overlapping locus v_g.
- Let $s_{iv_g} = d + 1$ if the subject has more than one type of lesion overlapping locus v_g.

Once the subjects have been assigned to subgroups according to their lesions in this manner, use the rank-based Kruskal-Wallis test to compare the RNA transcript expression levels across the subgroups. The Kruskal-Wallis test is chosen because it is a robust rank-based method. For each transcript g and its annotated loci v_g, this yields a p-value p_{gv_g}. To address multiple testing, use this set of p-values to compute q-values as described above.

6.2 ALEX Analysis of T-ALL Example

We applied ALEX to the T-ALL example data set. Table 4 summarizes the results for selected genes. Notably, we find compelling evidence ($q < 0.10 \Leftrightarrow -\log_{10} q > 1$) that RNA expression significantly associates with DNA lesions for the genes TAL1, CDKN2A, PTEN, MYB, and ETV6 ($-\log_{10} q > 1$), but DNA lesions and RNA expression of AHI1 and NOTCH1 are not significantly associated with mutation status after multiplicity adjustment ($-\log_{10} q < 1$). Figure 4 shows the DNA lesion and RNA expression data of AHI1 and PHF6 with side-by-side median-centered waterfall plots. For AHI1, gains spanning a portion of the gene were observed in the leukemia of several subjects but were not associated with an increase of RNA expression relative to that of tumors with no genomic abnormality. This suggests that partial gains of AHI1 do not impact tumor biology via copy-number-driven increase in RNA expression. A handful of subjects had mutations coupled with (mostly) gain or (a few with) loss and RNA expression that was typically greater than that of the abnormality-free cohort. However, the sample size is very small, so it is difficult to make a statistically robust evaluation of this observation. In contrast, RNA expression of PHF6 in tumors with loss, mutation, or both loss and mutation in that gene is much less than that of tumors with no abnormality.

Table 4
ALEX results for selected genes

	TAL1	CDKN2A	PTEN	MYB	ETV6	AHI1	NOTCH1
$-\log_{10} q$	15.0	12.1	4.7	2.5	1.4	0.7	0.5
Median expression no lesions	1.2	1.0	5.4	6.0	4.1	2.5	5.1
Median expression loss		0.0	4.7	4.9	3.9	2.0	
Median expression gain			6.1	6.5	3.8	2.3	5.0
Median expression mutation		4.6	5.3	5.6	5.0	2.8	5.0
Median expression fusion	3.6	3.2		6.3		3.1	
Median expression multiple lesion types		2.4	5.1	7.0	4.5	3.4	5.3

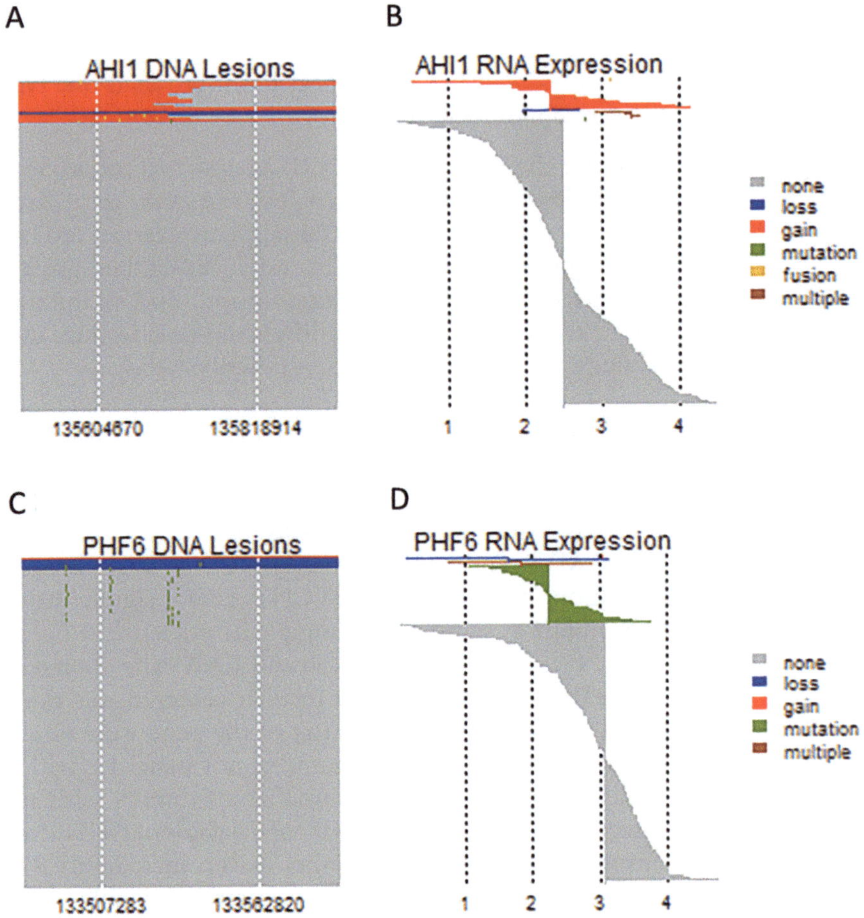

Fig. 4 Association of DNA lesions with RNA expression for AHI1 and PHF6. Panel (**a**) shows the lesions affecting the locus of AHI1 in the tumor of each subject (row) as rectangles or tick marks of the colors indicated in the color legend to the right. Panel (**b**) shows the $\log_2$-scale RNA expression of the same subjects in side-by-side group-specific median-centered waterfall plots (the vertical line in each waterfall plot is the group median). Panels (**c**) and (**d**) provide analogous information for PHF6

This suggests that PHF6 may be a T-ALL tumor suppressor, consistent with recent results from animal model studies [8]. These results illustrate some ways that an ALEX analysis can help deduce the biological roles of the DNA lesions in tumor biology.

7 PROMISE

7.1 Background

It is very important to identify molecular omic features that associate with prognosis. This knowledge can be used to personalize the intensity of therapy for future patients, select new agents for future clinical trials, or develop new agents to target poor prognostic features. In oncology, there are several metrics of outcome, including the response of the tumor to the first course of therapy, time from diagnosis or initiation of therapy until a failure event (such as relapse or death), and time until death. In pediatric oncology, sample sizes are limited, and several diseases have a good long-term prognosis. Thus, statistical power for identifying features that associate with one outcome metric (such as fitting a one-predictor Cox model for each omic variable) can have very limited statistical power. Typically, the outcome metrics are strongly correlated. For example, patients with a poor initial response to therapy are likely to have brief failure-free survival and short overall survival. Leveraging knowledge about the relationships among clinical outcomes in a statistically efficient and rigorous manner can help more reliably identify prognostic genes.

Therefore, attempts have been made to identify genes that associate with better prognosis according to all outcome metrics or associate with worse prognosis according to all outcome metrics. Genes that consistently associate with better prognosis or worse prognosis across multiple clinical outcome metrics are intuitively more likely to be clinically useful discoveries than genes that do not show consistent prognostic associations. From a clinical research perspective, it is most sensible to prioritize evaluation of therapies that increase expression of genes that consistently associate with better outcomes according to multiple metrics and/or decrease expression of genes that consistently associate with worse outcomes according to multiple metrics than to target genes that lack such a consistency in prognostic associations. For example, it would be difficult to determine whether it would be best to try to therapeutically increase or decrease expression of a gene that associated with better response but worse survival.

Frequently, a Venn diagram is used to narrow a candidate gene list by identifying genes that show a significant association with multiple clinical outcomes. This can be an effective method to narrow down a list of candidate genes to a set for more detailed study. However, it can sorely lack statistical power, especially in rare disease settings that typically involve small cohorts with few events. Each outcome association analysis will have limited power, so the overlap may be very small or even nonexistent unless the criteria for significance are so greatly relaxed that the validity of the results is questionable.

7.2 Definition of PROMISE

Projection onto the most interesting statistical evidence (PROMISE) is an integrative analysis method developed to increase statistical power to more directly identify genes consistently associated with better clinical outcomes or consistently associated with worse clinical outcomes according to multiple metrics [3]. PROMISE was developed and initially applied to associate microarray RNA expression data with multiple clinical and pharmacologic outcomes. It may be used to associate any form of high-dimensional omic data with multiple outcomes or phenotypes. Here, for simplicity of the narrative, we describe PROMISE in terms of associating the RNA expression of each of $j = 1, \ldots, g$ genes with each of $c = 1, \ldots, C$ clinical outcomes. PROMISE computes a statistic s_{jc} that measures the association of each gene j with each clinical outcome c. Additionally, for each gene j, PROMISE computes

$$\tilde{s}_{j\cdot} = \frac{1}{d}\sum_{c=1}^{d} \delta_c s_{jc}$$

as an overall measure of whether the expression of gene g has a *beneficial pattern of association* or *detrimental pattern of association* with the clinical outcomes $c = 1\ldots, C$. In this expression, δ_c is the sign of s_{jc} that indicates greater expression of j associates with better clinical outcomes. For example, if outcome c is overall survival and s_{jc} is the regression coefficient estimate (log hazard ratio) of a proportional hazards regression model, then δ_c is negative because a negative proportional hazard model coefficient indicates that an increase in expression associates with better overall survival. Thus, a large positive value for $\tilde{s}_{j\cdot}$ is evidence of a *beneficial pattern of* association in which greater expression of gene j associates with better prognosis as measured by clinical outcomes $c = 1\ldots, d$. Similarly, a large negative value for $\tilde{s}_{j\cdot}$ is evidence of a *detrimental pattern of association* in which greater expression of gene j associates with worse prognosis as measured by clinical outcomes $c = 1\ldots, d$.

7.3 Significance Determination of PROMISE

The statistical significance of $\tilde{s}_{j\cdot}$ is determined by permutation of the assignment of the gene expression data to the clinical outcome data. The permutation keeps the clinical outcome data of each subject intact and also keeps the gene expression profile of each subject intact. In this way, the correlation among clinical outcomes is retained, the correlation among the expression of genes is retained, but any correlation of the clinical outcomes with gene expression is broken. The observed statistic $\tilde{s}_{0j\cdot}$ is computed for the original data set, and a series of permutation statistics $\tilde{s}^{\star}_{b^{\star}j\cdot}$ is computed for a series of permuted data sets indexed by $b^{\star} = 1, \ldots, B^{\star}_j$. The permutation p-value is computed as

$$\widehat{p}_j = \frac{1}{B^{\star}_j}\sum_{b^{\star}=1}^{B^{\star}_j} \mathrm{I}\left(|s_{b^{\star}j\cdot}| \geq |s_{0j\cdot}|\right)$$

under the assumption that $E_0(s_{j\cdot}) = 0$, that is, the expected value of the statistic $s_{j\cdot}$ is zero under the null hypothesis.

It is common practice to use a fixed number of permutations $B^{\star}_j = B^{\star}$ for all genes. However, it is possible to compute a statistically sound p-value in a much more computationally efficient way with an *adaptive permutation* that performs a different *random* number of permutations for each gene [9]. The adaptive permutation procedure computes statistics for a series of permuted data sets until obtaining a minimum number $B^{\star}_{\min}$ permutation statistics with $|s_{b^{\star}j\cdot}| \geq |s_{0j\cdot}|$, or a maximum $B^{\star}_{\max}$ permutation statistics have been computed. The adaptive permutation procedure dramatically reduces computing time by performing a very large number of permutations only for the genes that are very statistically significant. For example, only $B^{\star}_{\min}$ permutations will be performed for each gene j with $s_{0j\cdot} = 0$ (yielding $\widehat{p}_j = 1$), whereas $B^{\star}_{\max}$ permutations will be performed for each gene j for which an extremely small proportion of permuted data sets yields a permutation test statistic of greater magnitude than does the original data set. Thus, the greatest computing effort is invested in computing the most statistically significant p-values and while minimal computing effort is spent on computing large p-values. This makes sense because it is usually not of interest to distinguish between genes with $p = 0.4$ or $p = 0.6$, but it is of great interest to distinguish between $p = 10^{-3}$ and $p = 10^{-6}$. The adaptive permutation procedure is described in greater detail [9].

7.4 PROMISE Analysis of Gene Expression in T-ALL

We applied the PROMISE procedure to evaluate the association of RNA-seq gene expression with minimal residual disease (MRD), event-free survival (EFS), and overall survival (OS) in the pediatric T-cell leukemia data set [1]. We treated MRD as a quantitative variable (percentage of bone marrow cells determined to be leukemic by flow cytometry) and computed its Spearman correlation with expression for each gene. We computed Jung's rank-based statistic to associate censored event-time data with gene expression [10] on a correlation-like scale [3] ranging from −1 (greater expression correlates perfectly with better outcome) to +1 (the opposite). The rank-based statistic is a natural extension of a nonparametric statistic of that associates a continuous quantitative variable with a censored survival outcome [11]. With these statistics to measure association of MRD (clinical outcome $c = 1$), EFS (outcome $c = 2$), and OS ($c = 3$) with expression of each gene, the direction coefficients were set as $\delta_1 = \delta_2 = \delta_3 = -1$. With this definition, a positive value of $s_{j\cdot}$ indicated gene j showed a beneficial pattern of association (increased expression associates with better outcomes) and a negative value of $s_{j\cdot}$ indicated a negative pattern of association (increased expression associates with worse clinical outcomes).

Figure 5 shows a three-dimensional scatterplot of the statistics associating each gene's expression with MRD, EFS, and OS. Table 5 and Fig. 6 show the results of genes with the most statistically significant results in the PROMISE analysis. Figure 7 shows how PROMISE more effectively enriches for genes with biologically concordant associations than does Venn diagram analysis. Intriguingly, LYL1 is the only gene with $\widehat{p} \leq 10^{-5}$ and $q \leq 10^{-5}$.

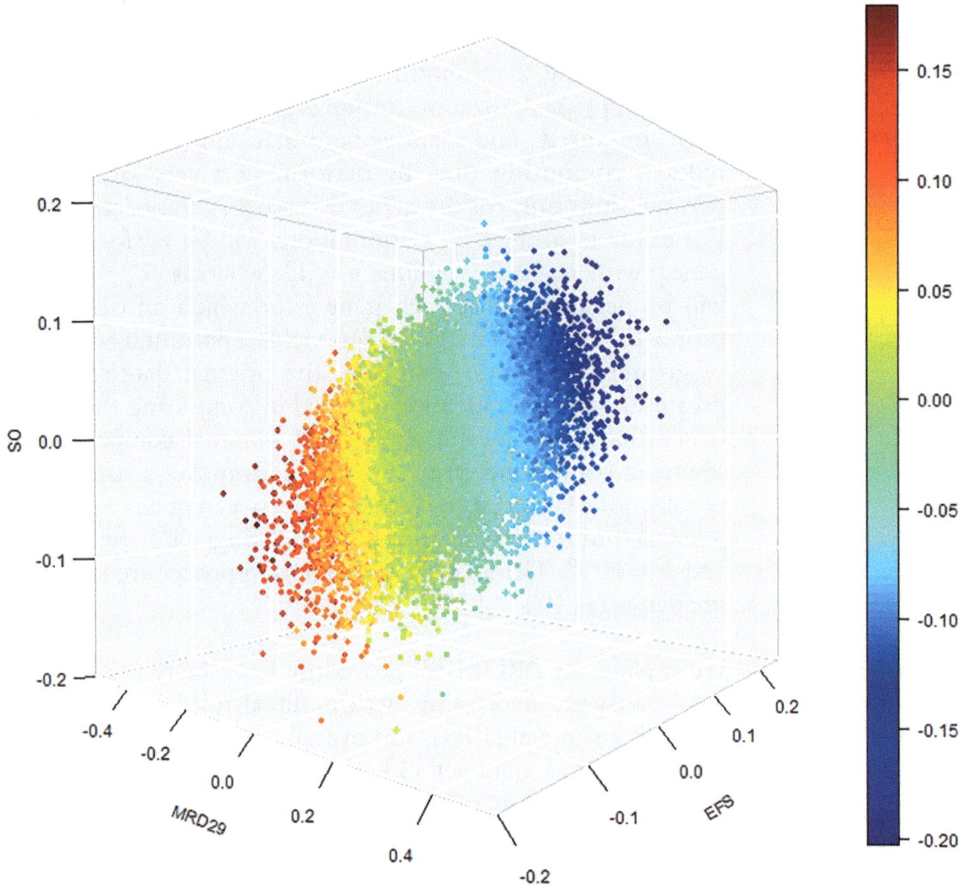

Fig. 5 Three-dimensional scatterplot of the correlation of each gene's expression with MRD, EFS, and OS. Each point in the three-dimensional scatterplot represents the correlation of one gene's RNA expression with MRD, EFS, and OS. The colors of the points correspond to the value of the PROMISE statistic as indicated by the color bar legend to the right

LYL1 is lymphoblastic leukemia-derived sequence 1. The top hit by PROMISE is a gene that is named after the disease under study. Finding this biological positive control as the top hit is a compelling example of how PROMISE can facilitate meaningful biological discovery.

Figure 8 shows that LYL1 expression has a statistically significant detrimental pattern of association with MRD, OS, and EFS ($\tilde{s}_{j.} = -0.173$ $p < 10^{-5}$; $q < 10^{-5}$). LYL1 expression has a very strong positive correlation with the quantitative level of MRD ($r = 0.363$; $p < 10^{-5}$; $q < 10^{-5}$). Also, greater diagnostic expression of LYL1 nonsignificantly associates with worse EFS

Table 5
The correlation of each gene with MRD, EFS, and OS as well as its PROMISE statistic, *p*-value, and *q*-value

Gene	MRD	EFS	OS	PROMISE	*p*	*q*
LYL1	0.363	0.079	0.078	−0.173	$<10^{-5}$	$<10^{-5}$
ADCY7	0.371	0.138	0.099	−0.203	0.00001	0.02854
CXXC5	0.463	0.008	0.085	−0.185	0.00001	0.02854
FUT7	0.408	0.052	0.090	−0.184	0.00001	0.02854
ESYT2	0.392	0.068	0.088	−0.182	0.00001	0.02854
PRSS45	0.336	0.098	0.096	−0.177	0.00001	0.02854
CTSW	0.475	0.014	0.081	−0.190	0.00002	0.03951
C6orf195	0.271	0.105	0.136	−0.171	0.00002	0.03951
HSH2D	0.509	0.008	0.016	−0.178	0.00003	0.03951
RPS6KA2	0.291	0.119	0.107	−0.172	0.00003	0.03951
C3orf52	−0.469	0.011	−0.041	0.166	0.00003	0.03951
PRSS46	0.387	0.045	0.048	−0.160	0.00003	0.03951
OR2G2	0.297	0.075	0.062	−0.145	0.00003	0.03951
RALGDS	0.471	0.012	0.033	−0.172	0.00004	0.04281
CCDC102A	0.252	0.163	0.091	−0.169	0.00004	0.04281
OR2G3	0.282	0.061	0.061	−0.135	0.00004	0.04281
NCKAP1	−0.396	−0.062	−0.080	0.179	0.00005	0.04467
CLIP2	0.448	0.053	0.031	−0.178	0.00005	0.04467

($r = 0.079$; $p = 0.19978$; $q = 1$) and OS ($r = 0.078$; $p = 0.20747$; $q = 1$).

8 Discussion

Integration of multiple forms of omic data and multiple clinical outcome metrics is a powerful approach to gaining valuable insights into the biological mechanisms that drive the development and prognosis of cancers and other catastrophic diseases. Integrated analysis can greatly increase statistical power to find biologically coherent associations among variables and thereby improve scientific understanding and accelerate clinical progress. The advantages of integrated analysis are especially valuable in rare disease settings with limited sample sizes such as pediatric oncology research.

The GRIN, ALEX, and PROMISE methods have been valuable tools in our molecular pediatric oncology research program.

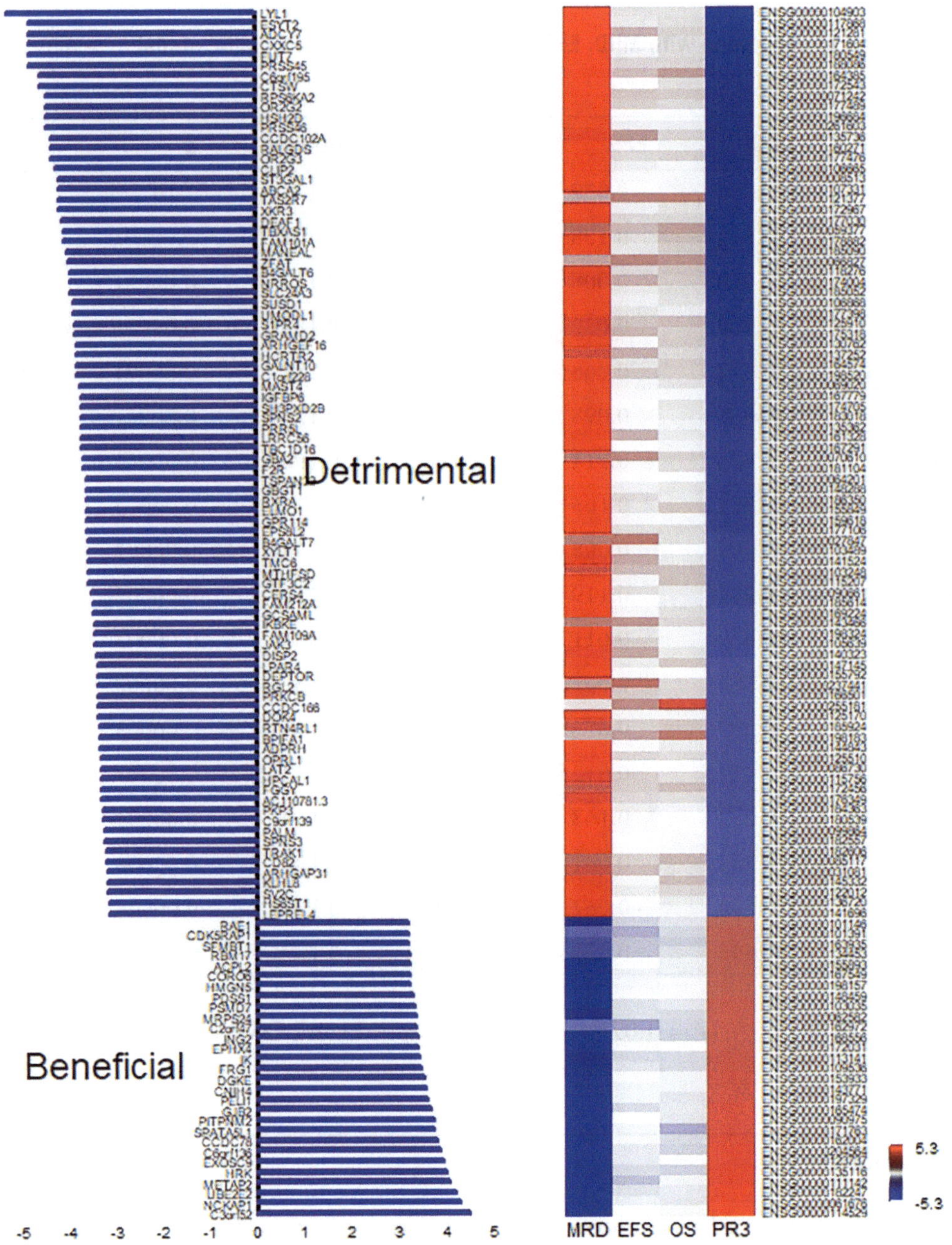

Fig. 6 PROMISE significance barplot. The barplot shows the sign of the PROMISE statistic and the magnitude of the PROMISE $-\log_{10}q$ value for the most significant genes. The heat map shows the sign and statistical significance of the individual endpoint correlation statistics according to the color legend to the bottom right

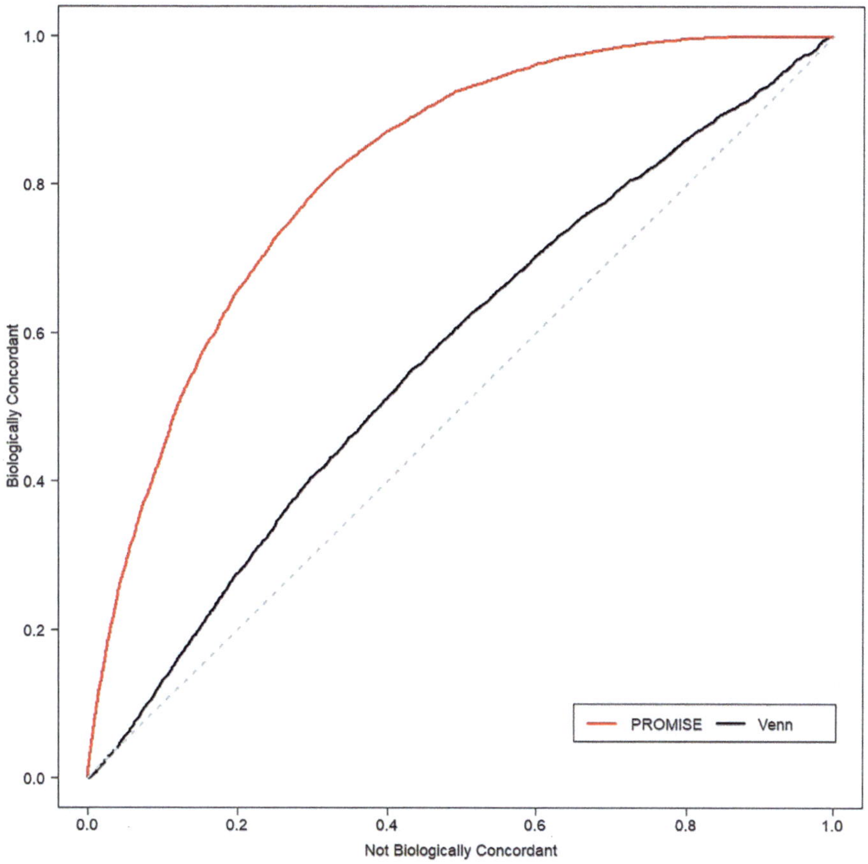

Fig. 7 Biological concordance receiver operating curve plot. The plot shows the proportion of genes with biologically concordant correlations with the three clinical outcomes (greater expression associates with better prognosis by all three outcome measures or associates with worse prognosis by all three outcome measures) versus the proportion of genes with biologically discordant correlations with the three clinical outcomes. The black curve shows these proportions for Venn diagram overlap as a function of a common statistical significance (p-value) threshold used for the three clinical outcomes' correlations. The red curve shows these proportions as a function of the statistical significance of the PROMISE statistic

The GRIN method has been used in high-impact studies of leukemia, brain tumors, solid tumors, and a pediatric pan-cancer study. ALEX complements GRIN by elucidating the transcriptional impacts of genomic lesions. PROMISE has combined with canonical correlation [12] to discover prognostic methylation-transcriptional profiles [13]. In particular, the discovery that DNMT3B demethylation and overexpression associated with greater total genome-wide methylation and worse prognosis provided the scientific rationale to evaluate the demethylating agents azacitidine and decitabine in the AML16 clinical trial (https://clinicaltrials.gov/ct2/show/NCT03164057), which is still ongoing at the time of this writing.

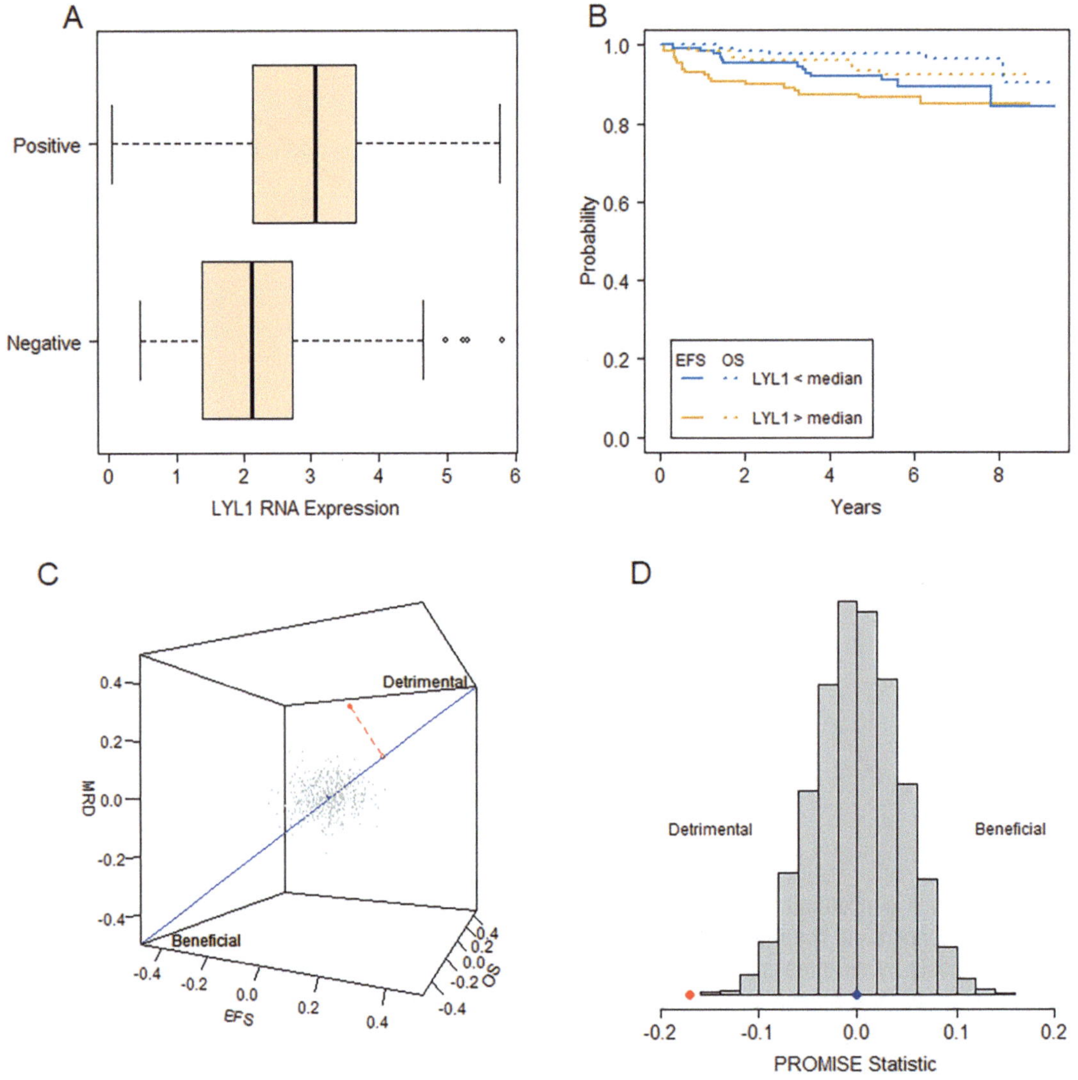

Fig. 8 PROMISE analysis of LYL1 expression. Panel (**a**) shows a boxplot of $\log_2$ LYL1 RNA expression by the presence of MRD. Panel (**b**) shows Kaplan-Meier estimates of EFS and OS by median-dichotomized LYL1 expression. Panel (**c**) shows the correlation of LYL1 expression with MRD, EFS, and OS as a solid red point in three-dimensional space and the correlations obtained in permuted data sets as gray points. The blue point is the origin (all three correlations equal zero). The blue line shows the most interesting statistical evidence for increased expression being detrimental to beneficial by the three clinical outcome metrics. The dashed red line shows the projection of the observed correlation onto the most interesting statistical evidence. The projection is shown as a hollow red point on that line. Panel (**d**) shows a histogram of the permutation projection statistics with a red dot for the observed projection statistic and a blue dot for the origin

The GRIN, ALEX, and PROMISE methods may also be generalized and extended to solve new challenges that present with new opportunities in the field of cancer multi-omics. It would be interesting and may prove scientifically valuable to explore methods to identify epigenetic abnormalities and incorporate them into the GRIN model framework so that epigenetic abnormalities and genomic abnormalities could be evaluated in an integrated manner. Additionally, there may be some utility in extending the GRIN null model to allow for random variability in lesion size and the possibility of lesions occurring on chromosomes other than the ones on which they are observed. The PROMISE method may be extended by considering other resampling methods (such as bootstrapping) in place of permutation and relaxing the requirement that association estimate vectors be projected onto a specific line of most interest. We are currently exploring those extensions of the PROMISE method.

References

1. Liu Y, Easton J, Shao Y, Maciaszek J, Wang Z, Wilkinson MR, McCastlain K, Edmonson M, Pounds SB, Shi L, Zhou X, Ma X, Sioson E, Li Y, Rusch M, Gupta P, Pei D, Cheng C, Smith MA, Auvil JG, Gerhard DS, Relling MV, Winick NJ, Carroll AJ, Heerema NA, Raetz E, Devidas M, Willman CL, Harvey RC, Carroll WL, Dunsmore KP, Winter SS, Wood BL, Sorrentino BP, Downing JR, Loh ML, Hunger SP, Zhang J, Mullighan CG (2017) The genomic landscape of pediatric and young adult T-lineage acute lymphoblastic leukemia. Nat Genet 49(8):1211–1218. https://doi.org/10.1038/ng.3909
2. Pounds S, Cheng C, Li S, Liu Z, Zhang J, Mullighan C (2013) A genomic random interval model for statistical analysis of genomic lesion data. Bioinformatics 29(17): 2088–2095. https://doi.org/10.1093/bioinformatics/btt372
3. Pounds S, Cheng C, Cao X, Crews KR, Plunkett W, Gandhi V, Rubnitz J, Ribeiro RC, Downing JR, Lamba J (2009) PROMISE: a tool to identify genomic features with a specific biologically interesting pattern of associations with multiple endpoint variables. Bioinformatics 25(16):2013–2019. https://doi.org/10.1093/bioinformatics/btp357
4. Winter SS, Dunsmore KP, Devidas M, Eisenberg N, Asselin BL, Wood BL, Leonard Rn MS, Murphy J, Gastier-Foster JM, Carroll AJ, Heerema NA, Loh ML, Raetz EA, Winick NJ, Carroll WL, Hunger SP (2015) Safe integration of nelarabine into intensive chemotherapy in newly diagnosed T-cell acute lymphoblastic leukemia: Children's Oncology Group Study AALL0434. Pediatr Blood Cancer 62(7):1176–1183. https://doi.org/10.1002/pbc.25470
5. Storey JD (2002) A direct approach to false discovery rates. J R Statist Soc B 64:479–498
6. Pounds S, Cheng C (2006) Robust estimation of the false discovery rate. Bioinformatics 22(16):1979–1987. https://doi.org/10.1093/bioinformatics/btl328
7. Berger GCaRL (2021) Statistical inference. Cengage Learning
8. Wendorff AA, Quinn SA, Rashkovan M, Madubata CJ, Ambesi-Impiombato A, Litzow MR, Tallman MS, Paietta E, Paganin M, Basso G, Gastier-Foster JM, Loh ML, Rabadan R, Van Vlierberghe P, Ferrando AA (2019) Phf6 loss enhances HSC self-renewal driving tumor initiation and leukemia stem cell activity in T-ALL. Cancer Discov 9(3): 436–451. https://doi.org/10.1158/2159-8290.CD-18-1005
9. Pounds S, Cao X, Cheng C, Yang JJ, Campana D, Pui CH, Evans WE, Relling MV (2011) Integrated analysis of pharmacologic, clinical and SNP microarray data using Projection Onto the Most Interesting Statistical Evidence with Adaptive Permutation Testing. Int J Data Min Bioinform 5(2):143–157. https://doi.org/10.1504/IJDMB.2011.039174
10. Jung SH, Owzar K, George SL (2005) A multiple testing procedure to associate gene expression levels with survival. Stat Med 24(20):3077–3088. https://doi.org/10.1002/sim.2179
11. O'Brien PC (1978) A nonparametric test for association with censored data. Biometrics 34(2):243–250
12. Cao X, Crews KR, Downing J, Lamba J, Pounds SB (2016) CC-PROMISE effectively integrates two forms of molecular data with multiple biologically related endpoints. BMC Bioinf 17(Suppl 13):382. https://doi.org/10.1186/s12859-016-1217-0
13. Lamba JK, Cao X, Raimondi SC, Rafiee R, Downing JR, Lei S, Gruber T, Ribeiro RC,

Rubnitz JE, Pounds SB (2018) Integrated epigenetic and genetic analysis identifies markers of prognostic significance in pediatric acute myeloid leukemia. Oncotarget 9(42): 26711–26723. https://doi.org/10.18632/oncotarget.25475

14. Emmanuelle, Clappier Wendy, Cuccuini Anna, Kalota Antoine, Crinquette Jean-Michel, Cayuela Willem A., Dik Anton W., Langerak Bertrand, Montpellier Bertrand, Nadel Pierre, Walrafen Olivier, Delattre Alain, Aurias Thierry, Leblanc Hervé, Dombret Alan M., Gewirtz André, Baruchel François, Sigaux Jean, Soulier (2007) The C-MYB locus is involved in chromosomal translocation and genomic duplications in human T-cell acute leukemia (T-ALL) the translocation defining a new T-ALL subtype in very young children. Blood 110(4):1251–1261 https://doi.org/10.1182/blood-2006-12-064683
15. Xiaoyan, Jiang Yun, Zhao Wing-Yiu, Chan Suzanne, Vercauteren Emily, Pang Sean, Kennedy Frank, Nicolini Allen, Eaves Connie, Eaves (2004) Deregulated expression in Ph+ human leukemias of AHI-1 a gene activated by insertional mutagenesis in mouse models of leukemia. Blood 103(10):3897–3904 https://doi.org/10.1182/blood-2003-11-4026
16. X, Liu K, Rothe R, Yen C, Fruhstorfer T, Maetzig M, Chen D L, Forrest R K, Humphries X, Jiang (2017) A novel AHI-1–BCR-ABL–DNM2 complex regulates leukemic properties of primitive CML cells through enhanced cellular endocytosis and ROS-mediated autophagy. Leukemia 31(11):2376–2387 https://doi.org/10.1038/leu.2017.108
17. Hiroshi, Fujiwara Naomichi, Arima Tomoko, Hashimoto-Tamaoki Kakushi, Matsushita Hideo, Ohtsubo Kosei, Arimura Shiroh, Hidaka Chuwa, Tei (1999) Alteration of p16 (CDKN2) gene is associated with interleukin-2–induced tumor cell growth in adult T-cell leukemia. Experimental Hematology 27(6): 1004–1009 S0301472X99000351 https://doi.org/10.1016/S0301-472X(99)00035-1
18. Sabine, Strehl Karin, Nebral Margit, König Jochen, Harbott Herbert, Strobl Richard, Ratei Stephanie, Struski Bella, Bielorai Michel, Lessard Martin, Zimmermann Oskar A., Haas Shai, Izraeli (2008) (2008) Clinical Cancer Research 14(4):977–983 https://doi.org/10.1158/1078-0432.CCR-07-4022
19. S. M. N., Onnebo P., Rasighaemi J., Kumar C., Liongue A. C., Ward (2012) (2012) Alternative TEL-JAK2 fusions associated with T-cell acute lymphoblastic leukemia and atypical chronic myelogenous leukemia dissected in zebrafish. Haematologica 97(12):1895–1903 haematol.2012.064659 https://doi.org/10.3324/haematol.2012.064659
20. Stefan, Nagel Letizia, Venturini Corinna, Meyer Maren, Kaufmann Michaela, Scherr Hans G., Drexler Roderick A.F., MacLeod (2010) Multiple mechanisms induce ectopic expression of LYL1 in subsets of T-ALL cell lines. Leukemia Research 34(4):521–528 S0145212609002975 https://doi.org/10.1016/j.leukres.2009.06.020
21. AHyun, Choi Anuradha, Illendula John A., Pulikkan Justine E., Roderick Jessica, Tesell Jun, Yu Nicole, Hermance Lihua Julie, Zhu Lucio H., Castilla John H., Bushweller Michelle A., Kelliher (2017) RUNX1 is required for oncogenic Myb and Myc enhancer activity in T-cell acute lymphoblastic leukemia. Blood 130(15):1722–1733 https://doi.org/10.1182/blood-2017-03-775536
22. Idoya, Lahortiga Kim, De Keersmaecker Pieter, Van Vlierberghe Carlos, Graux Barbara, Cauwelier Frederic, Lambert Nicole, Mentens H Berna, Beverloo Rob, Pieters Frank, Speleman Maria D, Odero Marijke, Bauters Guy, Froyen Peter, Marynen Peter, Vandenberghe Iwona, Wlodarska Jules P P, Meijerink Jan, Cools (2007) Duplication of the MYB oncogene in T cell acute lymphoblastic leukemia. Nature Genetics 39(5):593–595 https://doi.org/10.1038/ng2025
23. Yong-Mei, Zhu Wei-Li, Zhao Jian-Fei, Fu Jing-Yi, Shi Qin, Pan Jiong, Hu Xiao-Dong, Gao Bing, Chen Jun-Min, Li Shu-Min, Xiong Long-Jun, Gu Jing-Yi, Tang Hui, Liang Hui, Jiang Yong-Quan, Xue Zhi-Xiang, Shen Zhu, Chen Sai-Juan, Chen (2006) (2006) Clinical Cancer Research 12(10):3043–3049 https://doi.org/10.1158/1078-0432.CCR-05-2832
24. Agnieszka A., Wendorff S. Aidan, Quinn Marissa, Rashkovan Chioma J., Madubata Alberto, Ambesi-Impiombato Mark R., Litzow Martin S., Tallman Elisabeth, Paietta Maddalena, Paganin Giuseppe, Basso Julie M., Gastier-Foster Mignon L., Loh Raul, Rabadan Pieter, Van Vlierberghe Adolfo A., Ferrando (2019) (2019) Cancer Discovery 9(3):436–451 https://doi.org/10.1158/2159-8290.CD-18-1005
25. Alejandro, Gutierrez Takaomi, Sanda Ruta, Grebliunaite Arkaitz, Carracedo Leonardo, Salmena Yebin, Ahn Suzanne, Dahlberg Donna, Neuberg Lisa A., Moreau Stuart S., Winter Richard, Larson Jianhua, Zhang Alexei, Protopopov Lynda, Chin Pier Paolo, Pandolfi Lewis B., Silverman Stephen P., Hunger Stephen E., Sallan A. Thomas, Look (2009) High frequency of PTEN PI3K and AKT

abnormalities in T-cell acute lymphoblastic leukemia. Blood 114(3):647–650 https://doi.org/10.1182/blood-2009-02-206722

26. P Y, Jotta M A, Ganazza A, Silva M B, Viana M J, da Silva L J G, Zambaldi J T, Barata S R, Brandalise J A, Yunes (2010) Negative prognostic impact of PTEN mutation in pediatric T-cell acute lymphoblastic leukemia. Leukemia 24(1):239–242 https://doi.org/10.1038/leu.2009.209
27. Amélie, Trinquand Aline, Tanguy-Schmidt Raouf, Ben Abdelali Jérôme, Lambert Kheira, Beldjord Etienne, Lengliné Noémie, De Gunzburg Dominique, Payet-Bornet Ludovic, Lhermitte Hossein, Mossafa Véronique, Lhéritier Jonathan, Bond Françoise, Huguet Agnès, Buzyn Thibaud, Leguay Jean-Yves, Cahn Xavier, Thomas Yves, Chalandon André, Delannoy Caroline, Bonmati Sebastien, Maury Bertrand, Nadel Elizabeth, Macintyre Norbert, Ifrah Hervé, Dombret Vahid, Asnafi (2013) Journal of Clinical Oncology 31 (34):4333–4342 https://doi.org/10.1200/JCO.2012.48.5292
28. Tze King, Tan Chujing, Zhang Takaomi, Sanda (2019) Oncogenic transcriptional program driven by TAL1 in T-cell acute lymphoblastic leukemia. International Journal of Hematology 109(1):5–17 https://doi.org/10.1007/s12185-018-2518-z
29. Britt M., Gustafsson Kristin, Mattsson Gordana, Bogdanovic Gustaf, Leijonhufvud Emma, Honkaniemi Kim, Ramme Anthony M., Ford (2018) Pediatric Blood & Cancer 65(11) e27310- https://doi.org/10.1002/pbc.27310
30. B, Patel Y, Kang K, Cui M, Litt M S J, Riberio C, Deng T, Salz S, Casada X, Fu Y, Qiu K, Zhao S, Huang (2014) Aberrant TAL1 activation is mediated by an interchromosomal interaction in human T-cell acute lymphoblastic leukemia. Leukemia 28(2):349–361 https://doi.org/10.1038/leu.2013.158
31. Nosaka, K., Maeda, M., Tamiya, S., Sakai, T., Mitsuya, H., & Matsuoka, M. (2000). Increasing methylation of the CDKN2A gene is associated with the progression of adult T-cell leukemia. Cancer research, 60(4), 1043–1048.
32. Baer, R. (1993, December). TAL1, TAL2 and LYL1: a family of basic helix-loop-helix proteins implicated in T cell acute leukaemia. In Seminars in cancer biology (Vol. 4, No. 6, pp. 341–347).
33. Erbilgin, Y., Sayitoglu, M., Hatirnaz, O., Dogru, O., Akcay, A., Tuysuz, G., ... & Ozbek, U. (2010). Prognostic significance of NOTCH1 and FBXW7 mutations in pediatric T-ALL. Disease markers, 28(6), 353–360.

INDEX

Brooke Fridley and Xuefeng Wang (eds.), *Statistical Genomics*, Methods in Molecular Biology, vol. 2629, https://doi.org/10.1007/978-1-0716-2986-4,

S

T

U

X

MIX
Papier aus verantwortungsvollen Quellen
Paper from responsible sources
FSC® C105338

FSC
www.fsc.org

If you have any concerns about our products,
you can contact us on
ProductSafety@springernature.com

In case Publisher is established outside the EU,
the EU authorized representative is:
Springer Nature Customer Service Center GmbH
Europaplatz 3, 69115 Heidelberg, Germany

Printed by Libri Plureos GmbH
in Hamburg, Germany